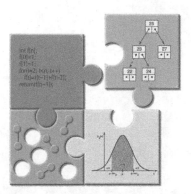

# BIG PRACTICAL GUIDE TO
# Computer
# Simulations
## Second Edition

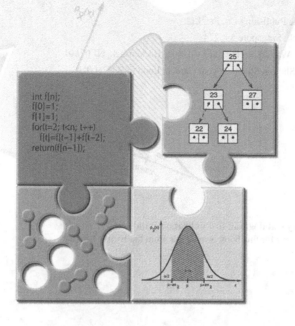

# BIG PRACTICAL GUIDE TO
# Computer
# Simulations
## Second Edition

## Alexander K Hartmann
University of Oldenburg, Germany

**World Scientific**

NEW JERSEY · LONDON · SINGAPORE · BEIJING · SHANGHAI · HONG KONG · TAIPEI · CHENNAI

*Published by*

World Scientific Publishing Co. Pte. Ltd.
5 Toh Tuck Link, Singapore 596224
*USA office:* 27 Warren Street, Suite 401-402, Hackensack, NJ 07601
*UK office:* 57 Shelton Street, Covent Garden, London WC2H 9HE

**British Library Cataloguing-in-Publication Data**
A catalogue record for this book is available from the British Library.

**BIG PRACTICAL GUIDE TO COMPUTER SIMULATIONS**
**2nd Edition**

ISBN 978-981-4571-76-0
ISBN 978-981-4571-77-7 (pbk)

Printed in Singapore

To my family

# Preface to the 2nd edition

After the first edition of this book was published, I received a lot of supporting comments from various people. Nevertheless, many readers recommended some further subjects to be included in the book, in particular related to new developments or tools. Furthermore, several misprints and little mistakes were reported by them. Also, for some paragraphs, I decided to rewrite them a bit or expand them by including more examples. Hence, I did not hesitate for long, when World Scientific asked me to write a second edition. You are holding the result of this effort in your hands.

The main extensions of the second edition are as follows:

- An introduction to the programming language *Python* has been added, see Sec. 2.3. This language is half-way between a script language (since it is interpreted) and a full high-level programming language (it offers, e.g. complex data structures, object-oriented elements and many libraries for various purposes). Thus, when using *Python*, on the one hand it is easy to write scripts for various tasks like data analysis. On the other hand, small to medium size simulation project can be implemented rapidly.

- Since simulation and analysis of *networks and graphs* has become more and more important in various fields of science, the Sec. 6.8 on graphs has been extended by a subsection (6.8.4) on the calculation of the *connected component*. This is one of the most-often performed analyses for graphs and the basis of many complex algorithms, e.g. to obtain shortest paths, diameters, or centrality measures.

- The Sec. 8.1.1 on how to *draw random numbers* for discrete random variables now includes the fastest available algorithm. It allows one to draw each random number according to the given distribution in constant time $O(1)$. This is much faster compared to the standard approach which takes $O(\log N)$, where $N$ is the number of the possible outcomes.

- The Sec. 8.2.4 covers the *rejection method*, which allows one to draw random numbers according to a given probability density, usually applied for non-invertible distribution functions. The section has been extended by the general approach to border the target distribution by arbitrary probability density functions. This enlarges strongly the range of distributions which can be realized by this approach.
  Furthermore, as an example for generating random points in higher dimensions, drawing points uniformly on the surface of *d-dimensional hyper sphere* is included now.

- In Chap. 8 on data analysis, in the Sec. 8.5 on hypothesis test and measuring correlations in data sets, a subsection on the *Receiver-operator characteristic* (ROC) has been included. The ROC is an advanced yet widespread approach which allows for the determination of optimal parameters for hypothesis tests.

- Furthermore a subsection (8.5.5) on the *principal-component analysis* has been added. This is the simplest approach to search for the most important "directions" in high-dimensional data.

- *Data clustering* methods are an important tool in the analysis of big data sets, since they allow for looking for structure in overwhelming amounts of data. This is a very sophisticated way of data analysis, much beyond the level of averages, histograms and curve fitting. Several approaches, ranging from the most-simple $k$-means method to the advanced hierarchical clustering algorithms are now explained extensively in Sec. 8.5.6.

- The Sec. 8.6.2 on *fitting* data using *gnuplot* has been extended by how to restrict ranges of fitting parameters (which is not possible by default) and how to fit functions to multiple sets of data simultaneously.

- In the first edition, *gnuplot* was only used to generate "quick and dirty" plots of data. In Sec. 9.2.1 it is now shown how one can generate publication-quality plots using *gnuplot*.

- The Sec. 9.2.4 on the ray tracer *Povray* has been extended by some paragraphs which explain how this ray tracer can be used to generated three-dimensional pictures of simulation snapshots. By using sequences of such figures, small movies can be generated easily.

- In the first edition, the Sec. 9.3.1 on LaTeX, a standard typesetting tool to generate high-quality documents as large as complete books (like this one), was a rather short introduction giving few examples. For the second editions this section has been expanded substantially, such that it now serves as a *complete introduction to LaTeX*, presenting the most-frequently used LaTeXelements in scientific publications, such that all necessary knowledge to generate standard to advanced manuscripts (from the typesetting point of view) is covered.

These new sections include and discuss, following the hands-on approach of the first edition, all necessary source codes such that the reader can apply the new knowledge to her or his case as quickly as possible. Thus, I hope that this new edition will allow you to learn all technical aspects of scientific computer simulations as fast as possible, ranging from the initial idea over the implementation, performing the simulations, doing the data analysis, up to the publication of your results. Thus, using this book will save you a lot of time which you can use to concentrate on the actual scientific problem you want to solve. Finally, since the new edition comprises almost all non-problem specific knowledge you need in computational science, I decided to rename the book to "Big Practical Guide to Computer Simulations".

This new edition would not have been possible without the help of many people. I am grateful to the following persons for communicating mistakes, making useful suggestions and providing extensions of the book: Pia Backmann, Jan Christoph Bernack, Gunnar Claussen, Timo Dewenter, Florian Effenberg, Pascal Fieth, Nikolai Gagunashvili, Hendrike Heidemann, Iwo Ilnicki, Simon Knowles, Karsten Looschen, Markus Manssen, Andreas Mohrs, Zacharais Njam Mokom, Oliver Melchert, Marc Mézard, Tudor Mitran, Christoph Norrenbrock, Tom Seren, Hendrik Schawe, Sagar Sinha, Verena Sterr, Sebastian von Ohr, and A. Peter Young. Furthermore, Oliver Melchert provided a Python script for data resampling which is included in the new edition.

*Alexander K. Hartmann*
Oldenburg, July 2014

# Preface to the 1st edition

You have decided to become an expert in computer simulations, congratulations! This is a good decision, because computational and simulation methods become more and more important in all areas of science, humanities, economy, engineering and mathematics. This book will help you a great deal in learning the most important basics which you need during all phases of a simulation research project, from the phases of program design, implementation, and debugging, through running the simulations, organizing the data and analyzing the results, to the final phase where you want to present and publish your results.

Note that nowadays thousands of different problems exist which are being investigated by computer simulations. One studies, for example, the diffusion of chemicals in soils, the folding of proteins in cells, the communication of neurons in the brain, the deformation of cars in accidents, the behavior of brokers working at the stock market, the evolution of the weather during the next days or weeks, the turbulent behavior of flowing water in a turbine, the movement of electrons in semiconductors, the patterns of words in languages, or the traffic of pedestrians in crowded shopping malls, to name only a few. Consequently, there are many algorithms to treat the variety of problems, for example, finite differences, finite elements, integration methods, matrix inversion, eigenvalue determination, equation solvers, molecular dynamics simulations, Monte Carlo methods, density functional approaches, graph algorithms, optimization methods, and so on. Which methods are suitable for your problems depends heavily on the problems you want to solve. Since there are way too many algorithms available for all these different problems, and because everybody needs usually a different one, these special-purpose algorithms are *not* covered in this book.

This book instead covers methods, techniques and algorithms, which you *always* have to apply, independent of the actual simulation research project you are considering. Here, *practical* aspects of conducting research via computer simulations are discussed. An overview is given towards the end of this preface. After reading this book, you only need some additional information about the specific project your are considering, usually provided in scientific papers, and maybe you need a second special-purpose book which you have to get. Then you are ready to start!

The book addresses people who have no or little experience with computer simulations. This book is in particular suited for students who want to start a project, like a PhD thesis, in the field of computer simulations. But also researchers who have conducted already some simulation projects may find a lot of the advanced material helpful. It is assumed that the reader is familiar with an operating system such as UNIX (e.g. Linux), a high-level programming language such as C, JAVA, Fortran or Pascal and has some experience with at least tiny software projects.

Throughout the book, because of the limited space, usually only short introductions to the specific areas are given, as "ready-to-use recipes". The material usually is presented here in a learning-by-example manner. Nevertheless, the material is extensive enough to provide a fundamental set of tools to perform all standard tasks when creating and performing simulations. In addition, references to more specialized literature are cited, allowing specific subjects to be studied more extensively. Most examples of code are in C/C++. Many examples, also solutions to exercises, are available on the website. This is indicated by a small **GET SOURCE CODE** box in the text. Also some freely available

| GET SOURCE CODE |
| --- |
| DIR: `c-programming` |
| FILE(S): `first.c` |

documentation is contained on the website. For details, see the appendix.[1]

Next, I give you some idea, how the book was realized. In fact, the work on this book started when I was doing simulations for obtaining my first university degree. I had to develop new algorithms and implement them. I had to do many large-scale simulations on a parallel computing cluster. The data had to be analyzed, usually in many different ways, and over-and-over again when new data became available. Finally, the results had to be presented and summarized in a scientific paper. These basic steps remained my main occupation during my PhD and during my first

---

[1]All supplementary materials can be downloaded from the website: http://www.worldscientific.com/r/9019-supp.

post-doc projects. I became more and more experienced and refined my approaches. I also improved my ways to work by reading books about software engineering, algorithms, data structures and data analysis, as well as by learning to use many programs. After the first-post doc years, I began to supervise students. Hence, I started to pass on my knowledge, trying to help other people to avoid many pitfalls and to devise highly efficient programs. Some of my experiences found their way to the last chapter of the book of Heiko Rieger and myself with the title "Optimization Algorithms in Physics". That chapter is in fact a very short version of the present book. Thus, it served as a seed of the this book and some material of the old chapter appears occasionally here.[2] During the years, I supervised more and more students, and even gave a university course on the practical aspects of computer simulations. This course contained a lot of new material compared to the book chapter and in fact it served as a seed for the present book. Nevertheless, the material I used for supervising students was still collected from several different sources, often not quite compatible with each other. I started to feel that it would help my work, and also increase the efficiency, if I wrote a full book about practical aspects of computer simulations, which should contain "all" needed material comprehensively. After two more years, I received an email from *World Scientific*, where my course web page was noticed. They asked me whether I would like to write a book, based on the course. Now it was not difficult to come to the decision indeed to realize the book. You are now holding the result in your hands.

Note that this book contains a very personal view of which tools are considered useful. Very often, I present several independent tools, such as tools for editing, compiling and analyzing programs rather than one all-purpose environment which usually contains a framework just integrating these basic tools. Nevertheless, most of the tools introduced are standard programs and available on all computer systems (for Microsoft operating systems they sometimes have other names).

Here, I give an overview over the contents of the book. First, a short introduction to C programming is given. Also related topics like macros, *make files* and shell scripts are touched. In the second chapter, the main ideas of software engineering are explained and several hints allowing the construction of efficient and reliable code are stated. In Chap. 4, three very

---

[2]taken from A.K. Hartmann and H. Rieger, *Optimization Algorithms in Physics*, pp. 293–357, 2002, Copyright Wiley-VCH Verlag GmbH & Co. KGaA. Reproduced with permission.

useful debugging tools are presented, which will help you to hunt down bugs in your programs quickly. In Chap. 5, a short primer on object-oriented software development is presented. In particular, it is shown that this kind of programming style can be achieved with standard procedural languages such as C as well, but also how C++ can be used. Next, basic types of algorithms and advanced data structures are explained. These can be used as auxiliary tools to create highly-professional, efficient simulation programs. In the subsequent chapter, the benefit of using libraries like the *Standard Template Library* and the *GNU Scientific Library* is explained and it is shown how you can build your own libraries. In Chap. 8, aspects of probability theory, random-number generation, data analysis, plotting data and curve fitting are covered. In the last chapter, an introduction to information retrieval and literature search in the Internet and to the preparation of presentations and publications is given.

I am indebted to all my colleagues for countless hours of joyful collaborations, which laid the foundations of this book. I am very grateful to Angelika Sievers for thoroughly reading all chapters of the book while checking for typos, language and grammar mistakes. Finally, I would like to thank Björn Ahrens, Luis Apolo, Bernd Burghardt, Niels Hoelzel, Magnus Jungsbluth, Reinhard Leidl, Oliver Melchert, Axel Schulz, Bruno Sciolla and Stefan Wolfsheimer for critically reading manuscript chapters and for giving me many useful comments which helped me to improve the book and to remove many typos.

*Alexander K. Hartmann*
Oldenburg, November 2008

Alexander Karl Hartmann is Professor of Computational Physics of the Department of Science at the University of Oldenburg (Germany). He obtained a M.S. in Computer Science at the University of Hagen and a M.S. in Physics at the University of Duisburg. After receiving his Ph.D. in Theoretical Physics at the University of Heidelberg, he worked at the University of Göttingen, the University of California in Santa Cruz and the Ecole Normale Supérieure in Paris. In 2007, he moved to Oldenburg. His research interests comprise computer simulations, disordered systems, combinatorial optimization, large deviations and Bioinformatics. Professor Hartmann is author or co-author of 8 books and more than 120 scientific publications.

# Contents

# Chapter 1

# Programming in C

Performing computer simulation in a sophisticated way always includes writing computer programs. Even if one builds upon an existing package, one usually has to perform minor or major modifications or write new subroutines for data analysis.

Several programming languages are commonly used in science. The historically first widespread high-level procedural programming language for this purpose is *Fortran*, which was developed in the 1950s. Thirty years ago, Fortran 77 was a standard in numerical computations, which has led to the development of many Fortran-based libraries like the NAG (*Numerical Algorithms Group* [Philipps (1987)] library. Since Fortran used to be a very inflexible programming language, other high-level structural programming languages like Pascal (and its extensions Modula-2 and Delphi) and C (and its object-oriented extension C++) were developed from the 1970s on. These languages allow for complex data types, dynamic memory allocation and modularization, which makes the treatment of large-scale simulations much easier. During the 1980s, the C programming language became a standard in particular for operating system programming. Hence, standard operating systems like UNIX (Linux) and Windows are written in C. This resulted in C being used in many other fields as well, in particular for scientific purposes. As a result, all standard numerical libraries are now available for C/C++ and nowadays C is the dominating language for writing simulation programs in science.

Note that Fortran caught up via the Fortran 90 and Fortran 95 standards, capable of basically everything C can do. Thus, Fortran is still heavily used, in particular because it allows for easier parallelization of codes related to linear and differential equations. Nevertheless, in this book we have to select one language to concentrate on. We select C (and C++ in

1

Chap. 5) because it offers, on the one hand, everything you can expect from a high-level programming language. On the other hand, C is relatively close to machine-level programming, which leads to a very high performance, but also makes insecure programming possible. Also, more recent developments like Java have a similar syntax to C and C++. C is also widespread outside the scientific community in business-related areas, a field where many people reading this book will work in after their studies. Therefore, a scientist benefits much from knowing C very well. In general, all high-level programming languages are equally powerful. Therefore, good proficiency in C allows to learn another programming language very quickly.

There are many additional programming languages used for specific purposes such as writing scripts (traditional shell script languages, Perl) or languages used within certain statistical and/or mathematical packages like *Mathematica, Maple, Matlab,* or *R*. If you want to learn these, you have to consult special literature.

Here, we start by explaining how to write simple C programs and what the most fundamental program ingredients are. This also includes advanced techniques like structures and self-defined data types. Note that we use a standardized form called *ANSI C*. Structuring programs is easier when using subroutines and functions, which are covered in the second section. Next, details for input-/output techniques are presented. In the fourth section of this chapter, more details concerning pointers and memory allocation are explained. In the last section, some examples for macro programming are given. C++ extensions related to object-oriented programming are covered in Chap. 5.

Before starting, you should be advised that learning to program is a matter of practice. Consequently, it will not be sufficient to just read the following sections. Instead, you should sit in front of a computer and try, use, and modify all given examples extensively!

As always in this book, most concepts are introduced via examples. Most standard statements and control structures are covered, but not explained in the most-general way. Nevertheless, it should be sufficient in 95% of all cases. For an exhaustive and complete description of C, you should consult specialized literature [Kernighan and Ritchie (1988)].

## 1.1 Basic C programs

It is assumed that you are familiar with your operating system, in particular with concepts like commands, files and directories, and that your are able to use a text editor. All examples given here are based on UNIX shell commands, but should be simple to convert to your favorite operating system. As always in this book, the basic functionalities are explained, such that fundamental applications are covered. For a complete reference to C programming, you should consult specialised books on this topic [Kernighan and Ritchie (1988)]. Note that the standard (so-called ANSI) C contains a complete set of commands such that programs for all tasks can be implemented. Nevertheless, for many standard applications, e.g. computing complex functions, solving equations, calculating integrals, or performing algorithms on complex data structures like graphs, there are extensions of the standard C, collected in so-called *libraries*. More about libraries and how to use them can be found in Chap. 7.

In this chapter, we start the introduction to C by presenting a very first C program, which you should enter in your favorite text editor. There you should save the file with the file

| GET SOURCE CODE |
| --- |
| DIR: `c-programming` |
| FILE(S): `first.c` |

name, say, `first.c`. The program looks as follows:

```
1   #include <stdio.h>
2
3   int main()
4   {
5     printf("My first program\n");
6     return(0);
7   }
```

The meaning of the different lines will be explained in a minute. First, you should *compile* the program. This means that in the current form the program cannot be executed by the computer. The compiler converts it into so-called *machine code*, which can be executed. Usually, the machine code resulting from your program will be put together with other already existing machine codes, e.g. for some standard tasks like input/output subroutines. This is the so-called *linking* state of the compilation performed by the *linker*.

Under UNIX/Linux, you can compile the program in a shell, provided the current directory of the shell is in the directory where the program is stored, via the command

```
cc -o first -Wall first.c
```

The main command is `cc`, which is an abbreviation of *C compiler*. In the above example, two *options* are used for the command `cc`:

-o (for ouput) gives the name the executable program will get, here `first`.

-Wall The W stands for *warning* and `all` means the compiler will report (almost) *all* strange things it observes, even if it is not an error in a strict sense. Since C is very sloppy about many rules, it lets you get away with a lot, which most of the times is not what you want to do.

The classical example is the comparison operator `==`, i.e. you have to write `a==b`, if you want to test whether the values stored in the variables `a` and `b` are equal. This is different from `a=b`, which assigns `a` the value of `b` (see below). Nevertheless, for conditional statements like `if(a==b)`, sometimes the programmer mistakenly writes `if(a=b)`, which is nevertheless a meaningful statement under some circumstances. Therefore, the compiler does not report this as an error, but with `-Wall`, the compiler warns you that you have used an assignment where a conditional statement is expected. Hence, you should use the highest warning level `-Wall` always, and ignore the warnings only, if you know exactly what you are doing. Note that there are several other warning options in C which are not turned on even when using `-Wall`, despite its name. For example, the option `-Wextra` (previously called just `-W`) turns on additional warnings. A complete list of all warning options can be found in the compiler documentation.

The last argument of the `cc` command is the file you actually want to compile. Note that, in principle, a program may consist of several files, which can all be passed via arguments to the compiler. By default, the C compiler will generate an executable. Consequently, the compiling process will consist of the actual compiling of the input source files, which generates a (hidden) *assembler code* for each input file. The assembler codes are translated to executable codes. Finally the different executable codes are *linked* into one program. If the option `-c` is given, the linking stage will be omitted. In this case several executable `.o` files are created. The C compiler, when called from the command line, offers a huge number of options, which are all described in the manual pages. Under UNIX/Linux, you can access the manual via the man command, i.e. by entering "man cc".

Now we come back to the structure of the program first.c. The first line *includes* a file. This means that the file stdio.h contains some definitions which the compiler should read before proceeding with the compilation process. In this case, stdio.h contains declarations of some built in standard functions for input/output i/o. The compiler needs these declarations, because it has to know, for example, how many parameters are passed to the function. The files containing declarations are usually called *header files*, hence the suffix is .h. In this example, the compiler needs the definition of the subroutine printf(), which means *print formatted* (see below). Note that the actual *definitions* of the subroutines, i.e. the code for the subroutine, is not contained in the header files, but in *libraries* (see page 17) and Sec. 7.1.

The main program of every C program is called main(), which comprises the remaining file of first.c. In fact, all codes which do something in C programs are functions, i.e. they take some arguments and return something. Hence, main() is also a function. Here, it takes zero arguments, indicated by the fact that nothing is written in the brackets () after the name (main) of the function in line three. Below, you will learn how to pass to main() some arguments via the command line. By default, the function main() returns an *integer* value, which is indicated by the int in front of the function name.

The definition of a function, i.e. the program code which is executed when the function is called, is contained inside the pair {...} of brackets (lines 4–7). Note that the opening and closing brackets are written below each other for readability, but this is not required by the compiler. You could have written everything in one line, or insert many empty lines, etc. Also, the position inside a line does not matter at all, e.g. your program could look like

```
#include <stdio.h>
int
main
(){
  printf(
"My first program\n");                    return(0);
              }
```

Certainly, the first version is much more readable than the second, so you should follow some formatting rules. This is discussed in Sec. 3.2. Note that concerning the `#include` *directive*, which is only a kind of command passed to the compiler but not a true C command, the syntax is more strict. Basically, each `#include` directive must be written in exactly one line (unless one explicitly uses several lines via the \ symbol at the end of a line). Also, only one directive is allowed per line. More information on the *compiler directives* and the related *macros* is given in Sec. 1.6.

The code describing the functionality of `main()` comprises the two lines 5–6 here. First, the `printf()` subroutine prints the string given as argument to the screen, i.e. here the string "My first program". The \n symbol is a formatting symbol. It indicates that after printing the string a *new line* is inserted. Consequently, any subsequent output, if also performed using `printf()` occurring during the execution of a program, will be printed starting at the next line. One can print many different things using `printf()`, more details are explained in Sec. 1.3. Note that formally the subroutine `printf()` is a *function*, since in C *all* subroutines are functions. Functions can have side effects, like printing out information or changing content of variables. Also, all functions calculate or *return* a value, which is often less important than the side effects, like in this case. The function call, together with the terminating semicolon, makes a *statement*. Each statement must close with a semicolon, even if it is the last statement of a function or the last statement of a block (see page 20) of statements.

In line 6, `return(0)` states the value the function returns. Here it is just 0, which means for `main()` by convention that everything was OK. In general, one can return arbitrary complex objects. Simple mathematical functions will return a real number, e.g. the square root of the argument, but also more complex objects like strings, arrays, lists or graphs are possible. More on functions in C is given in Sec. 1.2.

To actually execute the command, one types into the shell the name of the program, i.e.

```
first
```

Note that this assumes that the current directory is part of the search path contained in the environment variable PATH. If this is not the case, one has to type in "./first". As a result of the program execution, the message "My first program" will appear on the screen (inside the shell, just in the line after where you have called the program), the program will terminate and the shell will wait for your next command.

### 1.1.1  *Basic data types*

To perform non-trivial computations, a program must be able to store data. This can be done most conveniently using *variables*. In C, variables have to be *defined*, i.e. the *type* has to be given (see below on the syntax of a definition). A definition will also cause the compiler to reserve memory space for the variable, where its content can be stored. The type of a variable determines what kind of data can be stored in a variable and how many bytes of memory are needed to store the content of a variable. Here, we cover only the most important data types. The most fundamental type is

- `int`

  This type is used for integer numbers. Usually, four bytes are used to store integer numbers. Hence, they can be in the range $[-2^{31}, 2^{31} - 1]$. Alternatively you can use `unsigned int`, which allows for numbers in the range $[0, 2^{32} - 1]$

To define some variables of type `int`, one writes `int` followed by a comma-separated list of variable names and ended by a semicolon, like in

```
int num_rows, num_columns, num_entries;
```

which defines the three variables `num_rows`, `num_columns` and `num_entries`. Valid names of variables start with a letter {a...z,A,...,z} and may contain letters, digits {0,...,9}, and the underscore '_' character.[1] It is not allowed to used reserved key words. Forbidden are particularly statement key words like `for` or `while` as well as predefined and self-defined data types.

Initially, the values of variables are undefined (some compilers automatically assign zero, but you cannot rely on this). To actually assign a number, the C statement should contain the variable name, followed by an equal character '=' followed by a numeric constant for example, and terminated by a semicolon like in

```
num_rows = 10;
num_colums = 30;
```

Note that the spaces around the '=' are not necessary, but allow for better readability. To the right of the '=' symbol, arbitrary arithmetic expressions are allowed, which may contain variables as well. If a variable

---

[1]One can start variables also with the underscore '_' character, but this is also done by library subroutines, hence you should avoid an explicit use of the underscore.

appears in an expression, during the execution of the program the variable will be replaced by its current value. A valid arithmetic expression is

```
num_entries = num_rows * num_colums;
```

More on arithmetic expressions, including the introduction of the most common operators, will be presented in Sec. 1.1.2. Note that one can write numeric constants also in octal (number with leading 0) or in hexadecimal (number with leading 0x), e.g. the above statements could read as

```
num_rows = 012;
num_colums = 0x1e;
```

The value of one or several variables can be printed using the `printf` command used above. For this purpose, the *format string* passed to `printf` must contain *conversion specifications* for the variables, where the actual value will be inserted, when the string is printed. The variables (or any other arithmetic expression) have to be passed as additional arguments after the string, separated by commas. The most common conversion specification to print integers is "%d". The values of the variable can be printed in a C program for example with

```
printf("colums = %d, rows = %d, total size = %d\n",
       num_rows, num_colums, num_entries);
```

The '\n' at the end of the format string indicates again a *new line*. Using the above statements, the C program will print in this case

```
colums = 10, rows = 30, total size = 300
```

One can also assign variables right at the place where they are defined. This is called *initialization* and can be done e.g. using

```
int num_rows = 30;
```

This is even mandatory when defining a variable as being a *constant*. This means its content is not allowed to change. Syntactically, this is indicated by the qualifier `const` written in front of the type name, e.g.

```
const int num_rows = 30;
```

Hence, if you try to assign a value to `num_rows` somewhere else in the program, the compiler will report an error.

Other fundamental data types, which have similar fundamental properties and usage like `int`, include

- `short int`

  If you are sure that you do not need large numbers for specific variables, you can economize memory by using this data type. It consumes only two bytes, hence numbers in the range $[-2^{15}, 2^{15}-1]$ are feasible, while for `unsigned short int`, numbers in the range $[0, 2^{16}-1]$ can be used. Variables of these types are handled exactly like `int` variables.

- `char`

  This data type is used to store characters using the ASCII coding systems in one byte. Characters which are assigned have to be given in single quotes. When printing characters using `printf()`, the conversion specification `%c` can be used. The following three lines define a variable of type `char`, assign a character and print it.

  ```
  char first_char;
  first_char = 'a';
  printf("The first character is %c.\n", first_char);
  ```

  Note that you can also use `char` variables to store integers in the range $[-128, 127]$. Therefore, when you print the variable `first_char` using the `%d` conversion specification, you will get the ASCII code number of the contained character, e.g. the number 97 for the letter 'a' and the number 10 for '\n'. The code numbers might be machine and/or operating system dependent.

  Variables of the type `unsigned char` can store numbers in the range $[0, 255]$.

- `double` and `float`

  These are used to store floating-point numbers. Usually `float` needs four bytes, allowing for numbers up/down to $\pm 3.4 \times 10^{38}$, while `double` uses eight bytes allowing for much larger numbers $(\pm 9 \times 10^{307})$.

  When assigning `float` or `double` numbers, one can e.g. use the fixed point notation like in `-124.38` or a "scientific" notation in the form `-1.2438e2` which means $-1.2438 \times 10^2$ (equivalently you could write `-12.438e1` or `-1243.8e-1`). Note that the numeric constant 4 is of type `int`, while the constant 4.0 is of type `double`. Nevertheless, you can assign 4 to a floating-point variable, because types are converted implicitly if necessary, see Sec. 1.1.2.

When printing `float` or `double` numbers usually the conversion specifications `%f` for fixed point notation or `%e` for scientific notation are used.

Note that standard C does not provide support for complex numbers. This is no problem, because there are many freely available *libraries* which include all necessary operations for complex computations, see Chap. 7.

- `long int` and `long double`
  On some computer systems, more memory is provided for these data types. In these cases, it makes sense to use these data types if one requires a larger range of allowed values or a higher precision. The use of variables of these data types is the same as for `int` or `double`, respectively.

- addresses, also called *pointers*
  Generic addresses in the main memory can be stored by variables defined for example in the following way

  ```
  void *address1, *address2;
  ```

  Note that each variable must be preceded by a '`*`' character. The type `void` stands for "anything" in this case, although sometimes it stands for "nothing" (see Sec. 1.2). If one wants to specify that a pointer stores (or "points to") the address of a variable of a particular type, then one has to write the name of the data type instead, e.g.

  ```
  int *address3, *address4;
  ```

  One important difference from a `void` pointer is that `address3+1` refers to the memory address of `address3` plus the number of bytes needed to store an `int` value, while `address1+1` is the memory address of `address1` plus one byte.

  The actual address of a variable is obtained using the `&` operator (see also next section), e.g. for an integer variable `sum` via

  ```
  address1 = &sum;
  ```

The `printf()` conversion specification for pointers is '%p', the result is shown in hexadecimal. For example

GET SOURCE CODE

DIR: c-programming
FILE(S): address.c

```
printf("address = %p\n", address1);
```

could result in

```
address = 0xbffa0604
```

The value stored at the memory location where **address1** points to is obtained by writing **\*address1**. Consequently, the lines

```
int sum;
int *address1;

address1 = &sum;
sum = 12;
printf("%d \n", *address1);
```

will result in printing 12 when running.

A slightly advanced use of pointers can be found in exercise (2).

Note that you can determine the actual number of bytes used by any data type, also the self-defined ones (see Sec. 1.1.4), via the `size_of()` function within a C program. The

GET SOURCE CODE

DIR: c-programming
FILE(S): sizeof.c

name of the data type has to be passed to the function as argument. For example, the following small program `sizeof.c` will print the number of bytes used for each type:

```
#include <stdio.h>

int main()
{
  printf("int uses %d bytes\n", sizeof(int));
  printf("short int uses %d bytes\n", sizeof(short int));
  printf("char uses %d bytes\n", sizeof(char));
  printf("long int uses %d bytes\n", sizeof(long int));
  printf("double uses %d bytes\n", sizeof(double));
  printf("float uses %d bytes\n", sizeof(float));
  printf("long double uses %d bytes\n", sizeof(long double));
  printf("pointers use %d bytes\n", sizeof(void *));
  return(0);
}
```

Note that the format string of the `printf` statement, i.e. the first argument, contains a *conversion character* `%d`, which means that when printing the format string, the number passed as second argument will be printed as integer number in place of the conversion character. More details on the `printf` command and its format strings are explained in Sec. 1.3. Thus, the program may produce the following output:

```
int uses 4 bytes
short int uses 2 bytes
char uses 1 bytes
long int uses 4 bytes
double uses 8 bytes
float uses 4 bytes
long double uses 12 bytes
pointers use 4 bytes
```

Hence, the usage of `long int` does not increase the range compared to `int`, but so does `long double`. The `sizeof()` command will be in particular important when using dynamic memory management using pointers, see Sec. 1.4.

The precision of all standard C data types is fixed. Nevertheless, there are freely available libraries which allow for arbitrary precision including corresponding operators, see Chap. 7. The operators for the standard arithmetic data types are discussed in the next section.

### 1.1.2 *Arithmetic expressions*

Arithmetic expressions in C basically follow the same rules of traditional arithmetics. The standard operations addition (operator `+`), subtraction (`-`), multiplications (operator `*`), and division (`/`), as well as the usual precedence rules are defined for all numeric data types. Furthermore, it is possible to introduce explicit precedence by using brackets. The following expressions give some examples:

```
total_charge = -elementary_charge*(num_electrons-num_protons);
weighted_avg_score = (1.0*scoreA+2.0*(scoreB1+scoreB2))/num_tests;
```

When executing the program, the values which result in evaluating the expressions to the right of the '`=`' character are assigned to the variables shown to the left of the '`=`' character. Note that an assignment is an expression itself, whose result is the value which is assigned. Therefore, the result of the assignment `x = 1.0` is 1.0 and can be assigned itself like in

```
y = x = 1.0;
```

Nevertheless, such constructions are not used often and should usually be avoided for clarity.

Note that for integer variables and constants, the division operation results just in the quotient, being an integer again, e.g. 5/3 results in a value 1. This is even the case if 5/3 is assigned to a variable of type double! The remainder of an integer division can be obtained using the modulus operator (%), e.g. 5%3 results in 2. Note that the modulus operator is not allowed for one or two double or float operands. Nevertheless, in general, if a binary operator combines objects of different types, the value having lower resolution will be converted internally to the higher resolution. Consequently, the type of an expression will be always of the highest resolution. On the other hand, when assigning the result of an expression to a variable of lower resolution (or when passing parameters to functions as in Sec. 1.2), the result will also be automatically converted. Therefore, when assigning a floating-point result to an integer variable, the value will be truncated of any fractional part. One can also perform explicit type conversions by using a *cast*. This means one writes the desired type in () brackets left of a constant, variable or expression, e.g. in

```
void *addressA;
int *addressB;

addressB = (int *) addressA+1;
```

addressB will point 4 bytes behind addressA, while without the cast it would point only one byte behind addressA.

As pointed out in the last section, *address refers to the content of the variable where address points to. Hence, one can change the content of this variable also by using *address on the left side of an assignment, like in

```
int sum;
int *address1;

sum = 12;
address1 = &sum;
*address1 = -1;
printf("%d \n", sum);
```

which will result in printing the value -1 when the program is executed.

The value of `address1` will not be changed, but will still be the address of variable `sum`.

It occurs very often that a variable is modified relative to its current value. Hence, the variable occurs on the left side of an assignment as well as on the right side within an arithmetic expression. For the simplest cases, where the variable is incremented, decremented, multiplied with, or divided by some value, a special short syntax exists (as well as for the modulus operator and for the bitwise operations presented below). For example,

```
counter = counter + 3;
```

can be replaced by

```
counter += 3;
```

Consequently, the general form is

$$\langle variable \rangle \ \langle operator \rangle = \langle expression \rangle$$

where $\langle operator \rangle$ can be any of

```
+  -  *  /  %  &  |  ^  <<  >>
```

The last five operators are for bitwise operations which are discussed below. On the left side, $\langle variable \rangle$ stands for an arbitrary variable to which the value is assigned, an element of an array or of a structure, see Sec. 1.1.4. The $\langle expression \rangle$ on the right side can be any valid arithmetic expression, e.g. it may contain several operators, nested formulas, and calls to functions. Standard mathematical functions are introduced at the end of this section.

There are special unary operators `++` and `--` which increase and decrease, respectively, a variable by one, e.g. `counter++`. It can be applied to any variable of numeric or pointer type.

| GET SOURCE CODE |
| --- |
| DIR: `c-programming` |
| FILE(S): `increment.c` |

In addition to the change of the variable, this represents an expression which evaluates to the value of the variable *before* the operator is applied or *after* the operator is applied, depending on whether the operator is written *behind* (postincrement) or *in front* (preincrement) of the variable. For example, compiling and running

```
int counter1, counter2, counter3, counter4;

counter1 = 3;
counter2 = counter1++;
```

```
counter3 = ++counter1;
counter4 = --counter1;
printf("%d %d %d %d\n", counter1, counter2, counter3, counter4);
```

will result in

4 3 5 4

Since the computer uses a binary representation of everything, in particular any natural number $n$ can be interpreted as sequence $a_k a_{k-1} \ldots a_1 a_0$ of 0s and 1s when writing the number in binary representation $n = \sum_i 2^i a_i$. For instance, the number 201 is written as binary sequence (highest order bits at the left) 11001001. For these sequences, there are a couple of operators which act directly on the bits. We start with operators involving two arguments.

& Calculates a bitwise AND of the two operands. For each pair $(a, b)$ of bits, the & operation is defined as shown in the table on the right: The result is only 1 if $a$ AND $b$ are 1. Hence, for the numbers 201 (binary 11001001) and 158 (binary 10011110) one will obtain from 201 & 158 the result 136 (binary 10001000).

| $a$ | $b$ | $a\&b$ |
|-----|-----|--------|
| 0 | 0 | 0 |
| 0 | 1 | 0 |
| 1 | 0 | 0 |
| 1 | 1 | 1 |

| Calculates a bitwise OR of the two operands, defined as shown in the table on the right: The result is 1 if $a$ OR $b$ are 1. Therefore, for the numbers 201 and 158 one will obtain the result 223 (binary 11011111).

| $a$ | $b$ | $a|b$ |
|-----|-----|-------|
| 0 | 0 | 0 |
| 0 | 1 | 1 |
| 1 | 0 | 1 |
| 1 | 1 | 1 |

^ Calculates a bitwise XOR of the two operands, defined as shown in the table on the right: The result is 1 if $a$ OR $b$ is 1 but not both. Hence, for the numbers 201 and 158 one will obtain the result 87 (binary 01010111).

| $a$ | $b$ | $a\hat{\ }b$ |
|-----|-----|--------------|
| 0 | 0 | 0 |
| 0 | 1 | 1 |
| 1 | 0 | 1 |
| 1 | 1 | 0 |

<< The expression **seq** << **n** evaluates to a shift of the sequence of bits stored in the variable **seq** (or obtained from an expression in general) by **n** positions to the left, which is equivalent to a multiplication with $2^n$. Thus, if **seq** contains 51 (binary 00110011), **seq** << 2 will result in 204 (binary 11001100). Note that if the number overflows, i.e. gets too large relative to the type of the

variable or expression, the higher order bits will be erased. Consequently, if seq is of type unsigned char, seq << 4 will result in 48 (binary 00110000).

>> Likewise, the expression seq >> n evaluates to a shift of the sequence of bits stored in the variable seq by n positions to the right, which is equivalent to an integer division by $2^n$.

Furthermore, ~ is a unary inversion operator, which just replaces all 1s by 0s and all 0s by 1s. Therefore, the result depends on the standard number of bits, because all leading 0s will be set to one. For example, variable seq containing the value 12 (binary 1100, when omitting leading 0s as usual), ~seq will result in (4 bytes = 32 bits for int) 4294967283 (binary 11111111111111111111111111110011).

For floating point arithmetic, there are many predefined functions like trigonometric functions, exponentials/logarithms, absolute value, Bessel functions, Gamma function, error function, or maximum function. The declarations of these functions are contained in the header file math.h. Thus, this file must be included, when using these functions. When compiling without -Wall, note that the compiler will not complain, if math.h is not included and this may

| GET SOURCE CODE |
| --- |
| DIR: c-programming |
| FILE(S): mathtest.c |

yield incorrect results. Furthermore, math.h contains definitions of useful constants like the value of $\pi$ or the Euler number $e$. The following simple program mathtest.c illustrates how to use mathematical functions.

```
1   #include <stdio.h>
2   #include <math.h>
3
4   int main()
5   {
6     double z= 0.25*M_PI;
7     printf("%f %f %f\n", sin(z), cos(z), tan(z));
8     printf("%f %f \n", asin(sqrt(2.0)/2.0)/M_PI, acos(0.0)/M_PI);
9     printf("%f %f %f %f\n", pow(M_E, 1.5),exp(1.5), log(1.0), log(M_E));
10    printf("%f %f %f\n", fabs(-3.4), floor(-3.4), floor(3.4));
11    printf("%f %f\n", fmax(1.3, 2.67), fmin(1.3, 2.67));
12    printf("%e %e %e %e\n", erf(0.0), erf(1.0), erf(2.0), erf(5.0));
13    printf("%f %f %f %f\n", tgamma(4.0), tgamma(4.5),
14                           tgamma(5.0), exp(gamma(5.0)) );
15    printf("%f %f %f\n", j0(2.0), j1(2.0), jn(1.0,2.0));
16    return(0);
17  }
```

When compiling this program, one needs the additional option -lm, which means that the library for mathematical functions will be linked. In this library, precompiled code for all of these functions is contained and those which are needed will be included in the final program.[2] You can compile the program using

```
cc -o mathtest -Wall mathtest.c -lm
```

In line 6 of the program, the predefined constant for $\pi$ (3.14159265358979323846) is used. Other important constants are the Euler number (M_E = 2.7182818284590452354) and $\sqrt{2}$ (M_SQRT2 = 1.41421356237309504880). Lines 7–8 illustrate the usage of trigonometric functions and their inverse functions. Note that the arguments have to be in radians, i.e. units of $2\pi$. Line 9 shows how to use the general power $a^x$ pow() and the exponential $e^x$ exp() functions. In line 10, the function fabs() for calculating the absolute value of a number, and floor() for rounding downwards are shown. Line 11 introduces the usage of the functions fmax() and fmin() for calculating the maximum and the minimum of two numbers, respectively. In line 12, four values of the error function $\mathrm{erf}(x) = 2/\sqrt{\pi} \int_0^x e^{-y^2}\, dy$ are obtained. In line 13, sample usages of the gamma function are shown. Note that tgamma() is the gamma function $\Gamma(x)$ while gamma() represents $\ln \Gamma(x)$. Remember that for integer values $n$ one has $\Gamma(n) = (n-1)! = \prod_{k=1}^{n-1} k$. Finally, in line 15, the Bessel function of the first kind is shown. j($n,x$) calculates the Bessel function of order $n$, while $j0$ and $j1$ are shortcuts for jn(0, . ) and jn(1, . ), respectively. Hence, when running the program, you will get

```
0.707107 0.707107 1.000000
0.250000 0.500000
4.481689 4.481689 0.000000 1.000000
3.400000 -4.000000 3.000000
2.670000 1.300000
0.000000e+00 8.427008e-01 9.953223e-01 1.000000e-00
6.000000 11.631728 24.000000 24.000000
0.223891 0.576725 0.576725
```

These functions are documented on the corresponding man pages, e.g. man sin. More complicated functions, like Airy functions, elliptic func-

---

[2]Note that *dynamical* linking is usually used, i.e. the corresponding code is only linked when running the program. This keeps the executables short. Nevertheless, explicit linking at compile time can be forced using the option -static.

tions, or spherical harmonics can be found in special libraries, as explained in Chap. 7.

Finally, there is a special operator ?, which is related to the if statement and is explained in the next section.

### 1.1.3    Control Statements

Using only the statements presented so far, a program would just consist of a linear sequence of operations, which does not allow for complex problems to be solved. In this section, statements are introduced which enable the execution flow of a program to be controlled.

The simplest statement is the if statement, which allows to select between two statements depending on the evaluation of a conditional expression. A simple example is

```
if(x <= 0)
  step_function = 0.0;
else
  step_function = 1.0;
```

where the variable step_function will be assigned the value 0.0 if x is smaller than or equal to zero, and step_function will get 1.0 if x is larger than zero.

The general form of an if statement is as follows:

```
if( ⟨condition⟩ )
    ⟨statement 1⟩
else
    ⟨statement 2⟩
```

During the execution, first the ⟨condition⟩ is checked. If the condition is *true*, then ⟨statement 1⟩ is executed. If the condition is *false*, then ⟨statement 2⟩ is executed. These statements can be arbitrary, e.g. assignments as in the above example, calls to functions, another nested if statement, etc. Also, the type of the statements can be different, e.g. ⟨statement 1⟩ may be a call to a function and ⟨statement 2⟩ may be an assignment. Note that the else part can be absent. In this case, if the condition is *false*, the execution continues with the next statement after the if statement.

Basic conditional expressions can be formed by comparing (arithmetic) expressions using relational operators. In C we have

```
==  !=  <  <=  >  >=
```

Consequently, an expression of the form ⟨*expr1*⟩ ⟨*OP*⟩ ⟨*expr2*⟩, will be evaluated for ⟨*OP*⟩ being == to *true* if both expressions are equal, while for the operator != represents the test for non-equality. For the case ⟨*OP*⟩ being <=, *true* results if ⟨*expr1*⟩ is smaller than or equal to ⟨*expr2*⟩, etc. Note that the relational operators have lower precedence than the standard arithmetic operators, meaning they will be evaluated last. This means that `counter < limit + 1` is equivalent to `counter < (limit + 1)`.

C uses the integer 0 to represent a logical *false* and *all* other values to represent *true*. Therefore, for the comparison `counter=!n_max` one could write instead `counter-n_max`. This results in exactly the same behavior, since this expression becomes zero if `counter` is equal to `n_max`. In fact, C does not formally distinguish between arbitrary expressions and conditional expressions. Thus, wherever a condition is expected, any expression can appear in a C program. Nevertheless, to make a program more readable, which usually leads to less programming bugs, one should use conditional operators, if conditions are possible and meaningful.

A common mistake, as mentioned on page 4, is to use = instead of == when intending a comparison. Nevertheless, a statement like `if(a = b+1)` is valid C syntax, because an assignment is also an expression itself, where the result is the value which is assigned. Hence, in this case the result of the expression will be `b+1` which in C represents *false* if `b` equals -1 and *true* for all other values of `b`. Usually, an assignment inside an `if` statement is not intended by the program author. Hence, you should use the compiler option `-Wall`, which makes the compiler print out a warning, when it encounters such a case.

Compound conditional expressions can be formed by grouping them using braces and combining them using the logical OR operator ||, the logical AND operator &&, and the logical NOT operator !, e.g.

```
(counter>=10) && (counter<=50) && !(counter % 2==0)
```

which evaluates to *true* if the value of `counter` is inside the interval [10, 50] and not an even number, i.e. odd.[3] The NOT operator has higher precedence than the AND/OR operators. Note that a sequence of || or && operators is evaluated from left to right. This evaluation stops when the result of the full expression is determined, e.g. after the

---

[3]Instead of `!(counter % 2==0)` one could also write `(counter % 2!=0)` or `(counter % 2==1)` or right away `(counter % 2)`.

first occurrence of *true* for a sequence of || operators. This can be dangerous when using, inside a compound conditional expression, operators which change variables, like the post-increment operator ++, as used in (num_particles > 100)&&(num_run++ > 100). In this case, it depends on the result of the first condition whether the post-increment operator in the second condition is performed or not. Consequently, in most cases such constructions should be avoided for clarity, even if they work as intended.

Recall the first example of this section, where a value is assigned to a single variable, which is different within the different branches of the statement. For this special usage of the if statement, there exists the *conditional operator* ? as shortcut. For the above example the resulting *conditional expression* statement reads

```
step_function = (x<=0) ? 0.0 : 1.0;
```

The general format of the conditional expression is

$$\langle condition \rangle \ ? \ \langle expression \ 1 \rangle : \langle expression \ 2 \rangle$$

Hence, the ? operator is ternary. The result of the conditional expression is ⟨*expression 1*⟩ if the ⟨*condition*⟩ evaluates to true, else the result is ⟨*expression 2*⟩. The conditional expression can be used where a standard expression can be used.

According to the general form of the if statement shown above, it might appear that only one statement can be put into each of the two branches. This restriction can be overcome by grouping any number of statements into a *block*, which is created by embracing the statements by { ... } braces, e.g.

```
1   if(step % delta_measurement == 0)
2   {
3     num_measurements++;
4     sum += energy;
5     sum_squared += energy*energy;
6     sum_cubed += energy*energy*energy;
7     sum_quad += energy*energy*energy*energy;
8   }
```

This piece of code is from a simulation program, where the (somehow calculated) energy is recorded, if the number of steps is a multiple of delta_measurement. To obtain estimations for the average energy and a confidence interval ("error bar") as well as higher moments, one calculates running sums of the energy (line 4), the energy squared (line 5), the third (line 6) and the fourth power (line 7). The average energy will be sum_energy/num_measurements. For details on the calculation of simple statistical properties, see Sec. 8.3.

Note that it is also possible to define variables inside a block. For traditional C, the variable definition must appear at the beginning of the block, as it must appear at the beginning of main(), and in fact at the beginning of any function.[4] Using a block variable squared_energy, the above example would read

```
1   if(step % delta_measurement == 0)
2   {
3     double squared_energy = energy*energy;
4
5     num_measurements++;
6     sum += energy;
7     sum_squared += squared_energy;
8     sum_cubed += squared_energy*energy;
9     sum_quad += squared_energy*squared_energy;
10  }
```

In this way, the number of multiplications is reduced from five to three. Note that such block variables are only visible and accessible within the block where they are defined. The scope of variables will be explained in more detail in Sec. 1.2.

Note that in case of nesting if...else statements, sometimes one needs blocks to avoid ambiguities, e.g.

```
if(x==1)
   if(y==1)
     printf("case 11\n");
   else
     printf("case 1X\n");
```

is equivalent to

---

[4]This is different from C++, the object-oriented extension of C, and according the modern C standard, where variables can be defined everywhere. Nevertheless, the clarity of a program is increased, if all definitions of a block are collected in one place.

```
if(x==1)
{
  if(y==1)
    printf("case 11\n");
  else
    printf("case 1X\n");
}
```

which is different from

```
if(x==1)
{
  if(y==1)
    printf("case 11\n");
}
else
  printf("case 1X\n");
```

Therefore, one needs the brackets, if the `else` part should explicitly belong to the first `if` statement. In any case, if several `if` statements are nested, brackets always help to clarify their meaning.

Next, we consider the `for` loop, which enables one statement or a block of statements to be executed several times. The following piece of code calculates the sum of the squares of numbers from 1 to `n_max`, with `n_max=100` here:

```
1    int counter, sum, n_max;
2
3    sum = 0; n_max = 100;
4    for(counter = 1; counter <= n_max; counter++)
5      sum += counter*counter;
```

The `for` command (line 4) controls how often the addition in line 4 is executed, with `counter` getting the values $1, 2, 3, \ldots, 99, 100$, one after the other. Consequently, `sum` will contain the value 338350, when the loop is finished. The general form of a `for` loop is as follows:

> `for(` ⟨*init expression*⟩; ⟨*condition*⟩; ⟨*increment expression*⟩ `)`
> ⟨*body statement*⟩

During the execution, first the ⟨*init expression*⟩ is evaluated. Next, the ⟨*condition*⟩ is checked. If the condition is *true*, then the ⟨*body statement*⟩ is executed, next the ⟨*increment expression*⟩ evaluated. This completes the execution of one iteration of the loop. Then, once again the condition is tested to check whether it is still *true*. If it is, the loop is continued in the same way. This is continued until the ⟨*condition*⟩ becomes *false*. Then the execution of the `for` loop is finished. Note that it is not guaranteed that a loop finishes at some time, e.g. the following loop will run forever:

```
for(counter = 1; counter > 0; counter++)
  sum += counter*counter;
```

The ⟨*init expression*⟩, the ⟨*condition*⟩, and the ⟨*increment expression*⟩ are usually arithmetic expressions or assignments. Hence, instead of `counter++` one could write `counter = counter +1`. Due to its flexibility, C allows for even more general forms, e.g. one could use `printf("%d\n", counter++)` as ⟨*increment expression*⟩, which would result in printing the value of `counter` after each iteration of the loop. As you see, C allows for very compact code. Nevertheless, this often makes the program listing somehow hard to read, hence error-prone. Thus, it is preferable to write programs which might be a bit longer but whose meaning is obvious. Therefore, putting the `printf` statement also in the body of the loop in this case would lead to simpler-to-understand code. In this case one also has to use {...} braces to group a list of statements into one block.

In principle, one can use quite general constructions to perform a loop, so one could use also variables of type `double` as iteration counters like in

```
double integral, delta, x;

integral = 0.0; delta = 0.01;
for(x=0.0; x <=M_PI; x += delta)
  integral += delta*sin(x);
```

There are another two types of loops, the `while` loop and the `do ...while` loop. The former one has the following general form:

```
while( ⟨condition⟩ )
    ⟨statement⟩
```

The above ⟨*statement*⟩ is executed as long as the condition is true. Consequently, if one likes to perform a loop similar to a `for` loop, one has to put the initialization before the while loop and one has to include the increment in the ⟨*statement*⟩, usually a block of several statements, like in the following example:

```
1    int counter, sum, n_max;
2
3    sum = 0; n_max = 100;
4    counter = 1;
5    while(counter <= n_max)
6    {
7      sum += counter*counter;
8      counter++;
9    }
```

Very similar is the do ... while loop, which has the following general form

```
do
    ⟨statement⟩
while( ⟨condition⟩ )
```

The difference from the while loop is that the ⟨*statement*⟩ (again it may be a {...} block of statements) is executed at least once. Only after each iteration of the loop, the ⟨*condition*⟩ is tested. The iteration of the loop continues as long as the ⟨*condition*⟩ is *true*.

There are three different statements, which allow to exit at any time the execution of a block belonging to a loop. First, the **break** statement will immediately stop the execution of the loop without completing the block. Hence, the execution of the program continues with the first statement after the loop statement. Second, the **continue** statement will also terminate the current iteration of the loop, but the full loop is continued as normal, i.e., with the ⟨*increment statement*⟩ in case of a **loop** or with the next evaluation of the ⟨*condition*⟩ in case of a **while** or a **do...while** loop. Thus, if the ⟨*condition*⟩ allows, the loop will be continued with the next iteration. If several loops are nested, the **break** and **continue** statements act only on the innermost loop. Third, if a **return** statement is encountered, not only the current loop will be terminated, but the whole function where the loop is contained in will be exited.

Note that there is a fourth possibility. The C language contains a **goto** statement which allows to jump everywhere in the program. Possible jump targets can be identified by *labels*. Since the use of **goto** statements makes a program hard to understand ("spaghetti code"), one should not use it. One can always get along well without **goto** and no details are presented here.

Finally, the **switch** statement should be mentioned. It can be used when the program should branch into one of a set of several statements or blocks, and where the branch selected depends only on the value of one given expression. As an example, one could have a simulation with several different atom types, and depending on the atom type different energy functions are used. In principle, one could use a set of nested if...else statements, but in some cases the code is clearer using a switch. Since the switch statement is not absolutely necessary, we do not go into details here and refer the reader to the literature.

### 1.1.4 *Complex data types*

So far, only single variables have been introduced to store information. Here, complex data structures will be explained which enable the programmer to implement vectors, matrices, higher order tensors, strings, and data structures where elements of possibly different data types are grouped together to form one joint data type.

To define e.g. a vector intensity, which may contain the results of 100 measurements during a simulation, one can write

```
double intensity[100];
```

In C, such a vector is called an *array*. By this definition, the variable intensity is defined, i.e., also enough memory will be *allocated* during execution of the program to hold the 100 *elements* of the array. These 100 elements are stored consecutively in the memory, hence a chunk of 100*sizeof(double) bytes will be used to store the array. Note that the memory, which is allocated, may contain *any* content, i.e. it must be regarded as undefined. Although some compilers initialize all allocated memory with 0s, your program should perform always an explicit initialization. The allocated memory is available until the execution of the *block*, where the array is defined, is finished. This holds for all variables defined inside a block, including variables holding just a single element. More details on this so-called *scope* of variables can be found in Sec. 1.2.

One can access the i'th element of the array by writing intensity[i], i.e. the *index* (also called subscript) is written in [] brackets after the name of the vector. In C, array indices start with 0, hence here the array elements intensity[0], intensity[1], ..., intensity[99] are available. In the following example the average of the values in the array is calculated . Note that some part of the code is not shown, as indicated by a *comment* in the program (line 4). In C, comments are indicated by embracing the comment text in '/*' and '*/'.

```
1    double avg_intensity;
2    int i;
3
4    /* ... "some" more code to generate data ... */
5
6    avg_intensity = 0.0;
7    for(i=0; i<100; i++)
8      avg_intensity += intensity[i];
9    avg_intensity /= 100;
```

The index can be any expression which results in an integer, so one could also access every third measurement only, if needed, starting at index 1, by writing `intensity[3*i+1]`.

Technically, a one-dimensional array is equal to a pointer which points to the beginning of the memory chunk: `intensity` contains the address of the beginning of the chunk, while `intensity[0]` contains the content stored in the first element of chunk, which is equivalent to `*intensity`. The second element can be accessed via `intensity[1]` or via `*(intensity+1)`. More details about this equivalence will be given in Sec. 1.4.

Note that during the execution of the program it is *not* checked whether one accesses (reads or writes) elements *outside* the reserved range. Consequently, one could easily write `intensity[100] = 0.9;`, which is in fact a mistake that will *not* be detected by the compiler. Often, this will not do any harm during the execution of the program, because it means that a memory location will be accessed which is located just ahead the chunk that has been reserved for the `intensity` array. The basic reason is that memory is not allocated consecutively. Thus, very often the part just ahead this chunk will not be used to store other variables and consequently the program will not be affected. But *if* some other data is stored just ahead the chunk, e.g. when almost all the main memory is used, some other data will be overwritten by assigning a value to `intensity[100]`. In this case, the program *might* crash, or it might just give strange results. Therefore, different runs of the same program with the same input parameters might sometimes crash, sometimes give wrong results, and sometimes even do fine. Since such kind of bugs are obviously hard to detect, there are special tools to find them, so-called *memory checkers*. For details on these *very* useful tools, see Sec. 4.3.

The general form of an array definition is as follows:

⟨*type*⟩ ⟨*variable name*⟩ [⟨*size*⟩];

The ⟨*type*⟩ can be any predefined as well as self-defined data types, see below. For example, an array with 10 elements of the type `short int` is defined as follows:

```
short int flags[10];
```

It is also possible to mix the definition of single variables and arrays (and other variables based on the same basic type) in one line, e.g.

```
double avg_intensity, variance, intensity[100];
```

For old-fashioned standard C (before the standard ISO C99), the size of an array has to be known by compile time. Therefore, for compilers supporting only this type of C, it is not possible to write `int vector[size]` where `size` is a variable. In this case, if one wants to use arrays, where the size is not known at compile time, i.e. *variable-sized arrays*, one can always use the techniques of dynamic memory management as described in Sec. 1.4. This is the safest approach and also the most flexible. On the other hand, up-to-date compilers, like the *gnu C compiler* `gcc` [Loukides and Oram (1996)], support the definition of variable-sized arrays. Nevertheless, in case you use them, you might face a problem in rare cases, when you want to port your program to another system with old compilers. Furthermore, it is not possible to resize the array after it has been created, in contrast to arrays created using dynamic memory management.

As mentioned above, you cannot rely on arrays being initialized automatically upon creation. On the other hand, it is possible, like for single-valued variables, to combine the definition with an initialization, e.g.

```
int atom_weight[6] = {1,6,7,8,15,16};
```

Note that the length of the arrays can be larger than the number of elements given for initialization. If no array size is given, the size of the array will be automatically equal to the number of elements provided, like in

```
int atom_weight[] = {1,6,7,8,15,16};
```

where 6 elements will be allocated for `atom_weight`.

A very special type of array is one which has the basic type `char`. It is used to store *strings* and can be defined e.g. via

```
char atom_name[100];
```

One can assign values like for other types of arrays. Note that the last element of a string should contain a terminating 0. Hence, one could write:

```
atom_name[0] = 'c';
atom_name[1] = 'a';
atom_name[2] = 'r';
atom_name[3] = 'b';
atom_name[4] = 'o';
atom_name[5] = 'n';
atom_name[6] = 0;
```

This is quite space consuming. A simpler form exists, in case one performs an assignment together with the definition, i.e. an initialization:

```
char atom_name[100] = {"carbon"};
```

Consequently, it is not necessary to assign one element after the other. To indicate string constants, double quotes are used, in contrast to single quotes for single characters. The terminating 0 is not given explicitly.

To print a string using `printf()`, the conversion specification `%s` can be used. In this case, one only needs to pass the name of the array, not all variables one after the other. The printing system will automatically print the string character after character until the terminating 0 is reached, hence

```
printf("%s\n", atom_name);
```

will result in

```
carbon
```

being printed on the screen.

There are many auxiliary functions which are very useful for string processing. The declaration of these functions is contained in `string.h`, which must be included at the beginning of your program, if you use any of these functions. Here, only few examples are given: `strcpy` ("string copy") will copy the content of the string passed as second parameter, including the terminating 0, to the string passed as first parameter, like in

```
char atom_name1[100] = {"carbon"}, atom_name2[100];

strcpy(atom_name2, atom_name1);
```

Therefore, the content of `atom_name2` will be also `"carbon"`. Note that there is no check for the length of strings. Hence, if the second string until the terminating 0 is longer than the length of chunk reserved for the first string, one will write beyond the boundaries of the arrays. This may result in unpredicted behavior, as discussed above, and should be avoided (see Sec. 4.3). If you are not sure about the lengths of the strings, you could use `strncpy`, which has an additional third argument that states the maximum number of characters to be copied.

Note also that the above assignment *cannot* be performed by writing `atom_name2 = atom_name1`, which might appear natural. In fact, writing this in a program will result in a compiling error.[5]

Strings can be compared using `strcmp()`. A call `strcmp(s1,s2)`, will return 0 if the strings are equal, a negative value (usually -1) if `s1` is lexicographically smaller than `s2` and a positive value if `s1` is lexicographically larger than `s2`.

Another useful function is `sprintf()`, which allows a programmer to assemble strings easily. It works similar to `printf()`. As a difference, the result is not printed to the screen, but to the string which is passed as additional first argument. This is in particular useful when assembling file names, usually to write out simulation results, where the file names somehow contain some program parameters to distinguish different runs. This could look like

```
sprintf(file_name1, "sim_N%d_T%3.2f_id%d.out",
        num_particles, temperature, run_id);
```

Note that `%3.2f` means that a floating point number with at least 3 digits, 2 among them after the comma, is printed. More details on the use of `printf()` and related commands, including these conversion specifications, can be found in Sec. 1.3. Hence, if `num_particles = 10`, `temperature = 1.3` and `run_id = 15`, the resulting file name will be

```
sim_N10_T1.30_id15.out
```

---

[5]This is different when explicitly using pointers instead of arrays, i.e. `char *atom_name1, *atom_name2`. In this case one must explicitly allocate memory, see Sec. 1.4. Then one *could* in principle perform the given assignment, but no string will be copied, just both pointers will point to the same string. Afterward, a change of the content of `atom_name1` would also change the content of `atom_name2` since both are the same. Also, if some memory had been allocated for `atom_name2`, then it will be "lost" after this assignment, in case no other pointer refers to it. This would be a real bug.

This file name can be used to create a file where one can write some data to, see again Sec. 1.3 on file creation and input/output.

Also, strings can be converted back to numbers. The "inverse" function to `sprintf()` is `sscanf()` and will be described in Sec. 1.3. For the special case a string `s` contains an integer number, i.e. the digits of the corresponding number, one can obtain the number as `int` value via the `atoi()` function:

```
number = atoi(s);
```

For more details on string functions, please refer to the documentation, e.g. `man string` will give an overview over many useful string functions.

Matrices or tensors can be implemented using multi-dimensional arrays. E.g a matrix with elements of type `double` can be defined via

```
double mat1[10][20];
```

The `i,j`'th element of the matrix `mat1` is accessed using `mat1[i][j]`, again arbitrary `int` expressions can be used instead of `i` and `j`. Note that again no automatic check for array boundaries is performed. Assuming that `vec1` and `vec2` are `double` arrays of lengths 20 and 10, respectively, the following piece of code multiplies the matrix `mat1` with the vector `vec1` and stores the result in the vector `vec2`, which is finally printed:

```
1    for(i=0; i<10; i++)
2    {
3      vec2[i] = 0.0;
4      for(j=0; j<20; j++)
5        vec2[i] += mat1[i][j]*vec1[j];
6    }
7    for(i=0; i<10; i++)
8      printf("%f\n", vec2[i]);
```

Technically, the compiler treats multi-dimensional arrays like a big one-dimensional array. Let's see how this works for the matrix: First, a big chunk of memory is allocated (again often uninitialized). Next, the array is stored in this chunck consecutively row by row.

Consequently, in this case, the matrix will be stored in a chunk of $10 \times 20$ `double` elements each. Therefore, `mat1` will contain the address where chunk starts, which is the same as `mat1[0]`. `mat[1]` contains the address where the part for the second row starts, while `mat[0][0]` is the content of the first element of the first row, and so on. The size of the allocated memory can be evaluated afterward at any position in the program using again the `sizeof()` function. For example, `sizeof(mat1)` will result in the total number of elements $20 \times 10 \times$ `sizeof(double)`, i.e. in 1,600 if `double` uses 8 bytes. The amount of memory reserved for one row can be obtained via `sizeof(mat1[0])`.

Even higher-dimensional arrays can be defined like

```
double tensor1[10][20][10];
```

and accessed e.g. via `tensor1[1][2][3] = 5.0`. Further details are not necessary here. In exercise (3), permutation matrices are considered.

For arrays and matrices, several elements of the same type are put together. In many other cases, one would like to treat a set of elements of different types as single units. Consider as

| GET SOURCE CODE |
| --- |
| DIR: c-programming FILE(S): person.c |

toy example a simulation of a social system of people. Here, one wants to store, for example, the age, height, sex, marital status and city of residence for each person. In C, one can group these elements together using a *structure*. The elements are then called *members* of the structure. An example declaration of a structure for persons, including extensive comments, reads as follows:

```
struct person
{
    int         age;    /* age of person (years)              */
    double      height; /* height of person in meters         */
    short int   sex;    /* 0=male, 1=female                   */
    short int   status; /* 0=single,1=married,2=divorced,3=widowed */
    int         city;   /* cities are coded as integers 1,... */
};
```

It is possible for the members to have also more complex data types like arrays, other structures, or self-defined data types (see below).

To define variables, e.g. `p1`, `p2`, which are of type `struct person`, one writes

```
struct person p1, p2;
```

Alternatively, one could also list the variables to be defined directly after the closing bracket '}' of the structure declaration itself (and just before the final semicolon, which closes the declaration).

The members of a structure variable can be accessed by writing the variable name, a dot '.' and the name of the member. For instance, one could assign values to all members of p1 using

```
p1.age = 43;
p1.height = 1.73;
p1.sex = 0;
p1.status = 0;
p1.city = 23;
```

It is also possible to assign complete structures within one statement like in

```
p2 = p1;
```

Thus, p2.age will have value 43, p2.height will be 1.73, etc.

Now, struct person can be used as any existing type name. In this way, it is possible to define an array which contains all persons of the simulation:

```
int num_persons = 100;
struct person pers[num_persons];
```

The access to the different members of the elements of the pers array is obtained by combining the access methods for arrays with the access method for structures. For example, to initialize the marital status of all persons to 0 one could use:

```
for(p=0; p<num_persons; p++)
  pers[p].status = 0;
```

Please note that pers.status[p]=0 is *not* correct in *this* case. It would be correct, if pers contained as member an array status.

Sometimes, it is useful to declare new names for data types, in particular if they are of complicated form like pointers to functions (see Sec. 1.4). The typedef construct serves this purpose. The format is the same as that of a single variable definition, but preceded by the keyword typedef, i.e.:

```
typedef ⟨type⟩ ⟨type name⟩;
```

Consequently, the name given is now for a "new" type, not for a variable. For example, to define the new type name person_t for the person structure, one could write

```
typedef struct person person_t;
```

```
person_t p1, p2;
```

It is also possible to combine the structure declaration and the definition of the new name. In this case, one does not need to specify a name after the key word `struct`:

```
typedef struct
{
    int         age; /* age of person (years)                   */
    double      height; /* height of person in meters           */
    short int   sex; /* 0=male, 1=female                        */
    short int   status; /* 0=single,1=married,2=divorced,3=widowed */
    int         city; /* cities are coded as integers 1,...     */
} person_t;
```

The author recommends to use the suffix `_t` (or similar rules) for the names of self-defined types to make it obvious everywhere that a new type name is used.

It is even possible just to define via `typedef` new names for existing data types like `double`. This could be useful e.g. in a situation where one knows in advance that one needs a sophisticated data structure later on for some variables, like special types for extremely high-precision numbers. In this case, one might want to start the implementation for development purposes with standard `double` and only switch to the high-precision data type later on. Hence, one could use `typedef double precision_t` so that one would only have to change this `typedef` and not all occurrences of definitions of variables of the high-precision type later on.

C offers other data types, e.g. *enumerations* (where each variable can have only specific values), *bit fields* (for storing data bit wise) and *unions* (for allowing variables to be at different types), which are not covered here. These data types are useful for some applications but not absolutely necessary. Please refer to the literature for details.

Based on these complex data types, even more advanced data structures can be developed. They are often very useful for organizing simulation programs in a much better way, speeding up simulations considerably. Advanced data structures like lists, trees and graphs are discussed in Secs. 6.6 to 6.8.

## 1.2 Functions

Usually, there are many complex tasks, which have to be performed at different places in a simulation, e.g. evaluating mathematical functions or calculating statistical properties of a

| GET SOURCE CODE |
| --- |
| DIR: c-programming |
| FILE(S): step_fct.c |

given set of numbers. One could explicitly include code for the calculation each time, for example for the step function as shown on page 18. But this is a waste of programming effort and makes the code not well structured. It is better to move this task to a *subroutine*, which in C is always a *function*. A function for calculating the step function could look like this:

```
1   /************ step_function() ******************/
2   /** Mathematical step function:                **/
3   /** Returns 0 if argument is less or equal to 0 **/
4   /**         and 1 else                          **/
5   /**                                             **/
6   /** Parameters: (*) = return parameter          **/
7   /**           x: mathematical argument          **/
8   /*                                              **/
9   /** Returns:                                    **/
10  /**        step function value                  **/
11  /***********************************************/
12  double step_function(double x)
13  {
14    if(x <= 0.0)
15      return(0.0);
16    else
17      return(1.0);
18  }
```

As you see, the code of each function should be accompanied by a *function comment* (lines 1–11) which explains what the function does, what arguments are to be passed and what is returned. Although this is not necessary for the program to compile correctly, you should get used to writing function comments, and other comments, right from the beginning. *Whenever* I code a function, I start with the function comment! Medium or large simulation projects, which have more than few pages of source code, in particular if their lifetime is longer than some weeks, cannot be handled efficiently without a good practice of commenting. More about this and other aspects of *software engineering* can be found in Secs. 3.1 and 3.2. Note that all example source codes, which are available online for this book, carry extensive comments. Therefore, all functions will exhibit such a function comment. Nevertheless, to save space, the functions presented in the book will usually *not* show their respective comments.

In line 12, first the *return type* is declared, i.e. it will return a **double** value here. Next, the function name **step_function** is given. Next, enclosed by (...) brackets, the arguments to the function are listed. Here, one variable of type **double** is the argument, which can be addressed by the name x inside the function. Note that several arguments are possible. In this case one has to separate them by commas (see below). The main part of a function (here lines 13–18) is enclosed by {...} brackets. The main part consists of definitions and statements, as in the **main()** part of a full program. In fact, as you may have realized by now, the **main()** part of the program follows the same syntax rules like any other function, with the only difference that it always has to have the special name **main()**. Finally, the value returned by the function is determined via the **return()** statement. There can be several **return()** statements inside the main part of the function, but only one will be executed each time the function is called.

A function is called in other parts of the code by writing the function name and, in brackets, the list of arguments, quite the same as for predefined functions like **sin()** or **printf()**. The value returned by a function can be used as any result of an expression, e.g. printed or assigned to a value of suitable type. The main program for our example, which may contain e.g. a loop, printing out a few function values in the interval $[-1, 1]$, could look like

```
1   int main()
2   {
3     double x;
4
5     for(x=-1.0; x<=1.0; x+=0.1)      /* print step function in [-1,1] */
6       printf("Theta(%f)=%f\n", x, step_function(x));
7
8     return(0);
9   }
```

When running the program, the output looks like (only a part is shown here, as indicated by the dots)

```
Theta(-0.100000)=0.000000
Theta(-0.000000)=0.000000
Theta(0.100000)=1.000000
Theta(0.200000)=1.000000
```

To be able to use the function `step_function()` inside `main()`, the compiler has to "be aware" of it when it compiles the code for `main()`. There are four ways of achieving this:

(1) The full function definition of `step_function()` appears *before* `main()` in the source file. Then, when the compiler reaches the source code of `main()`, it will have processed already the part where `step_function()` is defined, hence the function is known to the compiler.

In this way, `step_function()` is a *global* object, hence it can also be called from other functions, not only `main()`.

(2) The full function definition comes *after* `main()` in the source file. To make the compiler aware of the existence of `step_function()` when it processes `main()`, one has to perform a declaration in advance, which is called *function prototype*. This means that in the source code before the definition of `main()`, the following line has to appear:

```
double step_function(double);
```

Note the closing semicolon at the end of the line. The function prototype carries all information which has to be known outside the function: the function name, the types and order of the arguments, and the type of the return argument.

All other functions which appear in the source code behind the function prototype of `step_function()` can use calls to `step_function()` in their code. Hence, if *all* function prototypes are listed at the beginning, each function is allowed to call each other function.

(3) Very often, the source code is distributed over many source files. This usually happens if your simulation program is large and can be decomposed into several independent modules. There might be, for example, one module for setting up the simulation, another for auxiliary functions, some for the main simulation functions, and so forth. In this case, the second possibility has to be used, too. This means that for each

source file where some function `func()` is to be used, a corresponding function prototype must be known, when it is compiled. Usually, the function prototype is not written in the file explicitly, but it is contained in a *header* file, just as the function prototypes for standard functions like `sin()` and `printf()`. This header file must be included using the `#include` directive, see Sec. 1.6. Note that for including your own header files, you have to use the form `#include "header.h"` instead of `#include <header.h>`, which is only used for predefined (system) header files.

In this case, where the function is not contained in a header file, the function prototype should be preceded by the key word `external`, which indicates that the function definition occurs outside the present file, e.g.

```
extern double step_function(double);
```

Now, the function `step_function()` can always be used after the corresponding header file has been made known by using `#include`. Usually, one puts all `#include` directives at the beginning of a source file, such that all definitions are known everywhere inside the source file.

(4) All first three possibilities allow `step_function()` to be used arbitrarily in other functions, provided that the functions are informed about `step_function()` via the definition itself or via a prototype. The reason is that `step_function()` is a *global* object.

Nevertheless, it is possible to put the definition of `step_function()` just inside `main()`, e.g. like in:

```
1    int main()
2    {
3      double x;
4      double step_function(double x)
5      {
6        if(x <= 0.0)
7          return(0.0);
8        else
9          return(1.0);
10     }
11
12     for(x=-1.0; x<=1.0; x+=0.1) /* print step function in [-1,1] */
13       printf("Theta(%f)=%f\n", x, step_function(x));
14     return(0);
15   }
```

In this case, `step_function()` can be only used *inside* `main()`. Now, `step_function()` is a *local* object. If you have another function in the source code which calls `step_function()`, the compiler will complain!

Thus, cases 1-3 provide *global* definitions of `step_function()`, while case 4 is a *local* definition. This distinction can also be made for other definitions like for variable definitions or type declarations. Nevertheless, type declarations are usually global. Function and variable definitions can be used quite often in the same context. Technically, in these two cases a name is connected to an address, either the address where the variable content is stored, or where the machine code for a function can be found.

The general form (according to ANSI standard) of a function definition is as follows:

⟨*return type*⟩ ⟨*function name*⟩ ( ⟨*arguments*⟩ )
{
    ⟨*main part of function*⟩
}

For the return type and the comma-separated list of arguments, variables of arbitrary type can be used. Later we will see examples where even functions are passed as arguments to functions, e.g. see Sec. 7.1. Note that the list of arguments can be empty, i.e. no arguments are passed. The ⟨*main part of function*⟩ can be any type of valid code, including definitions of variables, types and functions, and all kinds of statements. The variables defined here are local, as mentioned above. For the next example, we consider a function which takes an array of `double` numbers, and should return the minimum and maximum numbers. The function should work for arrays of arbitrary number of elements, hence one has to pass the array, in fact a pointer to the beginning of the array, and the number of elements. Since each function can return only one object,

| GET SOURCE CODE |
| --- |
| DIR: `c-programming`<br>FILE(S): `min_max.c` |

one can declare a structure with two elements, which should take the result, as in the following example:

```
1  typedef struct
2  {
3      double min;
4      double max;
5  } minmax_t;
6
```

```
7   minmax_t minmax(double x[], int num)
8   {
9     int i;
10    minmax_t mm;
11
12    mm.min = x[0];
13    mm.max = x[0];
14
15    for(i=1; i<num; i++)
16    {
17      if(x[i] < mm.min)
18        mm.min = x[i];
19      if(x[i] > mm.max)
20        mm.max = x[i];
21    }
22
23    return(mm);
24  }
```

The first argument of the function (line 7) shows that x is an array, but without specifying the size. Hence, the [] brackets are empty. Note that one does not need to pass the number of elements fitting into the array; it can be obtained using the `sizeof()` statement, see page 30. Nevertheless, it is more obvious what is going on when the number is passed explicitly. The function is also more flexible in this way, because one might not use all entries of a large array.

Note that `minmax()` contains, in lines 9 and 10, definitions of local variables. The actual determination of the minimum and maximum elements is done in lines 12–21. The result is returned in line 23.

There is also an "old" (non ANSI standard) form of a function definition, where inside the (...) brackets only the parameters are stated, but no types are given. In this old form, the types are listed separately behind the (...) brackets and before the {...} block, where the function's main code is contained, e.g.

```
minmax_t minmax(x, num)
double x[];
int num;
{
  ...
}
```

Sometimes you do not want a function to explicitly return something.

In this case, you can use the special type void. No return statement is necessary now. One can still use the **return** statement to leave a function, but it must be used without a return value. Without an explicit **return** statement, the function is just executed till the end of the block is reached.

Again, considering the minmax() example, instead of returning a structure of type minmax_t for returning two numbers, one can use a different approach: One can pass pointers (see page 10), which point to the variables where the return values should be stored. Using pointers enables the content of these variables to be changed. For the given example, this could look like:

```
1   void minmax2(double x[], int num, double *p_min, double *p_max)
2   {
3     int i;
4     double min_value, max_value;
5
6     min_value = x[0];
7     max_value = x[0];
8
9     for(i=1; i<num; i++)
10    {
11      if(x[i] < min_value)
12        min_value = x[i];
13      if(x[i] > max_value)
14        max_value = x[i];
15    }
16    *p_min = min_value;
17    *p_max = max_value;
18  }
```

When calling minmax2(), one must pass the addresses where the results should be stored, e.g. like in

```
minmax2(x, num, &x_min, &x_max);
```

Using pointers in this case is necessary because arguments are always passed *by value*. This means that the content of the argument variable is copied to a *local* variable of the function, i.e. which is accessible only inside the function, similar to the variables defined in the main part of the function. Consequently, if the value of an argument variable is changed inside the function, which is possible, it will not affect the content of the variable outside of the function. For example, consider the following variant of minmax():

```
1   minmax_t minmax(double x[], int num)
2   {
3     minmax_t mm;
4
5     mm.min = x[num-1];
6     mm.max = x[num-1];
7
8     while(num >=0)
9     {
10      if(x[num] < mm.min)
11        mm.min = x[num];
12      if(x[num] > mm.max)
13        mm.max = x[num];
14      num--;
15    }
16
17    return(mm);
18  }
```

Here, the argument num is also used as counter during the main loop (lines 8–15), hence it is changed (line 14) during the execution. Nevertheless, when the execution returns outside minmax(), the value which has been passed will still remain at its original value. The only way to change variables outside a function in C is to use pointers, as we have described above. Note again that passing an array is basically like passing a pointer. Thus, all changes to elements of an array will persist after the execution of the function has been terminated.[6] Nevertheless, one can quantify an argument, which is a pointer or an array, as being constant by writing const in front of the definition of the argument, e.g.

```
minmax_t minmax(const double x[], int num)
```

This means that no changes to the array x[] are allowed. This is sometimes useful to prevent programming errors, because if you know in advance that the content should not be altered and, when using const, accidentally include changes to this variable in the code, the compiler will complain.

As we have seen in the previous examples, it is possible to define variables inside a function, like i, min_value and max_value in minmax2(). Such variables are called *local* variables. The argument variables, which

---

[6]Note that for C++ there is also the possibility of passing variables *by reference*, which means that all changes to the argument will be effective also after the function has been finished. Technically, this is achieved by passing in fact pointers, but this is hidden to the programmer, who can use the argument variables like any local variables.

hold copies of the actual arguments passed to the function, are also local. These variables are accessible only *inside* the function where they are defined. This defines the *scope* of a variable. Local variables exist only during the time a function is executed: The memory for the variable will be reserved when the function is called and the variable will be "deleted", hence the memory freed again, when the call finishes. Thus, each time a function is called again, all local variables are created from scratch. As we have seen previously, one can have other local objects like functions. Structures and types can also be defined just locally. Note that the scope of a variable or any other object can also be restricted to any block enclosed in {... } brackets. Variables defined inside a block are accessible only within the block and they exist only during the execution of the block. Each time the block is executed again such a variable is defined from scratch. Thus, the variable has no memory of past executions of the block.

There is one exception, the so-called *static* local variables. They are identified by the key word **static** in front of the type. These variables are created upon the first call to the function, and they exist till the end of the execution of the program. Such variables are useful when giving functions some internal memory, e.g. to store a status variable which should be remembered each time the function is called. For static variables one gives usually an initialization together with the definition. This initialization is only performed during the first call to the function! For example, the following function counts how often it is called and returns this number each time it returns from execution:

| GET SOURCE CODE |
| --- |
| DIR: c-programming |
| FILE(S): staticvars.c |

```
1  int do_something(int n)
2  {
3    static int num_calls = 0;
4
5    printf("I got:%d\n", n);
6
7    return(++num_calls);
8  }
```

Another example for an application of static variables can be found in Sec. 8.2.1, where pseudo random number generators are treated.

On the other hand, there are *global* variables and other global objects, which are defined outside any function or block, hence on the level where **main()** is defined. These variables and other objects are accessible from everywhere![7] This makes it very convenient and is even necessary for the definitions of self-defined data types, which cannot be passed as arguments to functions. This may also appear convenient for variables, e.g. for a large-scale simulation consisting of many particles. Here, one could store their data in global variables, hence one does not have to pass these variables to all functions where they are used. Nevertheless, as discussed in Sec. 3.2, the use of global variables is very error-prone and also very inflexible. *Do not use global variables, whenever you can avoid it !!!*[8] For example, consider a function which calculates the average velocity among a set of moving particles. If you use global variables, you have to write a function for the velocity calculation explicitly stating the name of the array variable which is used to store the data, such as `particle[]`. Hence, if in your program you simulate also a second independent set of particles, say stored in `particleB[]`, you have to provide another function for the velocity calculation of `particleB[]`. This is inefficient, so it is better to have one single function, where the actual particles to be treated are passed as arguments. Extrapolating these considerations leads to the conclusion that all data should be passed via arguments to functions. Consequently, global variables are not necessary, even dangerous, and should be avoided.

Note that if you define a function `sub()` inside another function `func()`, all other variables and other objects defined in `func()` will be accessible from `sub()`, and hence are global relative to `sub()`. Furthermore, a local definition overrides any definitions which are global at the given point, as in the following example:

> **GET SOURCE CODE**
> DIR: `c-programming`
> FILE(S): `hide.c`

```
1  int number1, number2;      /* these are global variables */
2
3  void do_something()
4  {
5    int number1;              /* overides global definition */
6
7    number1 = 33;
8    printf("subr: n1=%d, n2=%d\n", number1, number2);
9    number2 = 44;
10   printf("subr: n1=%d, n2=%d\n", number1, number2);
11 }
```

---

[7]As an exception, a global variable which is declared `static` is only accessible inside the same source file. This may be useful when you want to use a name for a global variable, which is already or may be used as global variable elsewhere.

[8]This is basically always possible, except for some debugging purposes.

```
12  int main()
13  {
14    number1 = 11;
15    number2 = 22;
16    printf("main: n1=%d, n2=%d\n", number1, number2);
17    do_something();
18    printf("main: n1=%d, n2=%d\n", number1, number2);
19
20    return(0);
21  }
```

Inside do_something() only the global variable number2 is accessible, while number1 is purely local. Consequently, the assignment of number1 will not affect the global variable, but it does so for number2. Therefore, when running the program one gets

```
main: n1=11, n2=22
subr: n1=33, n2=22
subr: n1=33, n2=44
main: n1=11, n2=44
```

Nevertheless, one can always design simulation programs such that one does not need nested levels of variable scopes. Thus, for clarity they should be avoided unless absolutely necessary. There is a special compiler warning flag -Wshadow, which makes the compiler report when some local definition overrides a global definition. This flag is usually not included in the -Wall set of warnings.

More advanced programming techniques which are based on functions, like *recursion*, *divide-and-conquer*, and *backtracking* are discussed in Secs. 6.2 to 6.5.

## 1.3   Input/output

Since you want to know the results of your simulations, your program needs some output. The most basic command for this is printf(), which prints to the standard output, usually the screen, and which was already preliminarily discussed above. Usually, you write out some raw or intermediate results to files. For this purpose, fprintf() can be used. Below, we will also introduce the relevant commands for creating, opening, writing, closing, and deleting files. Typically, your intermediate results have to be read in again to be processed further. Hence, ways to read in data are discussed next. Also, simulation programs need command line arguments, which are discussed afterwards. Finally, an easy method to compress data files during the simulation is presented.

The `printf()` ("print formatted") command consists of a format string and some (possibly zero) expressions to be evaluated and then printed:

`printf(` ⟨*format strings*⟩, ⟨*argument 1*⟩, ⟨*argument 2*⟩, ...)

All arguments except the format string are optional. The format string may contain characters which are just printed as they are given. Furthermore, the format string determines how many of the optional arguments have to be given and it states how the values of these different arguments are to be printed. This is achieved using so-called *conversion specifications*, which are always preceded by the '%' character. Each conversion specification requires one argument, respectively. For example, a single integer is printed using %d, as shown in the following example:

```
int num_particles = 100;

printf("number of particles: %d\n",  num_particles)
```

which will result in

```
number of particles: 100
```

This format string also contains a character \n, which is an *escape sequence* and means that the printing after this output has been printed will continue in the next line. The most important escape sequences are

| | |
|---|---|
| \n | a new line |
| \t | a horizontal tabulator |
| \b | a backspace |
| \\ | a backslash '\' |
| \" | a double quote " |

There are different types o conversion specifications. The most important ones, in the context of simulations, are

| %d, %i | for integer numbers |
| %x, %X | hexadecimal representation of integers |
| %f | for floating point representation |
| %e, %E | scientific ("exponential") notation |
| %s | strings |
| %p | pointers (addresses). |

Note that one writes '%%' for printing a '%'. A complete list of conversion specifications can be found in the *manual* documentation of printf(), e.g. by typing man 3 printf under

| GET SOURCE CODE |
| --- |
| DIR: c-programming<br>FILE(S): printing.c |

a UNIX system, where the number 3 refers to the section ("Programmer's manual") since in Sec. 2 ("User commands") there is another printf shell command. Some of these conversion specifications appear in the following primitive example (note that the format string can be split into several pieces, like here):

```
int counter = 777;
double energy1 = 35678.99;
void *pointer = &energy1;
char name[100] = {"network"};

printf("After %d (hex:%x) iterations an energy of %f \n"
       "(%e, stored at %p) was obtained for %s\n",
       counter, counter, energy1, energy1, pointer, name);
```

When executing these lines of code, the following output will appear on the screen:

```
After 777 (hex:309) iterations an energy of 35678.990000
(3.567899e+04, stored at 0xbfaee7d0) was obtained for network
```

Although printf() can handle an arbitrary number of arguments, the number of arguments given must always match the number of conversion specifications in the format string. Otherwise the compiler will complain.

When printing just using the raw conversion specifications, e.g. the format of a floating-point number will always be the same. The predefined format can be modified by optional specifications. Details can be found again in the documentations (man pages). Here, we give only the most important examples. First, right after the '%' there can be a *flag*. Important flags are 0, to fill numbers with leading zeros, and - for left adjustment.

Next, the (minimum) field width can be stated, which is just an integer number. Next, one can optionally have a precision specifier, which is a dot '.' followed by another number (or .*, which means that the precision is given as next argument by an expression of type int). For floating point/ exponential numbers, this is the number of digits after the radix, while for integer numbers or strings it is the maximum field width. Finally, there can be a *length modification*. The most important one is 1 which stands for *long*, i.e. %ld will expect a long int as corresponding argument. The following simple statement gives some examples for the modifications of the output format:

```
printf("%06d, %4.3f, %-20s, %lf\n",
       45, 3.14159265358979, "Hello", 36.5);
```

which will result in the following output

```
000045, 3.142, Hello                 , 36.500000
```

Instead of printing the output to the screen (called standard output stdout), one can also print to files, see below, or to strings. The latter is done via the sprintf() function, which has the following format

sprintf( ⟨*target string*⟩, ⟨*format strings*⟩, ⟨*argument 1*⟩, ... )

Consequently, as example one can use this to concatenate strings

```
sprintf(name, "%s %s", first_name, family_name);
```

In the context of computer simulations, this function is quite useful to assemble parameter-sensitive file names, as shown in the example on page 29.

Printing to a file is more involved. First one must *open* the file. This is done using fopen(), which has the following format:

fopen( ⟨*file name string*⟩, ⟨*access mode string*⟩)

Thus, one must provide two argument strings. The first one contains the name of the file. This string may contain the path name relative to the working directory from which the

| GET SOURCE CODE |
| --- |
| DIR: c-programming<br>FILE(S): file_o.c |

program is started.[9] The second string states the access mode, which can

---

[9]This may be different for computer systems, where the jobs are submitted via queuing systems. There, sometimes all paths have to be specified relative to the home directory, or relative to some special scratch directories.

be "w" for *writing*, which means that a file will be created from scratch. Hence, if the file has existed previously, it will be deleted. Other important access modes are "a" for *appending* at the end of a file and "r" for *reading* a file. When the file is opened successfully, the function will return a *file pointer*, which points to an internal structure where all information is stored, which is needed by the operating system to access the file. Technically, the file is treated as a so-called *stream*, which makes an output to different media possible in a unified way. To open a file, e.g. "funcs.dat", for writing one can use

```
FILE *file_p;

file_p = fopen("funcs.dat", "w");
```

To actually write to the file, one can conveniently use the fprintf() function, which works exactly like the printf() function, except that the output is directed to a file. The format is similar to printf() except that the (additional) first argument must be a file pointer. Therefore, the general format is:

fprintf( ⟨*file pointer*⟩, ⟨*format strings*⟩, ⟨*argument 1*⟩, ... )

Assume that we want to write a four-column table containing some values of the functions $\sin(x), \cos(x), \exp(x)$ in the interval $[0, 2\pi]$. We write in the first line of the file a description of the following columns via

```
fprintf(file_p, "#    x      cos(x)    sin(x)    exp(x)\n");
```

Note that the first character '#' is recognized by most data analysis and plotting tools like **gnuplot** (see Secs. 8.4 and 8.6.2) as comment line. Hence, when you read in the file to postprocess the data, the first line will be ignored. Nevertheless, for your bookkeeping, you should always use such comment lines in your simulation output files. Having more information in the output files available will help you a lot in organizing and analyzing your results.

To write the actual data, one could use

```
for(x=0; x<=2*M_PI; x+=0.1)
  fprintf(file_p, "%f %f %f %f\n", x, sin(x), cos(x), exp(x));
```

There are some predefined file pointers like stdout (standard output) which writes to the screen. Hence, fprintf(stdout, ...) is equivalent

to printf(...). Also, there exists stderr, which is the standard output
for error messages. This is also usually directed to the screen in interactive
mode, but in case your programs runs in a special environment, like when
using a queuing system, stderr is usually different from stdout.

Note that there are several other C functions for writing to files. Examples are putc(), which writes single characters, and puts(), which writes
strings. Since they provide no functionality beyond fprintf(), they are
not discussed here.

Finally, a file has to be *closed* when the access is terminated. For this
purpose, the function fclose() has to be used, which expects as argument
a file pointer of the file to be closed, e.g.

```
fclose(file_p);
```

Once a file is closed, it can be accessed by other means, e.g. inspected via
an editor. Note that while a simulation is running, data which is written via
fprintf() will not be immediately forwarded to the file. Usually internal
buffers are used, and the data is output to the file blockwise, each time
the buffer is full. Nevertheless, emptying the buffer can be triggered within
a program via the function fflush(), which also takes a file pointer as
argument.

So far, we have discussed ways to output data from a program. On the
other hand, your simulation programs usually need some input as well. The
most direct way is to use interactive input, e.g. to type in some parameters
values on request. This can be accomplished in C using the getchar()
function, which just reads one character from the keyboard. Longer inputs
can be read in using multiple calls to getchar(). Nevertheless, simulation
programs are very often processed by batch queues on large-scale computing
facilities. These batch jobs start at some time which is unknown in advance,
hence you cannot sit in front of the screen, wait till your programs have
started, and then supply the necessary input. Nevertheless, some programs,
in particular for small systems and/or pedagogical purpose, may run interactively, quite often using a graphical user interface (GUI). Such interfaces
are beyond the scope of this book; also the way the user interface is pro-
grammed under C depends quite often on the programming environment.
Thus, the readers who are interested in simulations having a comfortable
GUI should consider for example the JAVA programming language [JAVA].
Anyway, in the context of (large-scale) computer simulations, input to programs is either done via command-line arguments, which are treated below,

or via reading in files, which we discuss next.

Input files are either parameter files, which tell the simulation program what the system to be simulated looks like, how many iterations are to be performed, or at which temperature your system is simulated. These files may also describe the complete (initial) state of a simulation, like the coordinates of all particles or, in general, the degrees of freedom. Input files may have been generated by previous simulation runs, which you would like to continue. Furthermore, files where the simulation results are contained in may be used as input files for further analysis. In case you do not use standard tools like *gnuplot* (see Secs. 8.4 and 8.6.2), you have to teach your self-written analysis program how to read in data. To summarize, reading data files is an important task. How this is performed in C is explained next.

Similar to writing to a file, one also has to *open* it for reading. This is again done with the `fopen()` function, but now the access mode should be `"r"`. This will again provide a file pointer (a stream in general), which can be used to actually read the file. The file pointer points to a memory area where all necessary internal information about the file is stored, and also to a *current position*, which indicates where the reading continues. Just after opening the file, the current position is the beginning of the file. There are several possibilities to actually read in and process the data. Here, we will describe one approach in detail, probably the most general one. For this purpose, we use the function `fgets()` ("file get string"), which exactly reads the next unread line of the file, i.e. till the next new line '\n' character. The format of the function is as follows:

`fgets( `⟨*file name string*⟩`, `⟨*maximum length*⟩`, `⟨*access mode string*⟩` )`

Consequently, you have to pass a string s, where the line is stored to, the maximum number $n_{max}$ of allowed bytes, and a file pointer to `fgets()`. Note that the string will be terminated by a 0. Also, no more than $n_{max} - 1$ bytes (without the closing 0) will be transferred to the string. This prevents you from writing beyond the reserved range of memory. `fgets()` will return the string which is passed (i.e. a pointer to the first character of the string), if something was read in. If the end of the file is already reached, i.e. no additional line could be read in, then a NULL pointer is returned.

Now, once the string is stored in memory, it can be further processed in many different ways, e.g. by using string-processing functions. There are many ways to treat the input lines, which basically depends on your chosen file format. For example, one can directly test the values of the string elements. Hence, if you want to filter out comment lines, which start by a '#' character, you can test whether s[0] == #. If yes, then one can continue with the next line.

A very convenient way to process strings is available if they follow some format, e.g. in case they were generated using fprintf() or similar functions. In this case one can use sscanf() which is basically the inversion of sprintf(): One has to supply as parameters an input string, which is to be analyzed, and a format string which may contain printable characters, escape sequences, and conversion specifiers. Depending on the conversion specifiers, *pointers* to variables also have to be supplied, one for each conversion specifier indicating a value to be read in.[10] The string is analyzed by comparing it to the format string. Whenever a conversion specifier in the format string is encountered, the corresponding value is extracted from the input string and stored in the given variable. This processing continues until the full input string is analyzed, or until *the first* mismatch between format string and input string is detected. For example, if the string s contains "particles: 20 runs: 100", then the call to

```
int num_read;
int num_part = 0, num_runs = 0;

num_read= sscanf(s, "particles: %d runs: %d", &num_part, &num_runs)
```

will assign the value 20 to num_part and 100 to num_runs, and will return the value 2, which is assigned to num_read. If the format string contains a mismatch, then the processing stops. Thus

```
num_read= sscanf(s, "particles: %d Runs: %d", &num_part, &num_runs)
```

will only assign the first variable num_part, while num_runs remains at its initial value 0, and num_read will be 1. Therefore, if you are only interested in the number of particles anyway, you may also use

```
num_read = sscanf(s, "particles: %d", &num_part)
```

The conversion specifiers are the same as for printf(). There is one important point: When you print a variable of type float or double, you can use a conversion specifier %f in both cases. But when reading in

---

[10]There may also appear the conversion specifier modification '*' right behind the '%' character which results in skipping the corresponding value; hence, *no* pointer should be supplied for this item.

a floating point number, which you want to assign to a variable of type
double, you *must* use the conversion specifier %lf, otherwise the value
will not be assigned correctly! Furthermore, there are also the functions
scanf() and fscanf() which enable standard (keyboard) input or file input
to be scanned directly. Nevertheless, this often generates problems. Hence,
it is better to first read in a line via fgets() (or gets() for standard input)
and then use sscanf().

As we have seen, fgets() reports if the end of the file has been reached.
This can also be tested directly via feof(), which takes a file pointer as
argument. It will return true (value 1), if the end of the file has been
reached. Note that it will not report *true* if just the last line has been read
in, i.e., an attempt to read in the next, non-existing line is necessary.

As a complete toy example, we next present
a source code which reads in the four-column
file which we have printed above (page 48). The
program ignores all comment lines starting with

| GET SOURCE CODE |
| --- |
| DIR: c-programming |
| FILE(S): file_in.c |

a '#' character, and it just prints for each line the value in the first column
and the sum of the values in the other three columns, a task which may
occur for some data analysis problems[11]

```
1   int main(int argc, char**argv)
2   {
3     char line[1000];    /* string where one line of file is stored */
4     double val1, val2, val3, val4;           /* values from file */
5     int num_got;      /* how many where obtained from current line? */
6     FILE *file_p;                             /* file pointer */
7
8     file_p = fopen("funcs.dat", "r");    /* open file for reading */
9
10    while(!feof(file_p))      /* read until end of file is reached */
11    {
12      if(fgets(line, 999, file_p) == NULL)        /* read in line */
13        continue;                        /* nothing was read in */
14      if(line[0] == '#')                     /* comment line? */
15        continue;                              /* ignore */
16      num_got = sscanf(line, "%lf %lf %lf %lf",    /* get values */
17                     &val1, &val2, &val3, &val4);
```

---

[11]For this simple purpose, one should not write a program but use the *awk* tool. The
example is just used for pedagogical purposes here.

```
18      if(num_got != 4)                            /* everything OK ? */
19      {
20        fprintf(stderr, "wrong line format in line: %s ", line);
21        continue;
22      }
23      printf("%f result= %f\n", val1,
24              val2+val3+val4); /* process */
25    }
26    fclose(file_p);                             /* close file */
27
28    return(0);
29  }
```

First, the file has to be opened in line 8. Note that here the filename is hard-coded in the program. In general, one will need analysis programs which work for any files; hence, the file name has to be passed to the program. This is discussed below. Here, for the purpose of the example, the filename "funcs.dat" is sufficient. The file is processed in the main loop (lines 10–25), until the end of the file has been reached. First, the current line is read (line 12). Comment lines are ignored in lines 14 and 15. In lines 16–22, the content of the line is analyzed, and finally processed in lines 23–24.

Finally, note that a very useful string processing function for analyzing input files is strstr(), which is able to locate given string patterns inside other strings. This is useful for reading in poorly structured input files, where different values are identified by key words at arbitrary positions.

So far, we have just linearly read in a file. Sometimes it is necessary to read a file several times. This can happen when you first want to count how many input lines the file contains, e.g. to set up enough local storage space dynamically (see Sec. 1.4), and then actually read in the data in a second sweep.[12] For this purpose, the rewind() function can be used, which sets the internal position pointer back to the beginning of the file. One can even navigate completely freely inside a file. For this purpose, functions like fseek() and ftell() are available. For more details on these functions, please refer to the documentation.

As mentioned above, one can use files to pass simulation parameters to a program. This is in particular useful if many parameters are available, and if one wants to archive the values of these parameters for the different runs. This is helpful if one has to perform many different runs without getting lost in all these data. Note that organizing large-scale simulations in a

---

[12]Alternatively, one could read in the data in one sweep, but dynamically extend the local storage space, if necessary.

useful way is an active area of research called *Computational Provenance* [Comp. Sci. Eng. (2008)]. A possible parameter file, e.g. for the Molecular Dynamics simulation[13] of a system of gas particles, could look like this

```
num_particles      = 512
temperature        = 37.3
number_steps       = 10000
step_size (fs)     = 1.2
box_size           = 10
appendix           = A67
save_config        = no
```

In real applications one could have many more simulation parameters, which for example describe different particle types and the coefficients for different force fields. Such a parameter file can be read in, as mentioned above, most conveniently using the `strstr()` function; we do not go into details here. Nevertheless, for many applications it is sufficient to pass simulation parameters as arguments when invoking the program, e.g. like

```
arguments 512 37.3
```

where it is assumed that `arguments` is the name of the program, the first parameter (here 512) determines, say, the number of particles, and the second (37.3) determines the temperature of the system. Different arguments always have to be separated by spaces. For other parameters of `arguments`, the default values are taken, unless they are changed via options, as described below.

To be able to read the program arguments, one has to define the `main()` function as follows:

```
int main(int argc, char *argv[])
```

During the execution of `main()`, `argc` will contain the number of arguments including the program name, i.e. three in the above example. The array `argv` of strings contains in `argv[0]` the program name, in `argv[1]` the first argument (here 512), in `argv[2]` the second argument (here 37.7) and so on. Note that the arguments are stored in strings; hence, if they are to be interpreted as numbers, they have to be converted, e.g. using `atoi()`, which converts a string into an integer, or using `sscanf()` as described above.

---

[13] For a Molecular Dynamics simulation, you have formulas describing the forces between different particles (*force fields*). Using the forces you integrate the Newton's equation of motions to study the dynamic evolution of such a system. Molecular Dynamics simulations are widespread, such as to study the dynamics of proteins in cells.

For our example, we use an additional counter **argz**, to treat one argument after the other. This is useful in particular in case the program has additional options, see below. Here, the program arguments are treated via:[14]

```
GET SOURCE CODE
DIR: c-programming
FILE(S): arguments.c
```

```
int N;                              /* number of particles */
double temp;                              /* temperature */
int argz = 1;                   /* counter to treat arguments */

N = atoi(argv[argz++]);
sscanf(argv[argz++], "%lf", &temp);
```

Some simulation parameters are usually kept at their default values and they do not have to be passed as arguments each time. In this case, it is convenient to use program options to change the default values. Program options are usually indicated by a '-' character at the beginning, followed by some name, and maybe some additional values. A call to the example program including some options could look like

```
arguments -size 10.3 -appendix XX 100 3.7
```

Options and their accompanying values are stored in the **argv[]** strings like any other argument. Consequently, they can be processed using standard string manipulation. For instance, one can implement a loop, that is executed before the non-optional arguments are read, which is iterated as long as the "current" argument starts with a '-' character. Then, one can compare the "current" argument via **strcmp** to the different available options. If the option matches, one can take suitable actions, like setting a flag or assigning some parameter value. Finally, one should implement printing an error message in case the option passed as argument is not known to the program. For the example program, this could look like:

```
char appendix[1000];            /* to identify output file */
int do_save;                    /* save files at end of output */
int print_help;                    /* print help message ? */
double l;                       /* lateral size of system */
int num_steps;             /* how many MD steps are performed */

/** default values **/
l = 10; do_save = 0; appendix[0] = 0;
```

---

[14]For this program, no line numbers are given here because we discuss different parts of the example program in non-linear order.

```
num_steps = 10000; print_help = 0;

/** treat command line arguments **/
while((argz<argc)&&(argv[argz][0] == '-'))
{
  if(strcmp(argv[argz], "-steps") == 0)
    num_steps = atoi(argv[++argz]);
  else if(strcmp(argv[argz], "-save") == 0)
    do_save = 1;
  else if(strcmp(argv[argz], "-size") == 0)
    sscanf(argv[++argz], "%lf", &l);
  else if(strcmp(argv[argz], "-appendix") == 0)
    strcpy(appendix, argv[++argz]);
  else
  {
    fprintf(stderr, "unkown option: %s\n", argv[argz]);
    print_help = 1;
    break;
  }
  argz++;
}
```

Note that different types of arguments have to be treated independently. For example, for -save, one just sets a flag variable, while for the other options additional values have to be read in, i.e. an integer value (-steps), a floating point value, (-size) or a string (-appendix).

It is also recommendable to print the calling sequence of your program and maybe some additional important simulation parameters at the beginning of each line to your output (or log) file, each line preceded by the '#' comment symbol. This helps to reconstruct later on how the output file was generated, i.e. supports Computational Provenance (see page 54). For the current example it reads

```
char name_outfile[1000];              /* name of output file */
FILE *file_out;                 /* file pointer to output file */
    ...

sprintf(name_outfile, "md_N%d_T%3.2f%s.out",      /* file name */
        N, temp, appendix);
file_out = fopen(name_outfile, "w");

fprintf(file_out, "# calling sequence: ");         /* print args */
for(t=0; t<argc; t++)                  /* print command line arguments */
  fprintf(file_out,"%s ", argv[t]);
fprintf(file_out, "\n");
```

```
fprintf(file_out, "# params: num_steps=%d, size=%f, save=%d\n",
        num_steps, 1, do_save);
```

Before treating the non-optional program arguments, you should always test whether the number of supplied arguments is as expected. In case this is not true, your program can conveniently print out a simple error message which also explains the usage of your program. For the example above it looks as follows (this code appears just before the values N and temp are assigned):

```
if( print_help || (argc-argz) != 2)    /* not enough arguments ? */
{                                       /* print error/usage message */
  fprintf(stderr, "USAGE: %s {<options>} <num_part> <temp>\n",
          argv[0]);
  fprintf(stderr, "   options: -steps <n> : num. MD steps "
          "(d:%d)\n",  num_steps);
  fprintf(stderr, "      -size <l>: system size (d:%d)\n",
          1);
  fprintf(stderr, "      -save: save config at end\n");
  fprintf(stderr, "      -appendix <s>: for output file name\n");
  exit(1);
}
```

Finally, when your simulation is finished, you can close the output file. Quite often, simulation programs write out huge amounts of information. In this case, to save hard-disk space, you should use a compression tool like *gzip* to compress your output files. This can be done most conveniently right in your program. For this purpose you can use the system() library function, which takes as argument a string which may contain any command line, e.g. a shell command line under UNIX/Linux. For our example, we can compress the output file simply in the following way:

```
char command[1000];                    /* for system() calls */

  ...

fclose(file_out);                      /* close file */

sprintf(command, "gzip %s", name_outfile);   /* zip file */
system(command);
```

Using this compression, you also do not have to think much about how to save data very efficiently, since the zipping program takes care of this automatically. To unzip a file before reading it into your program is also quite simple, see exercise (4). The names of your compression tools, and the behavior of the `system()` function depend heavily on the programming environment and on the operating system. Hence, the application of this zipping might make your program less portable. Note also that the output file is only compressed after the simulation has terminated, hence when the output file is written completely. Therefore, it may occupy a lot of disk space before being compressed. If this poses a problem for you, you can also use on-the-fly zipping libraries like *zlib* [zlib], which works for all standard operating systems.

## 1.4   Pointers and dynamic memory management

More complex data types involve references between different objects. Consider, for example, a simulation of a social system, where you simulate a set of individuals, and you want to store for each individual the other individuals he/she knows. You could for example generate a big array with one entry for each individual. The entry for each individual carries a small array, which contains the indices of the other individuals he/she knows. On the other hand, if you assume that you have two big arrays, one for the males and one for the females, then you already have to distinguish whether an index refers to a female or a male. As you see, for more complex simulations, this might become even more cumbersome.

A more general and elegant approach is to use *pointers* to the objects you want to refer to. In this case it does not matter whether the objects are stored in the same or in different arrays, because pointers are basically memory addresses, as introduced in Sec. 1.1.1. Here, it is explained how dynamically changing data structures can be implemented via pointers and via memory management. We also show how pointers and arrays are related to each other. Using these basic ingredients, quite complex data structures as *lists*, *trees*, or *graphs* can be built. These advanced applications are discussed in Secs. 6.6 to 6.8.

First, the previously given information about pointers is summarized. A pointer **p** to a memory position, where a variable of type ⟨*type*⟩ is stored, is defined via

⟨*type*⟩ \*p;

Initially, there is no value assigned to **p** (maybe zero for some compilers). To make **p** pointing to some relevant location, there are two possibilities:

First, if var is a variable of type ⟨*type*⟩, one can write

```
p = &var;
```

Now, as you know already from page 10, p contains the address where var is stored. Assigning a value to var will not change p. Nevertheless, one can access or change the content of var via *p.

The second possibility is that one acquires some available memory, and let p point to it. This works via the malloc() ("memory allocation") function, which requires as argument the number of bytes to be allocated. It returns a pointer to "something" (i.e. of type void *), pointing to the reserved memory area:

```
void *malloc(⟨size⟩);
```

If the operating system cannot provide the requested memory, the special pointer value NULL is returned. Since the size of a type is determined using sizeof(), one can reserve a so far unused memory location where p can point to via:

```
p = (⟨type⟩ *) malloc(sizeof(⟨type⟩));
```

Note the cast in front of the malloc() statement. This memory allocation does *not* take place during compile time or right when the program is started. It just happens, when the malloc() function is executed in the program. For this reason, the processes connected to malloc() are called *dynamic memory management*.

More interesting usages of pointers are possible. You could for instance have some init() function in your simulation (taking, say, an integer as argument) which also reserves some memory (where integers are stored) and returns the pointer to this memory. The function prototype might look like:

```
int *init(int);
```

On the other hand, functions can be referenced through pointers as well. To define a pointer fct_p, which points to a function taking an integer as argument and returning an integer, one writes

```
int (* fct_p)(int);
```

which looks very similar to the prototype above; it just differs by the brackets. Here, the usage of `typedef` is convenient. If pointers to functions of this form are needed frequently in the program, one can define a new type name and use it as follows, for example:

```
typedef int (* fctptr_t)(int);

fctptr_t fct_p;
```

Note that the name of the new type does not come at the end of the statement. An example where pointers to functions are needed is given in exercise (6), where a simple integration subroutine for an arbitrary function is given. Another example you find in Sec. 8.3.4.

With the above use of `malloc`, memory is allocated that can store exactly one value of the type ⟨*type*⟩. One can reserve a chunk of memory for several elements by just asking for more bytes. For example, one can reserve a chunk to store 10 persons of the `person_t` (see page 32) type via

```
person_t *pp;

pp = (person_t *) malloc(10*sizeof(person_t));
```

This is essentially the same as reserving an array of size 10. Here, one can access the i'th element via `*(pp+i)`, e.g. like in

```
for(i=0; i<10; i++)
  (*(pp+i)).age = 2*i;
```

Note that `pp+i` means that to the address `pp` one adds i times the number of bytes required to store `person_t`, i.e., increasing `p` by one here means to add `sizeof(person_t)` bytes to the address. Conveniently, it is also possible to write `pp[i]` to access the i'th element; hence, one can write `pp[i].age` to access the `age` member. For pointers to structures, there is an alternative syntax to access the members. Instead of `(*pp).age`, one can write `pp->age`. Similarly, to access the member `age` of the i'th element, the loop above can also read

```
for(i=0; i<10; i++)
  (pp+i)->age = 2*i;
```

Consequently, one-dimensional arrays and pointers, together with dynamic memory allocation, are very similar. One important difference is that for an array the size of the array cannot be changed, after it has been defined, while this is possible when using dynamic memory management. For this purpose, the `realloc()` function exists, which expects two arguments: 1. the pointer to the memory area to be extended and 2. the new size of the array:

void \*realloc(⟨*pointer*⟩, ⟨*size*⟩);

The function returns a pointer to the new start of the memory chunk. This can be the same position as before the call. Nevertheless, it may sometimes be necessary to extend the memory chunk considerably. In this case it might happen that the operating system allocates a completely different part of the memory and copies the memory content from the old chunk to the beginning of the new chunk. In this case, the execution of the command might take some time, depending on the size of the memory area to be copied.

Note that one should pass only pointers to `realloc()` which have been allocated via `malloc()`, otherwise the behavior of the function is undefined. Also, it might happen that the available memory is not sufficient. In this case the special `NULL` pointer is returned.

If one wants to allocate a matrix `mat` dynamically, one has to do this in two steps. First, one has to allocate one array which will contain pointers to the beginning of each row, respectively. Now, within a loop, the memory areas for the different rows can be allocated. If `n_rows` and `n_cols` are the numbers of rows and columns, respectively, the code to allocate the matrix `mat` could look like

```
double **mat;
int n_rows = 10, n_cols = 10;
int i, j;

mat = (double **) malloc(n_rows*sizeof(double *));
for(i=0; i<n_rows; i++)
  mat[i] = (double *) malloc(n_cols*sizeof(double));
```

The access can be performed in the same way as for standard matrices; hence, one can write, e.g. to initialize all entries to 0:

```
for(i=0; i<n_rows; i++)
  for(j=0; j<n_cols; j++)
    mat[i][j] = 0;
```

Note that due to the way the dynamically allocated matrix `mat` is stored, the actual access is different from a standard matrix `smat`, which is defined via `double smat[n_rows][n_cols]`: A standard matrix is stored in one large chunk of memory, where each row follows the next. On the other hand, the different rows for the dynamically allocated matrix `mat` are usually stored in different places. Therefore, a direct access could look like `*(*(mat+i)+j)`. This also means that a dynamically allocated matrix requires slightly more memory, because in addition to the actual matrix elements one needs an array with pointers storing the addresses of the different rows. Nevertheless, dynamically allocated matrices are more flexible, because the size can be changed during runtime. It is also possible that different rows contain a different number of elements, which is not possible for standard arrays either. This is useful, for example, when one wants to store an array of strings with strings of different lengths.

Finally, all memory which has been reserved inside the program should be released, one says *freed* again. For this purpose, the `free()` function can be used, which takes as argument a pointer to a previously allocated chunk of memory, e.g. like in:

```
free(p);
```

Freeing the memory which has been allocated for a matrix is performed in two steps: First, all rows are freed, then the array containing the pointers to the rows is freed. For the matrix `mat` this could look like:

```
for(i=0; i<n_rows; i++)
  free(mat[i]);
free(mat);
```

Freeing of memory is particularly important, if the memory is used only inside some function and will not be accessed again after the execution of the function has terminated. If it is not freed, and if the function is called several times, then more and more memory is allocated during the execution of the program, which may cause the operating system to run out of memory. Such a bug is called a *memory leak* and should be avoided under all circumstances. A useful tool to detect memory leaks and other bugs connected with dynamic memory management is introduced in Sec. 4.3. These bugs are often hard to spot by hand.

Using dynamic memory management, arbitrarily complex data structures can be generated. More advanced examples like *lists*, *trees*, and *graphs* are discussed in Secs. 6.6 to 6.8.

## 1.5 Important C compiler options

Throughout this book, many C compiler options are explained which are useful or even necessary for some tasks. Here, they are summarized:

-o

sets the name of the output file, e.g. like in `cc -o first first.c`

-c

Just the compile process is performed, no final executable is linked. Each `.c` file will result in a corresponding `.o` file. This is useful in case the source code contains some functions which are used in several different programs. This process can be made automatic using *make* files, see Sec. 2.1.

-l

states the name of a library where some precompiled functions can be found, e.g. for mathematical functions (m) in `cc -o mathtest mathtest.c -lm`

-Wall

Switches on (almost) all compiler warnings, e.g. indicates if one uses '=' inside a condition instead of '=='.

-Wextra

Switches on additional compiler warnings, e.g. indicates if one compares variables of different types.

-Wshadow

When using this option, the compiler will warn if a local definition shadows a global definition, e.g. in case there are two variables of the same name (see page 44).

-g

Switches on support for debuggers, see Sec. 4.1.

-pg

Switches on support for profiling, i.e. for measuring where the program spends how much running time, see Sec. 4.4.

-O

Switches on optimization of the code. This means it will run faster. For example, the complier may "unroll" loops where the number of iterations is fixed. Also, it writes the code of a function inline instead of calling the function (can be forced via the `inline` directive at the beginning of a function definition). It can get rid of intermediate expressions/find common subexpression. Furthermore, it can decide to put variables, which are used frequently, into registers (which can also be forced in the source by the `register` storage class specifier, which has to be printed in front of the type name of a variable).

Different levels of optimization are available like -O0 (no optimization), -O1, -O2 and -O3. The higher the level, the more optimized your code is. The highest level may even alter the meaning of your code under some circumstances; hence, to use -O3 is dangerous. Here, -O2 is recommended.

The effect on the running time of the different optimization options can be tested in exercise (7).

-I

Normally, the compiler looks for include files in standard search paths as indicated by the operating system, and in the current directory. Using this option, additional search paths can be stated where include files of your own libraries (see Sec. 7.4) can be found, e.g. `cc -o prog prog.c -I$HOME/include`.

-L

Normally, the compiler (in fact the linker) looks for library (`.a`) files in standard search paths as indicated by the operating system, and in the current directory. Using this option, additional search paths can be stated, where your own ond other local libraries (see Sec. 7.4) can be found, for instance `cc -o prog prog.c -L$HOME/lib`.

-D

Defines a macro (see Sec. 1.6), as if it was defined via the `#define` directive in the source code, e.g. `cc -o program program.c -DUNIX=1`.

C compilers have many more options, some of them are machine-dependent. You should have at least a quick look once at the documentation of your compiler to see what options are available in principle.

## 1.6   Preprocessor directives and macros

The compile process can be controlled via *preprocessor directives*. The most important one is the `#include` directive, which makes the compiler read in the given file at the current position. In principle, arbitrary files can be included, but it is most useful to include header files which contain declarations of external functions, variables, and global data types, but no actual code. Note that there are two variants for including files:

- #include <header.h>

  This will include a standard header file **header.h** (which is to be replaced by a real file name) as provided by the operating system. Hence, the search for such files will usually take place in directories provided by the operating system. Basically, every program has to have a #include <stdio.h> at least. Other important standard header files to be included are **stdlib.h** and **math.h**.

  When using the −I compiling option, one can specify additional directories, where the compiler searches for header files.

- #include "header.h"

  This is used for your private and local header files. This means, the compiler will first look in the current directory for the header files. Only if they are not found there, it will look in the standard directories.

  More information on what you have to take into account when writing your own header files is given below.

The second important type of compiler directive is the definition of a macro. Macros are shortcuts for code sequences in programming languages. Their primary purpose is to allow computer programs to be written more quickly. But the main benefit comes from the fact that a more flexible software development becomes possible. By using macros appropriately, programs become better structured, more generally applicable, and less error-prone. Here, it is explained how macros are defined and used in C; a detailed introduction can be found in C textbooks such as Ref. [Kernighan and Ritchie (1988)]. Other high-level programming languages exhibit similar features.

In C a macro is constructed via the **#define** directive. Macros are processed in the preprocessing stage of the compiler. This directive has the form

#define  ⟨name⟩  ⟨definition⟩

Each definition must be on one line, without other definitions or directives. If the definition exceeds one line, each line except the last one has to be ended with the backslash '\' symbol. The simplest form of a macro is a constant, e.g.

#define  PI  3.1415926536

You can use the same sort of names for macros as for variables. It is convention to use only upper-case letters for macros. A macro can be deleted via the #undef directive.

When scanning the code, the preprocessor just replaces literally every occurrence of a macro by its definition. If you have, for example, the expression 2.0*PI*omega in your code, the preprocessor will convert it into 2.0*3.1415926536*omega. You can use macros also in the definition of other macros. But macros are not replaced in strings, i.e. printf("PI"); will print PI and not 3.1415926536 when the program is run.

It is possible to test for the (non)existence of macros using the #ifdef and #ifndef directives. This allows for conditional compiling or for platform-independent code, such as in

```
#ifdef UNIX
  ...
#endif
#ifdef MSDOS
  ...
#endif
```

Please note that it is possible to supply definitions of macros to the compiler via the -D option, e.g. cc -o program program.c -DUNIX=1. If a macro is used only for conditional #ifdef/#ifndef statements, an assignment like =1 can be omitted, i.e. -DUNIX is sufficient.

When programs are divided into several modules, or when library functions are used, the definitions of data types and functions are provided in header files (.h files), as mentioned above. It is important that each header file should be read by the compiler only once for each source code file, otherwise declarations will appear twice and the compiler will complain. When projects become more complex, many header files have to be managed, and it may become difficult to avoid multiple scanning of some header files. This can be prevented automatically by this simple construction using macros:

```
/** example .h file: myfile.h  **/

#ifndef _MYFILE_H_
#define _MYFILE_H_

    .... (rest of .h file)
    (may contain other #include directives)

#endif /* _MYFILE_H_ */
```

After the body of the header file has been read the first time during a compilation process, the macro _MYFILE_H_ is defined, thus the body will never be read again.

So far, macros are just constants. You will benefit from their full power when using macros with arguments. They are given in braces after the name of the macro, such as in

```
#define  MIN(x,y)  ( (x)<(y) ? (x):(y) )
```

You do not have to worry more than usual about the names you choose for the arguments, there cannot be a conflict with other variables of the same name, because they are replaced by the expression you provide when a macro is used, e.g. MIN(4*a, b-32) will be expanded to (4*a)<(b-32) ?  (4*a):(b-32).

The arguments are used in ( ) braces in the macro, because the comparison < must have the lowest priority, regardless of which operators are included in the expressions that are supplied as actual arguments. Furthermore, you should take care of unexpected side effects. Macros do not behave like functions. For example, when calling MIN(a++,b++) the variable a or b may be increased twice when the program is executed. It is usually better to use inline functions (or sometimes templates in C++) in such cases. But there are many applications of macros, which cannot be replaced by inline functions, like in the following example, which closes this section.

The example illustrates how a program can be written in a clear way using macros, making the program less error-prone, and furthermore allowing for a broad applicability. A system of Ising spins is considered, i.e., a lattice where at each site $i$ a particle $\sigma_i$ is placed. Each particle can have only two states $\sigma_i = \pm 1$. It is assumed that all lattice sites are numbered from 1 to $N$. This is different from C arrays, which start at index 0. The benefit of starting with index 1 for the sites is that, for many simulations of Ising systems, one needs the site 0 for additional technical reasons, see below. For the simplest version of the model only neighbors of spins are interacting. With a two-dimensional square lattice of size $N = L \times L$ a spin $i$, which is not at the boundary, interacts with spins $i + 1$ (+$x$-direction), $i - 1$ (−$x$-direction), $i + L$ (+$y$-direction), and $i - L$ (−$y$-direction). A spin at the boundary may interact with fewer neighbors when free boundary conditions are assumed. With *periodic boundary conditions* (pbc), all spins have exactly 4 neighbors. In this case, a spin at the boundary interacts

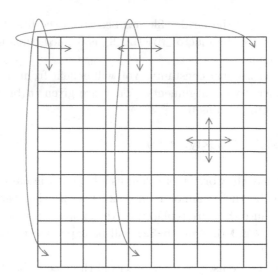

Fig. 1.1   A square lattice of size $10 \times 10$ with periodic boundary conditions. The arrows indicate the neighbors of the spins.

also with the nearest mirror images, i.e. with the sites that are neighbors, if you consider the system repeated in each direction. For a $10 \times 10$ system, spin 5, which is in the first row, interacts with spins $5 + 1 = 6$, $5 - 1 = 4$, $5 + 10 = 15$ and through the pbc with spin 95, see Fig. 1.1. The spin in the upper left corner, spin 1, interacts with spins $2, 11, 10$ and 91. In a program pbc can be realized by performing all calculations *modulo L* (for the $\pm x$-directions) and modulo $L^2$ (for the $\pm y$-directions).

This way of realizing the neighbor relations in a program has several disadvantages:

- You have to write the code for realization of the pbc everywhere where the neighbor of spins are accessed. This makes the source code larger and less clearly structured.
- When switching to free boundary conditions, you have to include further code to check whether a spin is at the boundary.
- Your code works only for one lattice type. If you want to extend the program to lattices of higher dimensions you have to rewrite the code or provide extra tests/calculations.
- Even more complicated would be an extension to different lattice structures such as triangle or face-center cubic. This would make the program look even more confusing.

An alternative is to write the program directly in a way it can cope with almost arbitrary lattice types. This can be achieved by setting up the neighbor relation in one special initialization subroutine (not discussed here) and storing it in an array next[]. Then, the code outside the subroutine remains the same for all lattice types and dimensions. Since the code should work for all possible lattice dimensions, the array next is one-dimensional. It is assumed that each site has num_n neighbors. Then the neighbors of site i can be stored in next[i*num_n], next[i*num_n+1],..., next[i*num_n+num_n-1]. Please note that the sites are numbered beginning with 1. This means, a system with N spins needs an array next of size (N+1)*num_n. When using free boundary conditions, missing neighbors can be set to 0. The access to the array can be made easier using a macro NEXT:

```
#define  NEXT(i,r)  next[(i)*num_n + r]
```

NEXT(i,r) contains the neighbor of spin i in direction r. For example, a quadratic system, r=0 is the $+x$-direction, r=1 the $-x$-direction, r=2 the $+y$-direction and r=3 the $-y$-direction. However, which convention you use depends on you, but you should make sure that they are consistent. For the case of a quadratic lattice, it is num_n=4. Please note that whenever the macro NEXT is used, there must be a variable  num_n defined, which stores the number of neighbors. You could include  num_n as a third parameter of the macro, but in this case a call of the macro looks slightly more confusing. Nevertheless, the way you define such a macro depends on your personal preferences.

The NEXT macro cannot be realized by an inline function, in case you want to set values directly like in NEXT(i,0)=i+1. Also, when using an inline function, you would have to include all parameters explicitly, i.e. num_n in the example. The last requirement could be circumvented by using global variables, but this is bad programming style as well.

When the system is an Ising spin glass, the sign and magnitude of the interaction may be different for each pair of spins. The interaction strengths can be stored in a way similar to the neighbor relation, e.g. in an array j[]. The access can be simplified via the macro J:

```
#define  J(i,r)  j[(i)*num_n + r]
```

A subroutine for calculating the energy $H = \sum_{\langle i,j \rangle} J_{ij} \sigma_i \sigma_j$ may look as follows (please note that the parameter N denotes the number of spins and the values of the spins are stored in the array sigma[]):

```
double spinglass_energy(int N, int num_n, int *next, int *j,
                        short int *sigma)
{
  double energy = 0.0;
  int i, r;                                         /* counters */

  for(i=1; i<=N; i++)              /* loop over all lattice sites */
    for(r=0; r<num_n; r++)               /* loop over all neighbors */
      energy += J(i,r)*sigma[i]*sigma[NEXT(i,r)];

  return(energy/2);    /* each pair has appeared twice in the sum */
}
```

The code for spinglass_energy() is very short and clear. It works for all kinds of lattices. Only the subroutine where the array next[] is set up has to be rewritten when implementing a different type of lattice. This is true for all kinds of code realizing such as Monte Carlo scheme or the calculation of a physical quantity. For free boundary conditions, additionally sigma[0]=0 must be assigned to be consistent with the convention that missing neighbors have the id 0. This is the reason why the spin site numbering starts with index 1 while C arrays start with index 0.

# Exercises

(solutions: can be downloaded from http://www.worldscientific.com/r/9019-supp)

(1) **Structure and its size**

Write a program which sets up using `typedef` a data structure for amino acids (the basic building blocks of proteins) and prints its size. The structure should contain: the following four elements:

| SOLUTION SOURCE CODE |
|---|
| DIR: `c-programming` |
| FILE(S): |
| `struct_size.c` |

- a `char` "type" of the amino acid
- an `int` "charge" (in units of the electron charge)
- a three-dimensional vector for the position
- a three-dimensional vector for the velcity

The main program should print the size of the data structure using the `sizeof()` command. Please compare with the sum of the individual sizes!

(2) **Pointer juggling**

Write a program which contains a `double` variable `value`, a pointer `p1` which should contain the address of `value` and a pointer `p2` which should contain the address of `p1`.

| SOLUTION SOURCE CODE |
|---|
| DIR: `c-programming` |
| FILE(S): `pointers.c` |

Perform the following steps

(a) Define `p1` and `p2` using suitable data types.
(b) Assign the desired values to `value`, `p1`, and `p2`.
(c) Print `value`, `p1` and `p2` using printf and suitable conversion specifiers. Print furthermore the content of `value` via just using `p1` and via just using `p2`, respectively.

(3) **Matrix permutation**

Write a program which permutes a given matrix `test` such that neighboring rows are exchanged. Use the multiplication of a suitably chosen permutation matrix `perm` (from the left) with `test`.

| SOLUTION SOURCE CODE |
|---|
| DIR: `c-programming` |
| FILE(S): `matrix.c` |

The program should

(a) contain the definition of three matrices, one is a n×n permutation matrix (n is an even number) `perm` and two (`test`, `result`) are n×m matrices,
(b) initialize `perm` such that neighboring rows are exchanged,
(c) initialize `test` in an arbitrary way such that different rows can be distinguished,
(d) calculate the matrix product `result = perm × test`

(e) print the matrix `result` row by row.

(4) **Online unzipping**

Write a program which prints a file, which is passed as argument. If the file is compressed, visible via the appendinx ".gz", the it should be uncompressed first.

> SOLUTION SOURCE CODE
> DIR: c-programming
> FILE(S): printzip.c

Requirements and hints:

(a) The filename should be passed as first argument.
(b) If no argument is given, a small error/usage message should be printed.
(c) Use `strstr()` to locate the substring ".gz" in the filename. If it is contained, the file should be decompressed via the help of the `system()` function.
(d) The file should be printed line by line.
(e) If the file was compressed before it was printed, it should be compressed again, before the program stops.

(5) **Bisection search**

Write a function that searches in an array 'number[]' of integers for a 'value' . The array is assumed to be sorted in increasing manner. Thus, you can apply bisection search.

> SOLUTION SOURCE CODE
> DIR: c-programming
> FILE(S): bisection.c

This works by keeping an interval [bottom,top] and testing the entry in the middle of the interval. If the value is found there, the search stops and the index of the current middle is returned. Otherwise, if the entry at the middle is larger than 'value' the search continues in the remaining part of the interval below the middle, and else above the middle. The function prototype reads as follows:

```
/********************** search() ********************/
/** Searches an int array sorted in increasing order **/
/** for the occurence of a certain 'value'            **/
/** PARAMETERS: (*)= return-parameter                 **/
/**        size: number of entries                    **/
/**      number: array                                **/
/**       value: to be found                          **/
/** RETURNS:                                          **/
/**     index of entry, -1 if not found               **/
/****************************************************/
int search(int size, int *number, int value)
```

Provide also a `main()` function which sets up an array with test numbers, e.g. the 100 numbers from 0 to 396 with step size 4, and let the function search for some test numbers, e.g. 120, 0, 396, 35 and 2000.

(6) **Integration of function**

Write a function which integrates a one-dimensional function $f(x)$ over the interval $[x_1, x_2]$ via the trapezoidal rule, i.e. calculates

```
SOLUTION SOURCE CODE
DIR: c-programming
FILE(S):
integration.c
```

$$I = \sum_{t=0}^{t_{\max}-1} \frac{f(x_1 + t\Delta) + f(x_1 + (t+1)\Delta)}{2}\Delta$$

$$= \frac{f(x_1)}{2}\Delta + \sum_{t=1}^{t_{\max}-1} f(x_1 + t\Delta)\Delta + \frac{f(x_2)}{2}\Delta$$

where $t_{\max}$ is the number of integration steps and $\Delta = (x_2 - x_1)/t_{\max}$. The function prototype reads as follows:

```
/******************** integrate() ********************/
/** Integrates a 1-dim function numerically using    **/
/** the trapezoidal rule                             **/
/** PARAMETERS: (*)= return-parameter                **/
/**          from: startpoint of interval            **/
/**            to: endpoint of interval              **/
/**     num_steps: number of integration steps       **/
/**             f: p. to function to be integrated    **/
/** RETURNS:                                          **/
/**      value of integral                           **/
/*****************************************************/
double integrate(double from, double to, int num_steps,
                 double (* f)(double))
```

Test the function in your `main()` function by integrating the `sin()` function defined in `math.h` over different intervals, for example $[0, \pi/2]$, $[0, \pi]$.

(7) **Optimizing code by compiler**

Get the program `optimization.c` which calculates for a set of numbers for all possible subsets of numbers the sum. This takes exponentially long in the number of numbers, hence it is a good testbed to see the effect of compiler optimization

```
GET SOURCE CODE
DIR: c-programming
FILE(S):
optimization.c
```

Compile the program via `cc -o optimization optimization.c -lm -Ox`, where x is 0, 1, 2, and 3. Measure the running time for executing the compile program via `time optimization` (note that the program does not print anything in the standard version).

# Chapter 2

# Scripts

Programmers's life can be made much easier when using scripts. The first task after writing the source code is to compile the code. For large software projects, containing many modules, this can be quite an effort if compiling is done by hand. Special scripts, called *Make files*, help a great deal to manage large software and other projects as explained in Sec. 2.1. Next, in Sec. 2.2, *shell scripts* are explained, which allow to perform many tasks automatically. Here, scripts for the *bash* (Unix/Linux) shell will be introduced. In fact, shell-scripting languages are like small programming languages. More sophisticated are extended script languages, like *Python*, which is introduced in Sec. 2.3. They allow to implement rather large projects in a compact way with small effort.

## 2.1  *Make* Files

If your software project grows larger, it will consist of several source-code files. Usually, there are many dependencies between the different files, e.g. a data type defined in one header file can be used in several modules. Consequently, when changing one of your source files, it may be necessary to recompile several parts of the program. In case you do not want to recompile your files every time by hand, you can transfer this task to the *make* tool which can be found on UNIX operating systems. A complete description of the abilities of *make* can be found in Ref. [Oram and Talbott (1991)]. You should look on the *man* page (type man make) or in the texinfo file [Texinfo] as well. Similar tools exists for other operating systems or software development environments. Please consult the manuals in case you are not working with a UNIX type of operating system.

The basic idea of *make* is that you keep a file which contains all dependencies between your source code files. Furthermore, it contains commands (e.g. the compiler command) which generate the resulting files called *targets*, i.e. the final program and/or object (.o) files. Each pair of dependencies and commands is called *rule*. The file containing all rules of a project is called *makefile*, usually it is named `Makefile` and should be placed in the directory where the source files are stored.

A rule can be coded by two lines of the form

$\langle target \rangle$ : $\langle sources \rangle$
<tab> $\langle command(s) \rangle$

The first line contains the dependencies and the second one the commands. The command line must begin with a tabulator symbol `<tab>`, which appears as a space in most editors. It is allowed to have several targets depending on the same sources. You can extend the lines with the backslash "\" at the end of each line. The command line is allowed to be left empty. An example of a dependency/command pair is

```
simulation.o: simulation.c simulation.h
<tab>    cc -c simulation.c
```

This means that the file `simulation.o` has to be compiled if either `simulation.c` or `simulation.h` have been changed. The *make* program is called by typing `make` on the command line of a UNIX shell. It uses the date and time of the last changes performed on each file, which is stored along with each file, to determine whether a rebuild of some targets is necessary. Each time at least one of the source files is newer than the corresponding target files, the commands given after the `<tab>` are executed. Specifically, the command is executed if the target file does not exist at all. In this special case, no source files have to be given after the colon in the first line of the rule.

It is also possible to generate meta rules which, e.g. tell how to treat all files which have a specific suffix. Standard rules, which tell how to treat files ending for example with .c, exist already, but can be changed for each file by stating a different rule. This subject is covered in the *man* page of *make*.

The make tool always tries to build only the first object of your *makefile*, unless enforced by the dependencies. Consequently, if you have to build several independent object files `object1, object2, object3`, the whole compiling must be toggled by the first rule, thus your *makefile* should read like this:

```
all: object1 object2 object3

object1:   <sources of object1>
<tab>      <command to generate object1>

object2: ...
<tab>      <command to generate object2>

object3: ...
<tab>      <command to generate object3>
```

Note that no command is given for the target `all`, because no commands to be executed are connected to it. It is just used to make sure that all objects on which it "depends" (artificially) are up to date. It is not necessary to separate different rules by blank lines. Here, they are just used for better readability. If you just want to rebuild e.g. `object3`, you can call `make object3`. This allows several independent targets to be combined into one *makefile*. When compiling programs via `make`, it is common to include the target "clean" in the *makefile* such that all objects files are removed when `make clean` is called. Thus, the next call of `make` (without further arguments) compiles the whole program again from scratch. The rule for 'clean' reads like

```
clean:
<tab>      rm -f *.o
```

Also, iterated dependencies are allowed, for example

```
object1: object2

object2: object3
<tab> ...

object3: ...
<tab> ...
```

The order of the rules is not important, except that *make* always starts with the first target. Please note that the *make* tool is not just intended to manage the software development process and toggle compile commands. Any project where some output files depend on some input files in an arbitrary way can be controlled. For example, you could control the setting of a book, where you have text-files, figures, a bibliography and an index as input files. The different chapters and finally the whole book are the target files.

Furthermore, it is possible to define variables, sometimes also called macros. They have the format

*variable*= *definition*

Also, variables belonging to your environment, like $HOME, can be referenced in the *makefile*. The value of a variable can be used, similar to shell variables, by placing a $ sign in front of the name of the variable, but you have to embrace the name by (...) or {...}. There are some special variables, e.g. $@ holds the name of the target in each corresponding command line; here no braces are necessary. The variable CC is predefined to hold the compiling command. You can change it by including for example

```
CC=gcc
```

in the *makefile*. In the command part of a rule the compiler is called via $(CC). Thus, you can change your compiler for the whole project very quickly by altering just one line of the *makefile*.

Finally, it will be shown what a typical *makefile* for a small software project might look like. The resulting program is called **simulation**. There are two additional modules init.c, run.c and the corresponding header .h files. In **datatypes.h** types are defined which are used in all modules. Additionally, an external precompiled object file **analysis.o** in the directory $HOME/lib is to be linked whose corresponding header file is assumed to be stored in $HOME/include. For init.o and run.o no commands are given. In this case **make** applies the predefined standard command for files having .o as suffix, which reads like

```
<tab>    $(CC) $(CFLAGS) -c $@
```

where the variable CFLAGS may contain options passed to the compiler and is initially empty. The *makefile* looks like this (please note that lines beginning with #' are comments:

```
#
# sample make file
#
OBJECTS=simulation.o init.o run.o
OBJECTSEXT=$(HOME)/lib/analysis.o
CC=gcc
```

```
CFLAGS=-g -Wall -I$(HOME)/include
LIBS=-lm

simulation: $(OBJECTS) $(OBJECTSEXT)
<tab>    $(CC) $(CFLAGS) -o $@ $(OBJECTS) $(OBJECTSEXT) $(LIBS)

$(OBJECTS): datatypes.h

clean:
<tab>    rm -f *.o
```

The first three lines are comments, then five variables OBJECTS, OBJECTSEXT, CC, CFLAGS, and LIBS are assigned. The final part of the *makefile* are the rules.

Please note that sometimes bugs are introduced if the *makefile* is incomplete. For example, consider a header file which is included in several code files, but this dependency is not mentioned in the *makefile*. Then, for example, if you change a data type in the header file, some of the code files might not be compiled again, especially those you did not change. Thus, the same object files can be treated with different formats in your program, yielding bugs which seem hard to explain. Hence, in case you encounter mysterious bugs, a **make clean** might help. However, most of the time, bugs which are hard to explain are due to errors in your memory management. How to track down those bugs is explained in Chap. 4.

The *make* tool exhibits many other features. For additional details, please consult the references given above.

## 2.2 Shell Scripts

Shell scripts are even more general tools than *make* files. They are in fact small programs, but they are usually not compiled, i.e. they are quickly written but they run slowly. Scripts can be used to perform many administration tasks like backing up data, installing software, or running simulation programs for many different parameters. Here, only an example concerning the last task is presented. For a general introduction to scripts, please refer to a book on your operating system like UNIX/Linux.

Assume that you have a simulation program called **coversim21** which calculates vertex covers of graphs, which are graph-theoretical objects.[1]

---

[1]For the definition of a graph see Sec. 6.8. A *vertex cover* of a graph $G = (V, E)$ is a subset $V' \subset V$ of vertices, such that each edge $\{i, j\} \in E$ is incident to at least one

Assume that you want to run the program for a fixed graph size L, for a fixed concentration c of the edges, average over **num** realizations, and write the results to a file, which contains a string **appendix** in its name to distinguish it from other output files. Furthermore, you want to iterate over different relative sizes x. Then you can use the following script **run.scr** (under UNIX/Linux, specifically the *bash* shell is used):

```
#!/bin/bash
L=$1
c=$2
num=$3
appendix=$4
shift
shift
shift
shift
for x
do
    ${HOME}/cover/coversim21 -mag  $L $c $x $num > \
                        mag_${c}_${x}${appendix}.out
done
```

The first line starting with "#" is a comment line, but it has a special meaning. It tells the operating system the language in which the script is written. In this case it is for the *bash* shell, the absolute pathname of the shell is given. Each UNIX shell is equipped with its own script language. Thus, you can use all commands which are allowed in the shell. There are also more elaborate script languages like *perl* or *Python*. The latter is covered in Sec. 2.3.

Scripts can have command line arguments, which are referred via $1, $2, $3 etc., the name of the script itself being stored in $0. Thus, in the lines 2 to 5, four variables are assigned. In general, you can use the arguments everywhere in the script directly, i.e., it is not necessary to store them in other variables. This is done here, because in the next four lines the arguments $1 to $4 are thrown away by four **shift** commands. Then, the argument which was on position five at the beginning is stored in the first argument, the initially sixth argument is now stored in the second one, and so on. Argument zero, containing the script name, is not affected by the shift.

Next, the script enters a loop, given by "for x; do ... done". This

---

vertex of $V'$, i.e. $i \in V'$ or $j \in V'$.

construction means that iteratively all remaining arguments are assigned to the variable "x" and each time the body of the loop is executed. In this case, the simulation is started via calling the program `coversim21` stored in the directory `${HOME}/cover/`with some parameters (where you do not have to care about their meaning here) and the output directed to a file. Please note that you can state the loop parameters explicitly like in "`for size in 10 20 40 80 160; do ... done`".

The above script can be called, for example, by

```
run.scr 100 0.5 1000 testA 0.20 0.22 0.24 0.26 0.28 0.30
```

which means that the graph size is 100, the fraction of edges is 0.5, the number of realizations per run is 1000, the string `testA` appears in the output file name, and the simulation is performed for the relative sizes 0.20, 0.22, 0.24, 0.26, 0.28, and 0.30.

## 2.3 *Python*

*Python* is an interpreted programming language, somehow half-way between a script language and a full high-level programming language. This makes it easy to write quickly programs, like for script languages, while allowing for rather large-scale projects.

Note that *Python* is in fact a full object-oriented language, so quite large projects are possible. Since the language is interpreted, the program execution is somehow slower compared to compiled languages like C or C++.[2] Thus, we concentrate on using *Python* for writing scripts related to data analysis. For an exhaustive coverage, the reader should consult the documentation on the *Python* web page [Python].

Since *Python* is interpreted, one can use the language interactively. Under UNIX/Linux, one can start it via entering `python` in a shell and enter *Python* commands. This is in particular useful when learning the language. The shell can be stopped via hitting <Ctrl>+<D> or typing `quit()`. Here we concentrate on providing the commands in *Python* script files, which usually end in the appendix `.py`. For a script `loop.py` you can either call *Python* with the name of the script file as argument, i.e.,

```
phython loop.py
```

---

[2]With the *Cython* language, which is very similar to *Python*, one can include external compiled C code to generate fast-running programs.

Alternatively, like in a shell script, you can write in the first line #! followed by the complete path name of the *Python* interpreter.[3] This is done in the following example, which we use to introduce some of the main *Python* concepts:

| GET SOURCE CODE |
| --- |
| DIR: scripts |
| FILE(S): loop.py |

```
1  #!/usr/bin/python
2
3  a=[12, 45, 98, 112, 114, 135, 167, 298, 312]
4  counter=0
5  for value in a:
6    counter=counter+value
7    print value
8  print "sum of", len(a), "elements:", counter
9  print range(7,12)
10 print a[1], a[2:6]
11 a.insert(3,77)
12 print a
13 a.extend([22, 8])
14 print a
15 a.sort()
16 print a
```

In line 3 of this script, a *list* is created, one of the basic data types in *Python*. A list is a linearly ordered sequence of objects, for details see Sec. 6.6. The members of the list are given in square brackets, separated by commas. *Python* allows several commands in a line, separated by semicolons. In contrast to C, a line does not have to be terminated by a semicolon, instead the line end identifies the end of a command as well. It is very important to note that in contrast to C, variables are not declared. This allows also for mixed types, i.e., the list may contain at the same time integers, floating points, strings, and even other lists as elements. Variables exist after they are created and initialized, until the end of the script or until the same variable name is used for something else. Thus, the variable `counter` initialized in line 4 will store integer variables. If one writes instead `counter=0.0` it would hold floating point variables.

In lines 5–7 a `for` loop is performed. A `for` loop is always iterated over the elements of a list (or a string). The body of the loop follows after the colon. In contrast to C, no brackets are used to define the body of the loop. Instead, the body of the loop is identified via indentation: The first line after the `for` statement belongs to the loop and must be indented by at

---

[3]You must make the script executable via `chmod +x loop.py` in this case.

least one space, here two spaces. All subsequent lines with the same level of indentation belong to the loop as well. Thus, line 8 does not belong to the loop. Note that if, for example, line 6 is not indented at all, or if line 7 is indented not by two spaces (current loop body) or zero spaces (level of the main program), an *IndentationError* is issued by the *Python* interpreter.

In line 6, an arithmetic statement is given. It works basically in the same way as in C, i.e., you can use numbers, variables, arithmetic operators, functions and (...) brackets. Note that ** is the power operator, i.e. 2**7 will result in 128. In line 7, a `print` statement is shown. In the most simple case, one can just give an expression after the command, e.g. a variable like here. The execution of the loop results simply in the elements of the list being printed:

```
12
45
98
112
114
135
167
298
312
```

In line 8, also *string* constants, identified by single or double quotes, are printed. Print statements can be formatted by giving several expressions separated by commas, resulting in the expressions being printed one after the other, separated by spaces. Therefore, the execution of this line yields

```
sum of 9 elements: 1293
```

Omitting one comma between two expression in a `print statement` would result in a *SyntaxError*. There are other ways for formatted printing: One can use format strings with '%' as formatting symbols, like in the C `printf()` function, or one can use a string method `format()`. Please refer to the *Python* documentation for these advanced approaches.

The remaining lines 9–16 serve to demonstrate more commands for creating and accessing lists. First, one can easily create lists of consecutive numbers with the `range()` command, see line 9. This is useful for standard loops, for example, `for i in range(3,10)` will loop from 3 to 9. If just one argument is given, the list will start at 0 and end one before the given argument. Also a third argument being the step size can be given, which is by default 1. Also negative step sizes are possible. All arguments are supposed to be integer numbers.

As visible on line 10, the elements of a list can be accessed by specifying the position, just like for an array in C. Parts of a list can be addressed by specifying ranges, called *slicing*. Note that also write access is possible in this way, i.e. a[1]=77 overwrites element 1 of the list. By assigning the empty list [] also sub lists can be removed, i.e. a[2:6]=[] removes elements 2, 3, 4 and 5 from the list. Elements can be inserted by using the **extend** *method*, see line 11. Thereby, the list is extended by appending all elements in the given argument list, [22, 8] The existence of methods signifies that *Python* is actually object oriented, the method is called for the list object. The object-oriented approach easily allows for more complex modifications, for example, to sort a list, see line 15. Other useful methods for lists are insert(i,x) which inserts element x at position i, append(x) which appends one single element x at the end of the list, index(x), which returns the index of the first occurrence of item x, and remove(x), which removes the first occurrence of item x from the list. Note that one could indeed use

```
a.append([22, 8])
```

but this results in a different list where the last element is itself a list consisting of the two numbers 22 and 8. If one wants to use the **append()** method resulting in both numbers to be added as single numbers, one has to use it twice:

```
a.append(22)
a.append(8)
```

So far, we have used the **for** statement to create a loop. More flexible is the while loop, which consists of the **while** statement, a condition, followed by a colon, and a block of commands, e.g.

```
counter = 3.5
while counter < 4.2:
  print counter*counter
  counter = counter + 0.1
```

Another standard way to direct the flow of the execution in a programming language is the **if** statement, detailed by the following example:

```
if reps < 10:
  reps = 10
  print "number of repetitions raised to 10"
```

```
elif reps > 1000:
  reps = 1000
  print "number of repetitions decreased to 1000"
else:
  print "number of repetitions:", reps
```

Note that again blocks of commands are defined via the level of indentation. The number of `elif` ("else if") branches can be zero, one or any higher number. The `else` branch can be omitted.

So far, we have learned about the data types numerical variables and lists. *Python* also offers strings, which have to be provided within single or double quotes, e.g.

| GET SOURCE CODE |
| --- |
| DIR: scripts<br>FILE(S): strings.py |

```
first_name = 'Donald E.'
family_name = "Knuth"
print family_name
print first_name
print len(first_name)
```

where the function `len()` returns the length of a string. Using a backslash \ one can include quotes in a string:

```
quote="The Beatles sang \"We all live in a yellow submarine\""
```

One can access single letters of a string using an array notation, or subsequences using *slicing*:

```
print first_name[3]
print first_name[0:3]
```

which will result in printing a and **Don**, respectively. The right limit of a slice is exclusive. One or both of the limits can be omitted. Note that strings are immutable, i.e. they cannot be changed. If you try an assignment like `first_name[3]='k'` you will get a *SyntaxError*. You can only "change" strings by creating new modified strings. The most important operators are the + operator for concatenation of strings and the * operator for repetition of strings. For example,

```
full_name = first_name + " " + family_name
print full_name

praise= "GREAT "
praise_tape = 10*praise
```

```
print praise_tape
```

will result in

```
Donald E. Knuth
GREAT GREAT GREAT GREAT GREAT GREAT GREAT GREAT GREAT GREAT
```

An even more sophisticated *Python* data type is a *dictionary*, which is like an array but allows for arbitrary objects as indices, called *keys* here. Technically, it is a collection of

| GET SOURCE CODE |
|---|
| DIR: scripts |
| FILE(S): dict.py |

key:value pairs, separated by commas and embraced by curled braces:

```
Z={'H':1, 'He':2, "Li":3}
print Z
print Z['He']
```

will result in

```
{'H': 1, 'Li': 3, 'He': 2}
2
```

Note that there is no inherent order of the elements or keys. For the above example, the keys are strings and the values are integers. Nevertheless, a single dictionary can hold keys or values of mixed type, e.g. one could for the same dictionary add

```
Z[55]=98
Z[78]='structure'
```

It is also possible to overwrite values by assigning a different value to the key, or to remove a key:value pair via the del command. For example,

```
Z[55]=[27,57]
del Z['He']
print Z
```

will result in:

```
{'H': 1, 'Li': 3, 78: 'structure', 55: [27, 57]}
```

It is also possible to extract all keys or all values from a dictionary by using the keys() or values() methods:

```
print Z.keys()
print Z.values()
```

result in

```
['H', 'Li', 78, 55]
[1, 3, 'structure', [27, 57]]
```

In this way one can iterate over all items of a dictionary:

```
for key in Z.keys():
    print "Z[", key, "]=", Z[key]
```

Alternatively, one can use the `iteritems()` method in the following way:

```
for key,value in Z.iteritems():
    print "Z[", key, "]=", value
```

Next, we consider a more complex maybe typical example for treatment of files via *Python*. Suppose you have two files with column-wise data and would like to merge them

| GET SOURCE CODE |
| --- |
| DIR: scripts |
| FILE(S): mix_files.py |

such that you extract, say from the first file columns one and two, as well as column one from the second file to generate a three-column file (under the assumption that the number of lines is equal in both files). This is achieved by the following script:

```
1  #!/usr/bin/python
2
3  # for getting command line arguments
4  import sys
5
6  # read file names from command line
7  file1=sys.argv[1]
8  file2=sys.argv[2]
9
10 # read in file 1
11 file = open(file1, "r")
12 line1=[]
13 for line in file:
14     if (line[0] != '#'):
15         line1.append(line)
16 file.close()
17
18 # read in file 2
19 file = open(file2, "r")
20 line2=[]
```

```
21  for line in file:
22     if (line[0] != '#'):
23        line2.append(line)
24  file.close()
25
26  # print line by line: 1st and 2nd col. of file1, 2nd col. of file2
27  counter=0
28  while counter<len(line2):
29     print line1[counter].split()[0], line1[counter].split()[1], \
30     line2[counter].split()[1]
31     counter=counter+1
```

In line 4 the "sys" module is imported, which allows to access command-line arguments passed to the string. Here, we want to pass two arguments corresponding to the two file names, see line 7 and 8. Note that sys.argv[0] contains the script name, as in the C programming language.

A file is opened via using the **open** command in line 11. The first argument of the command is the file name, the second the access mode. An empty list is created where the lines of the file shall be stored, see line 12. The files is read, line by line and stored in the list, while ignoring all lines starting with a '#', as visible in lines 13–15. After the file is read, it is closed, see line 16. The same procedure is performed for the second file, see lines 19–24.

Finally, the desired composition of the columns is printed in lines 27–31. Here the **split()** method is used, which generates for a given string a list of the space separated words, see lines 29,30.

Function definitions in *Python* are started with the keyword **def** followed by the function name, a comma-separated list of parameters in brackets and a colon. Optionally one can write

| GET SOURCE CODE |
| --- |
| DIR: scripts <br> FILE(S): functions.py |

an indented comment describing the function. In this case, the comment has to be enclosed by a pair of three double quotation marks """, signifying a so called *Python docstring*.

They actual function code is also indented. The function definition ends when the previous or a smaller indentation occurs. A function may return a value via the optional **return** statement, as in the following example:

```
def factorial(n):
    if n <= 1:
        return(1)
    else:
        return(n*factorial(n-1))
```

As usual in *Python*, no types are declared for parameters or for the return type. Thus, you can also pass or return any type of object, like lists, dictionaries or of self-defined classes. Simple variables like numbers are passed by value, i.e., changes to them will be effective only locally inside the function, e.g.

```
def change1(x):
    x=5
    print x

z=10
print z
change1(z)
print z
```

will result in

```
10
5
10
```

Also, it is not immediately possible to change global variables, since any assignment to a variable inside a function will create a local variable of this name. For the (bad-programming style) purpose of using global variables the **global** statement might be used, see the *Python* documentation.

If the objects passed to a function are mutable compound objects like lists or dictionaries, they can be changed inside a function in the usual way, e.g.

```
def change2(l):
    ll=len(l)
    l.append(ll)
list=[]
change2(list)
change2(list)
change2(list)
print list
```

will generate the list [0, 1, 2].

This section is closed by a somehow larger example, which performs a bootstrap resampling of a set of random numbers to obtain an error estimate for the variance of the set of random numbers. For the concept of bootstrapping, see Sec. 8.3.4.

| GET SOURCE CODE |
|---|
| DIR: scripts |
| FILE(S): |
| resampling.py |

We start with a function, which computes the empirical mean, the empirical standard deviation and the error bar of the mean (see Sec. 8.3.1) for a given set of numbers, stored in a list:

```
1   def basicStatistics(myList):
2       av=var=tiny=0.
3
4       # first pass to get mean
5       N  = len(myList)
6       av = float(sum(myList))/N
7       # second pass to get variance
8       for el in myList:
9           dum   = el - av
10          tiny += dum
11          var  += dum*dum
12      # correction step
13      var = (var - tiny*tiny/N)/(N-1)
14      sDev = sqrt(var)
15      sErr = sDev/sqrt(N)
16      return av, sDev, sErr
```

Here a very accurate two-pass algorithm [Press et al. (1995)] is used for the calculation of the variance. Note that in one single **return** statement a *tuple* of three values is returned, see line 16. The elements of a tuple can be accessed separately using the array notation, see the following function which picks the variance from the tuple. In the main program we will use the resampling approach to calculate the error bar of the variance:

```
def Var(x):
    return basicStatistics(x)[1]*basicStatistics(x)[1]
```

The actual resampling is performed in the following function.

```
1   def bootstrap(array,estimFunc,nBootSamp=128):
2     # estimate mean value from original array
3     origEstim=estimFunc(array)
4
5     ## resample data from original array
6     nMax=len(array)
7     h   = [0.0]*nBootSamp
8     bootSamp = [0.0]*nMax
9     for sample in range(nBootSamp):
10      for val in range(nMax):
11        bootSamp[val]=array[randint(0,nMax-1)]
12      h[sample]=estimFunc(bootSamp)
13
14    ## estimate error as std deviation of resampled values
15    resError = basicStatistics(h)[1]
16    return origEstim,resError
```

The function receives three arguments, the sample of data points stored in a list `array`, a function calculating the estimator which is applied to the data points, and the number `nBootSamp` of times a bootstrap sample is taken. As usual, one cannot infer from the arguments which "type" their are, this is only clarified by the way they are used in the body of the function, highlighting the "duck-typing" feature of dynamical programming languages such as *Python*. For example, in line 3 it is clear that `estimFunc` is a function. For the third argument, a default value of 128 is stated, i.e. if one calls `bootstrap` with only two arguments, the default value is used. In line 7, a list (used as an array) `h` of `nBootSamp` entries is initialized with zeros. In the same way, a list `bootSample` of `nMax` entries is initialized in line 8. The main resampling is performed in lines 9–12: `nBooSamp` times a bootstrap sample is taken, i.e., the list `bootSample` is filled with numbers drawn randomly and independently from the original data `array`. Here, the function `randint(0,nMax-1)` is used, which returns a uniform random integer in the range from 0 to `nMax-1`. Note that in the bootstrap sample, every data point may appear more than once. In line 12, the estimator is calculated for the current bootstrap sample and stored in the current entry of `h`. Finally, the variance of the bootstrap estimators (line 15) is the error bar for the actual estimator, which both are returned in line 16.

Note that in the sample program `resampling.py`, also the main function is included, which sets up a random data set, drawn from a uniform distribution, and calculates the variance together with a bootstrap error bar.

# Chapter 3

# Software Engineering

Performing simulations or other large-scale software projects is not only a matter of skillful coding. A lot depends on the organization of the design and the way of programming. Also the methods used for testing and the organization of actually running the simulations and performing the data analysis have a strong influence on the success of a project. This is in particular true, if several people are involved in the project. In this chapter, strategies for performing software projects are discussed. The corresponding computer-science field is called *software engineering*. An introduction to software engineering is given, for example, in Ref. [Sommerville (1989)], while a more practical approach is presented in Ref. [Kernighan and Pike (1999)].

First, a general introduction to software engineering is given (Sec. 3.1). In Sec. 3.2, some hints are given on how to structure the source-code documents efficiently, such that programming, debugging and documentation are facilitated. In the Sec. 3.3, the *subversion* tool is presented, which facilitates the management of any kind of projects, also software projects, in particular if several people cooperate.

## 3.1 How to manage a (simulation) project

When you are creating a program, you should start never immediately writing the code. In this way, only tiny software projects such as scripts can be completed successfully. Otherwise your code will probably be very inflexible and contain several hidden errors which are very hard to find. If several people are involved in a project, it is obvious that a considerable amount of planning is necessary.

But even when you are programming alone, the first step you should undertake is to sit down and think for a while. This will save you a lot of time and effort later on. For medium- to large-scale projects, planning and structuring is indispensable. There are many specialized books in the field of software engineering, see for example Refs. [Sommerville (1989); Ghezzi et al. (1991)]. Here, just the main steps are listed of how to create a sophisticated software development cycle. The following descriptions refer to the usual situation you find in computational science: One or a few people are involved in the project. How to manage the development of big programs involving many developers is explained in the above-mentioned expert literature.

### 3.1.1   *Definition of the problem and solution strategies*

Write down the problem you want to solve. Drawing diagrams is always helpful! Discuss your problem with others and tell them how you would like to solve it. In this context many questions may appear of which some examples are given here:

- What is the input you have to supply? In case you have only a few parameters, they can be passed to the program via command-line arguments or options. In other cases, especially when chemical systems are to be simulated, many parameters have to be controlled and it may be a good idea to use extra parameter files.
- Which results do you want to obtain and which quantities do you have to analyze? Very often it is useful to write the raw results of your simulations, such as the positions of all atoms or the orientations of all spins of your system, to a configuration file. All measurable quantities can be obtained by post-processing. Then, in case new questions arise, it is very easy to analyze the data again. When using configuration files, you should estimate the amount of data you generate. Is there enough space on your disk? It may be helpful, to include the compression of the data files directly in your programs (see page 57).
- Can you identify "objects" in your problem? Objects may be physical entities like atoms or molecules, but also internal structures like nodes in a tree or elements of tables. Considering the system and the program as a hierarchical collection of objects usually makes the problem easier to understand. More details on object-oriented development can be found in Chap. 5.

- Maybe the program will be extended later on? Usually a code is "never" finished. You should foresee later extensions of the program and set up everything in a way to be reused easily.
- Do you have existing programs available which can be included into the software project? If you have implemented your previous projects in the above-mentioned fashion, it is very likely that you can recycle some code. But this requires experience and is not very easy to achieve at the beginning. Over the years, however you will have a growing library of programs which enables you to finish future software projects much faster.
  Has somebody else created a program which you can reuse? Sometimes you can rely on external code like libraries. Examples are the *Standard Template Library* and the *GNU Scientific library* which are covered in Chap. 7.
- Which algorithms are known to solve your problem? Are you sure that you can solve the problem at all? Many other techniques have been invented already. You should always search the literature for existing solutions. How searches can be simplified by using electronic data bases, is covered more deeply in Chap. 9.
  Sometimes it is necessary to invent new methods. This part of a project may be the most time-consuming but also the most interesting task.

### 3.1.2 *Designing data structures*

Once you have identified the basic objects in your systems, you have to think about how to represent them in the code. Sometimes it is sufficient to define some *struct* types in C (or simple *classes* in C++). But usually you will have to design a large set of data structures, referencing each other in a complicated way.

A sophisticated design of the data structures will lead to better organized programs; usually they will even run faster. For example, consider a set of vertices of a graph. Then assume that you have several lists $L_i$ each containing elements referencing the vertices of degree $i$. Maybe the graph is altered in your program and thus the degrees of the vertices change. Then it is sometimes necessary to remove a vertex from one list and insert it into another one. In this case you will gain speed, when your vertex data structures also contain pointers to the positions where they are stored in the lists. Hence, removing and inserting vertices in the lists will take only a constant amount of time. Without these additional pointers, the insert and delete operations have to scan partially through the lists to locate the elements. This leads to a linear time complexity of these operations.

Again, you should perform the design of the data structures in a way, such that later extensions are done more easily. For example, when treating lattices of atoms, you should use data structures which are independent of the dimensionality of the system or even of the structure of the lattice; an example is given in Sec. 1.6.

When you are using external libraries, they usually have some data types included. The above-mentioned Standard Template Library has many predefined data types like arrays, stacks, lists or heaps. Furthermore, it is possible to combine the data types in complicated ways. For instance, you can define a stack of graphs having strings attached to the vertices.

### 3.1.3 *Defining small tasks*

After setting up the basic data types, you should think about which basic and complex operations, i.e. which functions in C/C++, you need to manipulate the objects of your simulation. Since you have already thought a lot about your problem, you have a good overview of which operations may occur. You should break down the ultimate task "perform simulation" into small subtasks. This means, you use a *top-down* approach in the design process. When implementing the design, it is usually not possible to write a program in a sequential way as one code. For the actual implementation, instead a *bottom-up* approach is recommended. This means, you should start with the most basic operations. Later on, you can use them to create more complicated operations. As always, you should define the subroutines in a way such that they can be applied in a flexible way and such that extensions are easy to perform.

But it is not necessary to identify all basic operations at the beginning. During the development of the code, new applications may arise, which lead to the need for further operations. Also, it may be required to change or extend the data structures. However, the more you think in advance, the less you need to change the program later on.

As an example, the problem of finding ground states is considered, i.e. configurations with the lowest energies. A model for magnetic alloys consisting of little magnetic moments called *spins* is used. The method is called *simulated annealing*, which is a special Monte Carlo technique. Some of the basic operations are:

- Set up the data structures for storing the realizations of the interactions and the spin configurations.
- Create a random realization of the interactions.
- Initialize a random spin configuration.
- Calculate the energy of a spin in the local field of its neighbors.
- Calculate the total energy of a system.
- Calculate the energy changes associated with a spin flip.
- Execute a Monte Carlo step.
- Execute a whole annealing run.
- Calculate the magnetization.
- Save a realization and the corresponding spin configurations in a file.

It is not necessary to define a corresponding subroutine for all operations. Sometimes they require only few lines in the code, like the calculation of the energy of one spin in the example above. In this case, such operations could be written directly in the code, or a macro (see Sec. 1.6) could be used.

### 3.1.4  *Distributing work*

In case several people are involved in a project, the next step is to split up the work between the coworkers. If several types of objects appear in the program design, a natural approach is to make everyone responsible for one or several types of objects and the related operations. The code should be broken up into several modules (i.e. source files), such that every module is written by only one person. This makes the implementation easier and also helps testing the code (see below). Nevertheless, the partitioning of the work requires much care, since quite often some modules or data types depend on others. For this reason, the actual implementation of a data type should be hidden. This means that all interactions should be performed through exactly defined interfaces which do not depend on the internal representation, see also Chap. 5 on object-oriented programming.

When several people are editing the same files, then you should use a *version control system* (VCS), see Sec. 3.3. It helps to keep track of all versions, in particular it supports several people performing changes in the same file at the same time. This can be quite painful without a VCS.

### 3.1.5  *Implementing the code*

With good preparation, the actual implementation becomes only a small part of the software development process. General style rules guarantee clear-structured codes as explained in Sec. 3.2. Following style rules helps a lot to understand a code several months after you, or someone else, has written it. You should use a different file, i.e. a different module, for each set of closely related data structures and functions; when using an object oriented language, you should define different classes (see Chap. 5). This rule should be obeyed for the case of a one-person project as well. Large software projects containing many modules are easily maintained via *make-files* (see Sec. 2.1).

Each function and each module should be tested separately, before combining many modules into one program. In the following, some general hints concerning testing are presented.

### 3.1.6  *Testing*

When performing tests on single subroutines, usually standard test cases are used. This is the reason why many errors become apparent much later. When modules have already been combined into one single program, errors are much harder to find. For this reason, you should always try to find also special and rare cases when testing a subroutine. Consider, for example, a procedure which inserts an element into a list. Then, not only inserting in the middle of the list, but also at the beginning, at the end and into an empty list must be tested. Also, it is strongly recommended to read your code carefully once again before considering it finished. In this way many bugs can be found easily which otherwise must be tracked down by intensive debugging.

The actual debugging of the code can be performed by placing print instructions at selected positions in the code. But this approach is quite time-consuming, because you have to modify and recompile your program several times. Therefore, it is advisable to use debugging tools like a *source-code debugger* and a program for checking the memory management. More information about these tools can be found in Chap. 4. But usually, you also need special operations which are not covered by an available tool. You should always write a procedure which prints out the current instance of the system that is simulated, such as the nodes and edges of a graph or the interaction constants of an disordered Ising system. This facilitates the types of tests, which are described in the following.

After the raw operation of the subroutines has been verified, more complex tests can be performed. When, for example, testing an optimization routine, you should compare the outcome of the calculation for a small system with the result which can be obtained by hand. This allows you to follow the execution of the program step by step. If the outcome differs from the expected result, the small size of the test system enables you much better to find the reason for the discrepancy. For each operation you should think about the expected outcome and compare it with the result originating from the running program.

Furthermore, it is very useful to compare the outcome of different methods applied to the same problem. For example, you know that there must be something wrong, in case an approximation method finds a better value than your "exact" algorithm. Sometimes analytical solutions are available, at least for special cases. Another approach is to use invariants. For example, when performing a Molecular Dynamics simulation of an atomic/molecular system (or a galaxy), energy and momentum must be conserved; only numerical rounding errors should appear. These quantities can be recorded very easily. If they change in time, there must be a bug in your code. In this case, usually the formulas for the energy and the force are not compatible or the integration subroutine has a bug.

You should test each procedure directly after writing it. Many developers have experienced that if the interval between implementation and tests is large, then the motivation for performing tests becomes very low, resulting in more undetected bugs.

The final stage of the testing process begins when several modules are integrated into one large running program. In the case where you are writing the code alone, not many surprises should appear, if you have performed many tests on the single-module level. If several people are involved in the project, many errors occur at this stage. But in any case, you should remember: There is probably no complex program which is free of bugs. Always remember the following important result from theoretical computer science [Lewis and Papadimitriou (1981)]: It is impossible to invent a general method, which can prove automatically that a given program obeys a given specification. Thus, all tests must be designed to match the current code.

In case a program is changed or extended several times, always keep the old versions. Quite commonly, new bugs are introduced by changing the code, even if the change is small. In that case, you can compare your new code with the older version. Please note that many editors only keep the second latest version as backup, so you have to take care of this problem yourself, unless you use a version control system, which keeps all older version automatically.

For C programmers, it is always advisable to apply the `-Wall` (warning level: all) option. Then several bugs already show up during the compiling process, for example, the common mistake to use '=' in comparisons instead of '==', or the access to uninitialized variables.[1]

In C++, some bugs can be detected by defining variables or parameter as `const`, when they are considered to stay unchanged in a block of code or subroutine. Here again, you will receive an error message already at compile stage, if attempts are made to alter the value of such a variable.

This part finishes with a warning: Never try to save time when performing tests. Bugs which appear later on are much harder to find and you will have to spend a lot more time than you have "saved" before.

### 3.1.7 *Writing documentation*

This part of the software development process is very often disregarded, especially in the context of scientific research, where no direct customers exist. But even if you are using your own code, you should write good documentation. It should consist of at least three parts:

- *Comments in the source code*: You should place comments at the beginning of each module, in front of each subroutine or each self-defined data structure, for blocks of the code and for selected lines. Additionally, meaningful names for the variables are crucial. Following these rules makes later changes and extension of the program much more straightforward. You will find more hints on how a good programming style can be achieved in Sec. 3.2.
- *On-line help*: You should include a short description of the program, its parameters and its options in the main program. It should be printed, when the program is called with the wrong number/form of the parameters, or when the option `-help` is passed. Even if you are the author of the program, it is quite hard to remember all options and usages.
- *External documentation*: This part of the documentation process is important, when you would like to make the program available to other users or when it grows really complex. Writing good instructions is

---

[1]But this is not true for some C++ compilers when combining with option -g.

really a hard job. When you remember how often you have complained about the instructions for a video recorder or a word processor, you will understand why there is a high demand for good authors of documentation in industry.

### 3.1.8  *Using the code*

Performing the actual simulations usually requires careful preparation. Several questions have to be considered, for example:

- How long will the different runs take? Either by analysing your algorithm you will know, whether your simulation is $\mathcal{O}(n)$, $\mathcal{O}(n^k)$, $\mathcal{O}(e^n)$ or something else. If you do not know, you should perform simulations of small systems and extrapolate to large system sizes.
- Often you have to average over different runs or over several realizations of a "disordered system". This occurs, for example, when simulating random graphs or alloys. The system sizes should in this case be chosen in a way that the number of samples is large enough to reduce the statistical fluctuations. It is better to have good statistics for a small system than bad statistics for a large system. If you are lucky, your model exhibits *self-averaging*. This means, the larger the sample, the less the number of samples can be. Nevertheless, usually the numerical effort grows stronger than the system size, so there will be a maximum system size which can be treated with satisfying accuracy. To estimate the accuracy, you should always calculate the statistical error bar $\sigma(A)$ for each quantity; see Chap. 8 on statistical analysis.
  A good rule of a thumb is that each sample should take no more than 10 minutes. When you have many computers and much time available, you can handle larger problems as well.
- Where to put the results? In many cases you have to investigate your model for different parameters. You should organize the directories where you put the data and the names of the files in such a way that the former results can be found quickly even years later. You should put a README file in each directory, explaining what it contains. Note that organizing large-scale simulations in a useful way is an active area of research called *computational provenance* [Comp. Sci. Eng. (2008)]. If you want to start a sequence of several simulations, you can write a short script, which calls your program with different parameters within a loop; see the example script in Sec. 2.2.

- The program should write some information about the ongoing processes into logfiles during each simulation. The logfile should state the version number of the program and the parameters which have been used to start the simulation in the first few lines. This facilitates a reconstruction of how the results have been obtained.

The steps listed here do not usually occur in linear order. It is quite common that after you have written a program and performed some simulations, you are not satisfied with the performance or new questions arise. Then you start to define new problems and the program will be extended. It may also be necessary to extend the data structures, when for instance, new attributes of the simulated models have to be included. It is also possible that a nasty bug is still hidden in the program, which is found later on during the actual simulations and becomes obvious by results which cannot be explained. In this case, changes cannot be circumvented either.

In other words, the software development process is a *cycle* which is traversed several times. As a consequence, when planning your code, you should always keep this in mind and set up everything in a flexible way, so that extensions and code recycling can be performed easily.

## 3.2 Programming Style

The code should be written in a style that enables the author, and other people as well, to understand and modify the program even years later. Here, some principles are stated briefly that you should follow. Just a general style of description is given. Everybody is free to choose his/her own style, as long as it is precise and consistent.

- Split your code into several modules. This has several advantages:
  - Performing changes, you have to recompile only the modules which have been edited. Otherwise, if everything is contained in a long file, the whole program has to be recompiled each time again.
  - Functions which are related to each other can be collected in single modules. It is much easier to navigate in several short files than in one large program.
  - Having been finished and tested, a module can be used for other projects. Thus, software reuse is facilitated.
  - Distributing the work among several people is impossible if everything is written into one file. Furthermore, you should use a version

control system (see Sec. 3.3) in case several people are involved. This helps to avoid uncontrolled editing.

- To keep your program logically structured, you should always put data structures and implementations of the operations in separate files. In C/C++, this means, you have to write the data structures in a header (.h) file and the code into a source code (.c/ .cpp) file.
- Try to find meaningful names for your variables and subroutines. Thus, during the programming process it is much easier to remember their meanings, which helps a lot in avoiding bugs. Additionally, it is not necessary to look up the meaning frequently. For local variables like loop counters, it is sufficient and more convenient to have short (e.g. one letter) names.

  In the beginning, this might seem to take additional time (e.g. writing 'kinetic_energy' for a variable instead of 'x10'). But several months after you have written the program, you will appreciate your effort, when you read the line

```
kinetic_energy += 0.5*atom[i].mass*atom[i].veloc*atom[i].veloc;
```

  instead of

```
x10 += 0.5*x34[i].a*x34[i].b*x34[i].b;
```

- You should use proper indentation of your lines. This helps a great deal in recognizing the structure of a program. Many bugs are caused by misaligned braces forming a block of code. Furthermore, you should place at most one command per line of code. The reader will probably agree that

```
for(i=0; i<number_nodes; i++)
{
  degree[i] = 0;
  for(j=0; j<number_nodes; j++)
    if(edge[i][j] > 0)
      degree[i]++;
}
```

  is much faster to understand than

```
for(i=0; i<number_nodes; i++) { degree[i] = 0; for(j=0;
j<number_nodes; j++)    if(edge[i][j] > 0)  degree[i]++; }
```

- Avoid jumping to other parts of a program via the "goto" command. This is bad style originating from programming in assembler or BASIC. In modern programming languages, there are corresponding commands for every logical programming construct. "Goto" commands make a program harder to understand and much harder to debug if it does not work as it should.

  In case you want to break out of a loop, you can use a while/until loop with a flag that indicates if the loop is to be stopped. In C, you can also use the commands `break` or `continue`.

- Do not use global variables. At first sight, the use of global variables may seem tempting: You do not have to care about parameters for subroutines; the variables are accessible everywhere and everywhere they have the same name. Programming is done much faster.

  But later on you will have a bad time: Many bugs are created by improper use of global variables. When you want to look up the definition of a variable you have to search the whole list of global variables instead of just checking the parameter list. Sometimes the range of validity of a global variable is overwritten by a local variable. Furthermore, software re-usage is almost impossible with global variables, because you always have to check *all* variables used in a module for conflicts and you are not allowed to employ the name for another object. When you want to pass an object to a subroutine via a global variable, you do not have the choice of how to name the object which is to be passed. Most important, when you have a look into a subroutine after some months, you cannot see immediately which objects are changed in the subroutine; instead, you will have to read the whole subroutine again. If you avoid this practice, you just have to look at the parameter list. Most annoying, when a renaming occurs, you have to change the name of a global variable everywhere in the whole program. Local variables can be changed with little effort.

- Finally, an issue of utmost importance: Do not be economical with comments in your source code! Most programs, which may appear logically structured when writing them, will be a source of great confusion when being read some weeks later. Every minute you spend on writing reasonable comments you will save several times over later on. You should consider different types of comments.

  - Module comments: At the beginning of each module you should state its name, what the module does, who wrote it and when it was written. It is a useful practice to include a version history, which lists the changes that have been performed. A module comment might look like this:

```
**********************************************************/
/*** Functions for spin glasses.                      ***/
/*** 1. loading and saving of configurations          ***/
/*** 2. initialization                                ***/
/*** 3. evaluation functions                          ***/
/***                                                  ***/
/*** A.K. Hartmann  January 1996                      ***/
/*** Version 1.8    09.10.2000                        ***/
/***                                                  ***/
/**********************************************************/

/*** Vers. History:                                   ***/
/*** 1.0 feof-check in lsg_load...() included 02.03.96 ***/
/*** 2.0 comment for cs2html added           12.05.96 ***/
/*** 3.0 lsg_load_bond_n() added             03.03.97 ***/
/*** 4.0 lsg_invert_plane() added            12.08.98 ***/
/*** 5.0 lsg_write_gen() added               15.09.98 ***/
/*** 6.0 lsg_energy_B_hom() added            20.11.98 ***/
/*** 7.0 lsg_frac_frust() added              03.07.00 ***/
/*** 7.1 use new call-form of llist.c library 04.07.00 ***/
/***       -> no memory leak (through copy data)      ***/
/*** 8.0 lsg_mc_T() added                    23.08.00 ***/
```

- Type comments: For each data type (a structure in C or class in C++) which you define in a header file, you should attach several lines of comments describing the data type's structure and its application. For a class definition, also the methods which are available should be described. Furthermore, for a structure, each element should be explained. A nice arrangement of the comments makes everything more readable. An example of what such a comment may look like can be seen in Chap. 5 for the data type histo_t.
- Function comments: For each function, its purpose, the meaning of the input and output variables and the preconditions which have to be fulfilled before calling must be stated. In case you are lazy and do not write a *man* page, a comment atop of a subroutine is the only source of information, if you want to use the subroutine later on in another program.

If you use some special mathematical methods or clever algorithms in the subroutine, you should always cite the source in the comment.

Later on, this facilitates the understanding of how the methods works.

The next example shows what the comment for a function may look like:

```
/*********************** mf_dinic1() ****************/
/** Calculated maximum flow using Dinics algorithm    **/
/** See: R.E.Tarjan, Data Structures and Network      **/
/** Algorithms, p.104f.                               **/
/**                                                   **/
/** PARAMETERS: (*)= return-parameter/altered var's   **/
/**         N: number of inner nodes (without s,t)    **/
/**       dim: dimension of lattice                   **/
/**      next: gives neighbors next[0..N][0..2*dim+1] **/
/**         c: capacities  c[0..N][0..2*dim+1]        **/
/**    (*) f: flow values f[0..N][0..2*dim+1]         **/
/** use_flow: 0-> flow set to zero before used.       **/
/**                                                   **/
/** RETURNS:                                          **/
/**          0 -> OK                                  **/
/****************************************************/
int mf_dinic1(int N, int dim, int *next, int *c,
              int *f, int use_flow)
```

– Block comments: You should divide each subroutine, unless it is very short, into several logical blocks. A rule of thumb is that no block should be longer than the number of lines you can display in your editor window. Within one or two lines you should explain what is done in the block. Example:

```
/* go through all nodes except source s and sink t in  */
/* reversed topological order and set capacities       */
for(t2=num_nodes-2; t2>0; t2--)
    ...
```

– Line comments: They are the lowest level comments. Since you are using (hopefully) meaningful names for data types, variables and subroutines, many lines should be self-explanatory. But in case the meaning is not obvious, you should add a small comment at the end of a line, for example:

```
C(t, SOURCE) = cap_s2t[t];     /* restore capacities */
```

Aligning all comments to the right makes a code easier to read. Please avoid unnecessary comments like

```
counter++;                              /* increase counter */
```

or unintelligible comments like

```
minimize_energy(spin, N, next, 5);    /* I try this one */
```

The line containing C(t, SOURCE) is an example of how to apply a macro. This subject is covered in Sec. 1.6.

Finally: The author of this book writes module, function and block comments *before* he starts to write the corresponding first line of the code. This forces him to think more clearly about the structure of the program. You should do the same.

## 3.3 Version management with *subversion*

All simulation projects will evolve over time. This is reflected by changes of documents like design descriptions, program source codes or papers where the simulation algorithms and results are going to be published. If you are just one person performing the project, i.e. one author, and if the project is not very complicated, you can just keep always the "current" files; hence, you do not perform explicit version management.

Nevertheless, you often need older versions, for example, to compare an improved algorithm to a previous version. This can happen as well if the latest version of your program behaves strangely, because you have entered a little bug in your code. Or, when writing your paper, you can restore a chapter which has accidentally been deleted.

To handle these cases, you could run your hand-made version management, for example, by naming your paper files config1v1.jpg, config1v2.jpg, config1v3.jpg and so on. In many cases, this works fine, but it is a bit waste of disk space, because each version is stored completely even if two successive versions differ only slightly.

In case several authors work on a project, hand-made version management becomes a bit more complicated. One standard way is to agree that only one author is allowed to edit each file at a time. One says, the file is *locked* by the current author. In practice, this means that you have mutually to agree who works next on which file. This is not bad, because you have to communicate anyway a lot to organize a project efficiently. This means, if an author wants to change a locked file, he has to wait with the changes until the file is unlocked again. Also, the lock is only virtual, i.e. it may happen accidentally that two person edit the same file. In this case, the two authors have to create a new latest version by merging the two personal files in some way.

Most of your work when using hand-made version management can be done automatically by a *version control system* (VCS). This means, the VCS stores *all* versions of the files. Furthermore, the VCS does the version counting for you. Each author can *update* at any time his or her local copy such that he or she has always the latest versions. One can perform changes to the files and *commit* the new version to the VCS, such that the other authors can use them as well. Also, one can obtain any older version or undo changes at any time.

Here, the *subversion* tool is presented. It is able to control every project which consists of a directory in your system. The directory may contain any types of files, also other directories or subdirectories, i.e. a full directory tree. This means that *subversion* is able to control the versions of arbitrary projects, such as program development, paper writing, or planning of large-scale birthday parties. Note that *subversion* does not store the different revisions completely. To save disk space, only the first version is stored. For later revisions, only the difference from the previous version is stored, respectively. Note that the version numbers, called *revisions* here, are given to the complete directory tree. This means, even if just one single file is changed, all files get virtually a new revision number. Internally, *subversion* stores by default the files in a *repository*, which is usually just a directory as well, containing files in custom file formats. As alternative, one can tell *subversion* explicitly to use a Berkeley data base system, but this is slower than the default manner.

By default, *subversion* does *not* perform locking. This means, several authors can change a file in parallel. Usually, these changes will not conflict with each other. In this case, *subversion* will automatically create a new revision, which contains the changes of all authors. Note that the term conflict applies only to the level of pure text comparison. Thus, if one author introduces in one part of the code a data type, which is different from what another author expects to use in his subroutine written elsewhere, *subversion* will not be able to create a working code, although the code will have no conflicts. If conflicts occur, for example, if two authors change the same paragraph of a text, then *subversion* will tell you about the conflict. Nevertheless, the authors have to resolve the conflict manually, usually by speaking to each other, which is a good idea anyway. Note that *subversion* is also able to lock documents, if the authors ask for this explicitly.

Next, the basic steps of using *subversion* are explained step by step. It is assumed that your system administrator has already installed a running version. Here, version 1.4 is discussed. Note that *subversion* is command-line based, but there exist also graphical front-ends.

To create a new repository, you have to use the program `svnadmin` with the command `create`. You have to specify a directory where the repository is to be created. Here, it is assumed that the directory `/home/hartmann/svn/` already exists (the commands were issued in the following example on the computer "Gene", the user `hartmann` having the current directory `se`):

```
Gene:se>svnadmin create /home/hartmann/svn/repos1
```

Then the subdirectory `repos1` will be created,[2] and initialized as empty repository. This means, some subdirectories and some files are created inside `repos1`.

To actually put something into the repository, you can *import* a complete directory tree. For this purpose, you use the program `svn`, which is the main workhorse of *subversion*, with the command `import`. The command is always the first argument passed to `svn`. You can use the `help` command to learn more about *subversion* and its commands, for example `svn help import`. As arguments for the import, you have to specify the directory where the files are located, a directory in the repository and a message via the option `-m`:

```
Gene:se>svn import list file:///home/hartmann/svn/repos1/list \
        -m "init repository"
Adding         list/list_main.c
Adding         list/list.h
Adding         list/list_remove_element.c
Adding         list/list_mergesort.c
Adding         list/list.c

Committed revision 1.
```

To specify the directory inside the repository, one has to use the URL syntax. Here, since the repository is stored locally, the URL prefix is `file://`. For repositories with distributed authors, the system administrator can activate remote access (see below). In this case, either standard URLs (`http://`, `https://`) or special *subversion* access (`svn://`, `svn+ssh://`) are feasible. Please have a look at the documentation for details.

---

[2]It may already exist as empty directory, which would be no harm.

In this example, the files (and all subdirectories recursively) of the directory list are put into a repository directory, which will be created automatically. Here, the repository directory has the same name as the imported directory, but they can have different names. It is also possible to specify the repository without subdirectory (file:///home/hartmann/svn/repos1/). On the other hand, you can also specify a longer pathname if necessary. All directories and subdirectories in the repository are automatically created, if they do not exist. When directly listing the content of a repository using standard operating system commands, you will not be able to see the files or directories you have put in. They are stored internally.

Note that if you do not specify the -m option, *subversion* will call an editor, where you have to enter a message.

An import command is issued only *once* for each top-level repository directory. For later changes you have to use the commands commit, add, or remove, see below. Remember always that *all* files and subdirectories of the given directory are imported. Hence, you should have a *clean* directory, when you perform an import. For a programming project, for example, this means there should be no object files .o or auxiliary files present.

With the list command, you can view the content of the repository or of subdirectories. Each time only one directory level is shown:[3]

```
Gene:se>svn list file:///home/hartmann/svn/repos1/
list/
Gene:se>svn list file:///home/hartmann/svn/repos1/list
list.c
list.h
list_main.c
list_mergesort.c
list_remove_element.c
```

Before you start editing some files the first time, you should first *checkout* the current revision of the project to create a *working copy*. This starts a so-called *working cycle*. This is done using the checkout command. As second argument, again the URL of the repository path has to be specified:

```
Gene:listtest>svn checkout file:///home/hartmann/svn/repos1/list
A    list/list_main.c
A    list/list.h
A    list/list_remove_element.c
A    list/list_mergesort.c
A    list/list.c
Checked out revision 1.
```

---

[3]Unless you use the option -R.

This will create a subdirectory in the current directory (`listtest` here), which contains the files (and possible other subdirectories recursively) contained in the given repository directory. Note that it is also possible to check out just subdirectories. Hence, if `list` contained a subdirectory `docs`, you could specify `file:///home/hartmann/svn/repos1/list/docs` as well. By default, *subversion* uses the base name as the name of the directory which is created; in this case, the subdirectory `list/docs` would be created. Also, one can give a third optional argument which specifies the path of the directory or directories being checked out; hence, specifying `ldocs` as forth argument would put all repository files and directories of `list/docs` into `./ldocs`. Please read the online help (`svn help checkout`) for more information.

In addition to copying the regular files from the repository, *subversion* will create a subdirectory `.svn` in all directories which are checked out. These directories contain all system information needed by *subversion* to administrate the working files. Please do not remove or edit the files in these directories.

You can check out any directory or subdirectory of the repository as often as you like. This means, you can have many working versions in parallel (if you do not get confused by this).

Now you may edit some files (say after changing the directory to `list`). This is done using the *emacs* editor here. Editing and other operations change the *status* of a file, which can be inspected using the *subversion* command `status`. You can perform these operations locally, which means that you do not have to enter the repository URL:

```
Gene:list>emacs list_main.c
Gene:list>svn status
M       list_main.c
```

The letter 'M' indicates that the file is *modified*. Other important status values are 'A' for a file or directory which has been added using `svn add` or copied using `svn copy`, 'D' for a file or directory which has been deleted using `svn delete`. If you use `svn move`, the old file or directory will be deleted and the new one created. Do *never* use operating system commands like `rm` or `mv` to remove files which are under control of *subversion*, because this would mess up the whole VCS. Finally, the status '?' is used for files or directories which are not under the control of *subversion*.

Once you are finished editing your files, you should finish a working cycle by *committing* the files back to the repository using the `commit` command. Again a message has to be passed, either via the option `-m` or interactively:

```
Gene:list>svn commit -m "exchange two main parts"
Sending        list_main.c
Transmitting file data .
Committed revision 2.
```

After the committing process is finished, *subversion* tells you that this has created a new revision, `revision 2` in this case.

By default, other authors are *not* informed about the new revision. But it is possible to use *hooks*. Hooks are Unix shell scripts (see Sec. 2.2) made known to *subversion*. They are run either before or after a *subversion* command is performed. These scripts are placed in the `hook` subdirectory of your repository. By default, there are no active scripts, but the directory contains some inactive template scripts, which can be used to implement your own hooks. For example, you may want a mail is to be sent to all authors informing them about the latest commit.

You can see the differences between the last updated revision and the files in your working directory using then command `diff`. Say, you perform further changes to `list:main.c`, then you get:

```
Gene:list>svn diff
Index: list_main.c
===================================================================
--- list_main.c (revision 2)
+++ list_main.c (working copy)
@@ -35,11 +35,11 @@
    list = insert_element(list, elem2, elem);
    print_list(list);

-   /*list = mergesort_list(list);
+   list = mergesort_list(list);
    printf("sort list\n");
-   print_list(list);*/
+   print_list(list);

-   printf("delete 6:\n");
+   /*printf("delete 6:\n");
    elem = search_info(list, 6);
```

```
   list = remove_element(list, elem);
   delete_element(elem);
@@ -49,7 +49,7 @@
   elem = list;
   list = remove_element(list, elem);
   delete_element(elem);
-  print_list(list);
+  print_list(list);*/

   return(0);
}
```

The output is in *diff* format. The lines of the old revision are marked by '-' symbols, the lines of the working copy by '+' symbols. Line numbers (where each block of empty lines is counted as one line) are given after the @@ symbols. For convenience, also some lines are shown which have not been changed.

If other authors performed changes to project files in the meantime, you can get the latest revision by entering **svn update** (having the working directory as current directory). This will update all files in your current working directory to the latest revision. It is possible that some files in the repository were changed which you have changed as well. In this case, *subversion* will automatically merge the file from the latest revision with the working file, if no *conflicts* from contradicting changes to the same text positions arise. You *must* perform an update, before you can commit a file which has been changed in the repository in the meantime:

```
Gene:list>svn commit -m "element 8"
Sending        list_main.c
svn: Commit failed (details follow):
svn: Out of date: '/list/list_main.c' in transaction '3-1'
```

This shows that you are forced to perform an update. If no conflict arises, everything will be fine. If a conflict arises, in some versions of *subversion* you have the opportunity to resolve the conflicts interactively. We consider the most general case, where you resolve the conflicts "by hand" (which is the default in our example):

```
Gene:list>svn update
C   list_main.c
Gene:list>ls list_main.c*
list_main.c  list_main.c.mine  list_main.c.r2  list_main.c.r3
```

Suppose that your changes are based on revision 2 (.r2), while the latest revsion is .r3. In this case, the update has created four files: list_main.c.r2 contains the revision your changes are based on, while list_main.c.r3 contains the latest revision. The file list_main.c.mine contains your working file which was named list_main.c, before the update was performed. Finally, list_main.c now contains a "mix" of your working file and the latest revision. The parts where conflicts have been found are indicated between extra lines of the form

```
<<<<<<< .mine
⟨text of your working file⟩
=======
⟨text of the latest reversion⟩
>>>>>>> .r3
```

Now you can edit the file list_main.c and create a consistent version. Usually, you have to speak to the author of the latest revision to resolve the conflicts most efficiently and to satisfy all authors. In the resulting file all auxiliary lines containing <<<<<<<, ======= and >>>>>>> should be removed. Then you have to tell *subversion* that you have resolved the conflicts using the resolved command. This will remove all auxiliary files. Then you can commit:

```
Gene:list>emacs list_main.c
Gene:list>svn resolved list_main.c
Resolved conflicted state of 'list_main.c'
Gene:list>ls
list.c  list.h  list_main.c  list_mergesort.c  list_remove_element.c
Gene:list>svn commit -m "element 8"
Sending        list_main.c
Transmitting file data .
Committed revision 4.
```

You can have a look at the past revision history using the command log. Also the messages describing the changes are shown, for example:

```
Gene:list>svn log
-------------------------------------------------------------------
r3 | hartmann | 2008-11-08 13:00:29 +0100 (Sa, 08 Nov 2008) | 1 line

element 7
-------------------------------------------------------------------
r2 | hartmann | 2008-11-08 12:48:05 +0100 (Sa, 08 Nov 2008) | 1 line

exchange two main parts
-------------------------------------------------------------------
r1 | hartmann | 2008-11-08 12:38:47 +0100 (Sa, 08 Nov 2008) | 1 line

init
-------------------------------------------------------------------
```

So far, all examples were for the case of one author or, if the access rights for the repository are set accordingly, also for several authors. In any case, the repository is accessed only locally.

Often, people from different places cooperate in a project. In this case, one must have access to the repository across the web. For the web access, your local system administrator has to install a server. There are currently two possibilities. The first one is to use *svnserve*, which is very fast. Practically, when setting up a repository, you have usually to use a special directory as parent directory; please ask your system administrator. After creating your repository, you should create or change the **passwd** file in the **conf** subdirectory of your repository, e.g.

```
[users]
alex = project1
bernd = project2
stefan = project3
```

This creates three "users" (not to be confused with the normal user accounts on your system) for the repository, which have the passwords **project1**, **project2** and **project3**, respectively. Note that the passwords are not encrypted; hence, you should not use very secret passwords for this purpose.

Furthermore, in the **svnserve.conf** file in the same **conf** subdirectory, you have to uncomment the following lines (i.e. remove the '#' symbol in front of them):

```
[general]
auth-access = write
password-db = passwd
realm = This repository contains ...
```

The first line contains a tag, the second states that only authenticated users have write access to the repository, i.e. only they can commit files. The third line states the password file as entered above. The fourth line should describe your repository. Here, you should put a meaningful short description, which will be shown when logging in. When using *svnserve*, you should use URLs starting with `svn://` for the commands instead of URLs beginning with `file://` as above, for example for the commands `checkout` or `list`:

`svn checkout svn://svn.physik.uni-oldenburg.de/hartmann/repos1`

All commands, where no URL is passed as argument, such as `update` and `commit`, work like in the single-user case.

Another access method to *subversion* repositories is based on the *Apache* server, in this case the URLs will begin with the standard `http://` prefix. Details about the server configuration with *Apache*, also more information for the *svnserve* case, can be found in the documentation [SVN].

# Chapter 4

# Debugging and Testing

In Chap. 3, the importance of thorough testing has already been stressed. Here, four useful tools are presented which significantly assist in the debugging process. Please note again that the tools run under UNIX/Linux operating systems. Similar programs are available for other operating systems as well. The tools covered here are *gdb*, a source-code debugger, *ddd*, a graphic front-end to *gdb*, *valgrind*, which finds bugs resulting from bad memory management, and finally *gprof*, which assists in finding running-time consuming parts of your program.

## 4.1  *gdb*

The *gdb* gnu debugger tool is a *source code debugger*. Using *gdb* you can watch the execution of your code "live". You can stop the program at arbitrarily chosen points by setting *breakpoints* at any lines or subroutines in the source code, inspect variables/data structures, change them and let the program continue (e.g. line by line). Here, some examples for the most basic operations are given, detailed instructions can be obtained within the program via the **help** command.

As an example of how to debug, please consider a program with the following little **main()** function. The program does not do anything really meaningful, it just allocates an array, fills

| GET SOURCE CODE |
| --- |
| DIR: debugging FILE(S): gdbtest.c |

it with numbers, calculates the sum of the number and prints the sum:

```
26  int main(int argc, char *argv[])
27  {
28      int t, *array, sum = 0;
29
30      array = (int *) malloc (100*sizeof(int));
31      for(t=0; t<100; t++)
32          array[t] = t;
33      for(t=0; t<100; t++)
34          sum += array[t];
35      printf("sum= %d\n", sum);
36      free(array);
37      return(0);
38  }
```

When compiling the code you have to include the option -g to allow debugging:

```
cc -o gdbtest -g gdbtest.c
```

The debugger is invoked using **gdb** ⟨program name⟩, i.e.

```
gdb gdbtest
```

Now you can enter commands in textual form. It is very useful to list the source code of the program via the `list` command, it is sufficient to enter just l. By default always ten lines around the *current* position are printed. Therefore, at the beginning, ten lines near the beginning of `main()` are shown[1] (the line with `(gdb)` shows the input, the other lines state the answer of the debugger):

```
(gdb) l
19
20          for (t=0; t<n; t++)
21              sum += array[t];
22          return(sum);
23      }
24
25
26      int main(int argc, char *argv[])
27      {
28          int t, *array, sum = 0;
(gdb)
```

---

[1] At the top you see some lines of another function of the program, which is discussed below.

When entering the command l again, the next ten lines are listed. Furthermore, you can refer to program lines of the code in the form list ⟨from⟩, ⟨to⟩ or to functions by typing list ⟨name of function⟩. More information can be obtained by typing **help list**.

To let the execution stop at a specific line, one can use the **break** command (abbreviation b). To stop the program *before* line 33 is executed, you enter

```
(gdb) b 33
Breakpoint 1 at 0x8048443: file gdbtest.c, line 33.
```

One can also give the name of a function when specifying a breakpoint, see **help break**. Breakpoints can be removed via the **delete** command. All current breakpoints are displayed by entering **info break**.

To start the execution of the program, you enter **run** or just **r**. Note that in case your program requires some arguments, you can set them via **set args** ⟨*argument list*⟩.

As requested before, the program will stop at line 33:

```
gdb) r
Starting program: /home/hartmann/book4/programs/debugging/gdbtest 100

Breakpoint 1, main (argc=2, argv=0xbfdfcc94) at gdbtest.c:33
33              for(t=0; t<100; t++)
```

Now you can inspect, for example, the content of variables via the **print** (or just **p**) command:

```
(gdb) p array
$1 = (int *) 0x8049680
(gdb) p array[99]
$2 = 99
```

To display the content of a variable permanently, the **display** command is available. You can change the content of variables via the **set** command

```
(gdb) set array[99]=98
```

You can continue the program at each stage by typing **next** (or just **n**), then just the next source-code line is executed:

```
(gdb) n
34                  sum += array[t];
```

Functions are regarded as one source-code line as well. If you want to debug the function in a step-wise manner as well, you have to enter the **step** command *before* the call to the function is performed. By entering **continue**, the execution is continued until the next breakpoint is encountered. Also a severe error or the end of the program will stop the execution. Please note that the output of the program being debugged appears in the *gdb* window as well:

```
(gdb) c
Continuing.
sum= 4949

Program exited normally.
```

As you can see, the final value (4949) the program prints is affected by the change of the variable **array[99]**.

Very useful is the possibility to include conditions with breakpoints. In this case, the breakpoint will come into action only if the condition is true. This is useful, for example, if one wants to inspect the state of the simulation after a certain, possibly large, number of iterations have been performed. This can be done using the command **condition**. The command can be applied to breakpoints which have already been defined. One must supply the break point number and the condition. For example:

```
(gdb) delete 1
(gdb) b 34
Breakpoint 2 at 0x8048477: file gdbtest.c, line 34.
(gdb) condition 2 (t==50)
(gdb) r
Starting program: /home/hartmann/book4/programs/debugging/gdbtest

Breakpoint 2, main (argc=1, argv=0xbff8fe34) at gdbtest.c:34
34                  sum += array[t];
(gdb) print t
$1 = 50
```

In case you are dealing with complex data structures, which is likely if your simulation program evolves over some time, printing the state of your objects using many applications of the **print** command is not very efficient. For this case, you should implement in the C code of your simulation program separate functions, which print program objects completely. Whenever the debugger has stopped, after a breakpoint or when using the `step` command, you can call any function included in your program with the `call` command. Here, this is illustrated with a simple checksum function, which does in this case the same as the above `main()` program. Such a function can be useful in case you want to check quickly, whether the value of a variable in an array has been changed:

```
int chksum(int n, int *array)
{
  int sum=0, t;
  for (t=0; t<n; t++)
    sum += array[t];
  return(sum);
}
```

For our example, the call could be performed as follows:

```
(gdb) call chksum(100, array)
$2 = 4950
```

Debugging is in particular useful when pointers are involved. Here, we consider the example of a linear list, which is a sequence of elements connected by pointers. The actual data

| GET SOURCE CODE |
| --- |
| DIR: `algorithms`<br>FILE(S): `list_error.c` |

structure is presented in Sec. 6.6. We assume that a list consisting of elements 3 5 7 10 6 has been built up, but for some reason, the connection is broken after the second element, see Fig. 4.1.

Fig. 4.1  A list consisting of five elements containing the integer numbers 3, 5, 7, 10 and 6. The list is represented by a pointer `list` which points to the first element. Each element contains a pointer to its successor (represented by arrows in the figure), except the last element, which contains a NULL pointer, represented by a filled circle. Due to a bug in the program, the connection to the successor of the second element is lost.

When now the program attempts to insert a new element after the third element, it will not work, see `list_error.c`. We now use the debugger to investigate the situation. For this purpose, a break point will be set in the main program, just before the corresponding function is called, and the program is started:

`break 42`

```
Breakpoint 1 at 0x4007c0: file list_error.c, line 42.
(gdb) run
Starting program: list_error
Breakpoint 1, main (argc=1, argv=0x7fffffffdcf8) at list_error.c:42
42          list = insert_element(list, elem3, elem2);
```

Now, for example, the content of the variable list can be examined:

```
(gdb) print list
$1 = (elem_t *) 0x601030
```

This information is not particularly useful. When using pointers, it is better to look at the content of the memory the pointer points to:

```
(gdb) print *list
$2 = {info = 3, next = 0x601010}
```

You can even go the the next (or further away elements), via the **next** field of the element:

```
(gdb) print *list->next
$3 = {info = 5, next = 0x0}
```

Here, you can see that the list does not continue, thus you know better for which bug you have to look for. Note, as mentioned before, it is very useful when your program contains functions to print complex data structures, which you can use within *gdb*, e.g.

```
(gdb) call print_list(list)
3 5
```

Almost any type of information about the debugged program can be obtained using the **info** commands, which was already introduced above to show the breakpoints (**info break**). For example, **info registers** shows the contents of all integer registers of the processor, while **info float** does the same for floating point registers. The command **info variables** display a long list of all know symbols, **info types** a list of all types and **info program** shows the current state of the execution. There are many more **info** commands, see **help info**.

The commands explained in this section are sufficient for most of the standard debugging tasks. Note that *gdb* offers many special commands, please have a look at the documentation [Loukides and Oram (1996)].

## 4.2   ddd

Some users may find graphical user interfaces more convenient. For this reason there exists a graphical front-end to the gdb, the *data display debugger (ddd)*. On UNIX operating systems it is just invoked by typing ddd (see also *man* page for options). Then a nice windows pops up, see Fig. 4.2. The lower part of the window is an ordinary *gdb* interface, several other windows are available. By typing file <*program*> you can load a program into the debugger. Then the source code is shown in the main window of the debugger. All *gdb* commands are available, the most important ones can be entered via menus or buttons using the mouse. For example, to set a breakpoint it is sufficient to place the cursor in a source-code line in the main *ddd* window and click on the *break* button. A good feature is that the content of a variable is shown when moving the mouse onto it. For more details, please consult the online help of *ddd*.

Fig. 4.2   The *data display debugger* (*ddd*). In the main window the source code is shown. Commands can be invoked via a mouse or by entering them into the lower part of the window.

## 4.3 Memory checker

Most program bugs are revealed by systematically running the program and cross-checking with the expected results. But other errors seem to appear in a rather irregular and unpredictable fashion. Sometimes a program runs without a problem, in other cases it crashes with a `Segmentation fault` at rather puzzling locations in the code. Very often a bad memory management is the cause of such a behavior. Writing beyond the boundaries of an array, reading uninitialized memory locations or addressing data which has been freed already, are the most common bugs of this class. Since the operating system organizes the memory in a different way each time a program is run, it is rather unpredictable whether these errors become apparent or not. Furthermore, it is very hard to track them down, because the effect of such errors becomes visible most of the time at program locations different from where the error has occurred.

As an example, we consider the case where the program writes beyond the boundary of an array. This array is stored in the memory area called *heap*,[2] where all allocated memory is taken from. If at the location behind the array another variable is stored, it may be overwritten in this case. Hence, the error becomes visible the next time the other variable is read. On the other hand, if the memory block behind the array is not used, the program may run without any problems this time. Unfortunately, the programmer is not able to influence the memory management directly.

To detect such types of nasty bugs, one can take advantage of *memory checkers*. Here, *valgrind* is considered, which is a very convenient tool and freely available. It works under UNIX and is included by just writing `valgrind` in front of the calling sequence of your program. No special compile commands or libraries are necessary. You can even debug standard programs like `ls`, try `valgrind ls`.

As an example, the program from Sec. 4.1 is considered, which is modified a bit; the memory block allocated for the array is now slightly too short (length 99 instead of 100):

| GET SOURCE CODE |
| --- |
| DIR: debugging<br>FILE(S): `memerror1.c` |

```
1   #include <stdio.h>
2   #include <stdlib.h>
3
```

---

[2]Not to be confused with the data structure *heap* introduced in Sec. 6.7.1.

```
4   int main(int argc, char *argv[])
5   {
6       int t, *array, sum = 0;
7
8       array = (int *) malloc (99*sizeof(int));
9       for(t=0; t<100; t++)
10          array[t] = t;
11      for(t=0; t<100; t++)
12          sum += array[t];
13      printf("sum= %d\n", sum);
14      free(array);
15      return(0);
16  }
```

The program is compiled via

```
cc -o memerror1 memerror1.c -g
```

We have compiled including -g, because the memory checker shall tell us where the bug appears and, in a second run, we want to transfer the control to the debugger automatically, if a bug appears.

Starting the program produces the following output, the program terminates normally:

```
sum= 4950
```

The result is correct, everything seems to be fine. Nevertheless, the array was used beyond its allocated regime. This can be seen by invoking valgrind in addition to the program:

```
1   [hartmann@comphy01 debugging]$ valgrind memerror1
2   Memcheck, a memory error detector.
3   Copyright (C) 2002-2005, and GNU GPL'd, by Julian Seward et al.
4   Using LibVEX rev 1575, a library for dynamic binary translation.
5   Copyright (C) 2004-2005, and GNU GPL'd, by OpenWorks LLP.
6   Using valgrind-3.1.1, a dynamic binary instrumentation framework.
7   Copyright (C) 2000-2005, and GNU GPL'd, by Julian Seward et al.
8   For more details, rerun with: -v
9
10  Invalid write of size 4
11      at 0x8048423: main (memerror1.c:10)
12    Address 0x402A1B4 is 0 bytes after a block of size 396 alloc'd
13      at 0x4004405: malloc (vg_replace_malloc.c:149)
14      by 0x80483FF: main (memerror1.c:8)
15
```

```
16    Invalid read of size 4
17       at 0x8048447: main (memerror1.c:12)
18    Address 0x402A1B4 is 0 bytes after a block of size 396 alloc'd
19       at 0x4004405: malloc (vg_replace_malloc.c:149)
20       by 0x80483FF: main (memerror1.c:8)
21    sum= 4950
22
23    ERROR SUMMARY: 2 errors from 2 contexts (suppressed: 13 from 1)
24    malloc/free: in use at exit: 0 bytes in 0 blocks.
25    malloc/free: 1 allocs, 1 frees, 396 bytes allocated.
26    For counts of detected errors, rerun with: -v
27    All heap blocks were freed -- no leaks are possible.
28    [hartmann@comphy01 debugging]$
```

Also the actual output of the program is printed (line 21 of the output). All other lines are generated by *valgrind*. Note that the first column shows always the ID (here ==16314==) of the process, which is omitted here to fit the output into the page. Two errors are reported, Invalid write of size 4 (line 10–14) and Invalid read of size 4 (line 16–20). Both errors consist of accesses to an array beyond the border. For each error both the location in the source code where the memory has been allocated (memerror1.c:8) and the location where the error occurred (memerror1.c:10 and memerror1.c:12) are given. In the end (line 23-27 of the output) *valgrind* gives a summary of the errors.

A very convenient option of *valgrind* is that one can tell it to transfer the control to the debugger once an error is detected. The debugger will show the program in the state of the execution, where *valgrind* just detected the error. This works by using the option --db-attach=yes. Once the execution reaches the error point, *valgrind* will ask you whether you really want to start the debugger. In the following example, 'y'+⟨RETURN⟩ was hit when the first error was encountered (line 16):

```
1    [hartmann@comphy01 debugging]$ valgrind --db-attach=yes memerror1
2    Memcheck, a memory error detector.
3    Copyright (C) 2002-2005, and GNU GPL'd, by Julian Seward et al.
4    Using LibVEX rev 1575, a library for dynamic binary translation.
5    Copyright (C) 2004-2005, and GNU GPL'd, by OpenWorks LLP.
6    Using valgrind-3.1.1, a dynamic binary instrumentation framework.
7    Copyright (C) 2000-2005, and GNU GPL'd, by Julian Seward et al.
8    For more details, rerun with: -v
9
10   Invalid write of size 4
11      at 0x8048423: main (memerror1.c:10)
```

```
12    Address 0x402A1B4 is 0 bytes after a block of size 396 alloc'd
13       at 0x4004405: malloc (vg_replace_malloc.c:149)
14       by 0x80483FF: main (memerror1.c:8)
15
16    ---- Attach to debugger ? --- [Return/N/n/Y/y/C/c] ---- y
17    starting debugger
18     starting debugger
19    GNU gdb Red Hat Linux (6.3.0.0-1.132.EL4rh)
20    Copyright 2004 Free Software Foundation, Inc.
21    This GDB was configured as "i386-redhat-linux-gnu"...
22
23    Attaching to program: /proc/16405/fd/1014, process 16405
24    0x08048423 in main (argc=1, argv=0xbec1a684) at memerror1.c:10
25    10                 array[t] = t;
26    (gdb) print t
27    $1 = 99
28    (gdb) quit
29    The program is running.  Quit anyway (and detach it)? (y or n) y
30    Detaching from program: /proc/16405/fd/1014, process 16405
31
32    Debugger has detached.  Valgrind regains control.  We continue.
33
34    Invalid read of size 4
35       at 0x8048447: main (memerror1.c:12)
36    Address 0x402A1B4 is 0 bytes after a block of size 396 alloc'd
37       at 0x4004405: malloc (vg_replace_malloc.c:149)
38       by 0x80483FF: main (memerror1.c:8)
39
40    ---- Attach to debugger ? --- [Return/N/n/Y/y/C/c] ---- n
```

After the second error was encountered 'n'+⟨RETURN⟩ was hit, and *valgrind* terminated as above (not shown here).

Other common types of errors are memory leaks. They appear when a previously used block of memory has been forgotten to be freed again. Assume that this happens in a function which is called frequently in a program. You can imagine that you will quickly run out of memory. This kind of errors are detected by *valgrind* as well. In the above example, no memory leak is present, as visible by lines 24–27 of the first example, culminating in `All heap blocks were freed -- no leaks are possible`.

Let us assume that the bug from above is removed and instead the `free(array)` command at the end of the program is omitted. After compiling and running the program again under *valgrind*, one obtains in the final part of the valgrind output

| GET SOURCE CODE |
| --- |
| DIR: debugging<br>FILE(S): memerror2.c |

```
[hartmann@comphy01 debugging]$ valgrind --db-attach=yes memerror1
   .....

searching for pointers to 1 not-freed blocks.
checked 51,748 bytes.

LEAK SUMMARY:
   definitely lost: 400 bytes in 1 blocks.
     possibly lost: 0 bytes in 0 blocks.
   still reachable: 0 bytes in 0 blocks.
        suppressed: 0 bytes in 0 blocks.
Use --leak-check=full to see details of leaked memory.
[hartmann@comphy01 debugging]$
```

Obviously, the memory leak has been found. You will obtain more information on the *valgrind* options using `valgrind -h`. Complete instructions are given at Ref. [valgrind].

A last advice: you should *always* (!) test a program with a memory checker, even if everything seems to be fine. The reason is that many memory faults occur only occasionally, depending on the actual circumstances of the run, such as memory usage. A memory checker on the other hand detects memory faults always, even if the program seems to run fine.

## 4.4   Profiling with *gprof*

If one wants to speed up a simulation, it is useful to analyze the run-time behavior of the program. The simplest way to do this is to measure the total running time, for instance for different problem sizes. This can be done easily under UNIX using the `time` command which can be followed by any other call of a program plus its arguments, e.g. the invocation of a simulation. After the execution is finished, `time` will report by default the total elapsed time (in seconds), the CPU time used by the process created by the command as well as the CPU time used by the operating system. In particular, if many processes are running in parallel, the total elapsed time might be much longer than the CPU time just consumed by the process of interest. Note that `time` (gnu version) may output some other non-timing-related information about a process, e.g. the amount of memory used and number of file I/O operations. These and many other output options can be specified via the `-f` option followed by a format string, which is given in a very similar format like the `printf` format string. More details can be found in the man page.

Although the overall timing behavior is already interesting, it is even better to know where a program spends most of its CPU time. This allows the program optimization effort to be concentrated on the most crucial parts of a program. To identify these parts, a *profiler* can be used. It basically measures for each function how much CPU time is spent there. Note that *valgrind*, which is presented in Sec. 4.3, is also capable of profiling. Here, the *gprof* tool is explained.

To understand how *gprof* operates, the following simple and short program is considered. It calculates for an array of numbers a[0]...a[n-1] two matrices `sum[][]` and

| GET SOURCE CODE |
| --- |
| DIR: **debugging**<br>FILE(S): **gproftest.c** |

`prod[][]` where `sum[i][j]` contains $\sum_{k=i}^{j} a[k]$ and `prod[i][j]` contains $\prod_{k=i}^{j} a[k]$. Note that this is achieved here in a very time-consuming way, just because we want to have an interesting analysis for *gprof* while keeping the example program short.[3] The program consists of two functions `calc_sum()` and `calc_prod()`, which do the main work. Both functions take as arguments the array of numbers and two indices which identify the subsequence for which the return value is calculated. Both functions return the calculated value. The C source code of the functions is as follows

---

[3]For an efficient calculation of sums and products of subsequences one should use a dynamic programming approach, see Sec. 6.4.

```
1   double calc_sum(double *a, int i, int j)
2   {
3     int t; double sum= 0.0;
4     for(t=i; t<=j ; t++)
5       sum += a[t];
6     return(sum);
7   }
8
9   double calc_prod(double *a, int i, int j)
10  {
11    int t; double prod= 1.0;
12    for(t=i; t<=j ; t++)
13      prod *= a[t];
14    return(prod);
15  }
```

The main work, to calculate all entries of the matrices sum[][] and prod[][], is done by the function all1(), which reads as follows

```
1   void all1(int n, double *a, double **sum, double **prod)
2   {
3     int i, j;
4
5     for(i=0; i<n; i++)
6       for(j=i; j<n; j++)
7       {
8         sum[i][j] = calc_sum(a, i, j);
9         prod[i][j] = calc_prod(a, i, j);
10      }
11  }
```

The main program is assumed to read in the size $n$ of the array, to initialize the array, to print the sums $\sum_{k=0}^{j} a[k]$ in advance and then call all1(). Thus, it contains the following three lines

```
for(t=0; t<n; t++)
  printf("%f\n", calc_sum(a,0,t));
all1(n, a, sum, prod);
```

To enable profiling, one has to use the option -pg when compiling. All necessary libraries will be linked automatically. Assuming that the test program is named gproftest.c, one can compile it via

```
gcc -o gproftest gproftest.c -pg
```

Now the program can be run as usual, passing $n$ as argument. In addition to the normal output, it will generate a *profile file* gmon.out, which contains the run-time statistics of the execution. If you want a different profile file name (for example, when collecting several profiles) you can rename the file after the execution of program is finished. The run-time overhead generated by the analysis is usually hardly measurable.

A human-readable analysis of the running time can be obtained by calling *gprof* with the name of the executable as first argument (a.out by default) and the name of the profile file as second argument (gmon.out by default). *gprof* allows for several options, which are not needed for the basic functionality. Information about the options can be obtained by reading the man page of *gprof*, i.e. via man gprof on UNIX systems. For the example program shown above, one obtains an analysis of the running times via

```
gprof gproftest gmon.out
```

The analysis of the running times is written to stdout. The head of the default output looks as follows (for running the program with the size of the array set to n=1000):

Flat profile:

Each sample counts as 0.01 seconds.

| %<br>time | cumulative<br>seconds | self<br>seconds | calls | self<br>s/call | total<br>s/call | name |
|---|---|---|---|---|---|---|
| 93.16 | 24.59 | 24.59 | 500500 | 0.00 | 0.00 | calc_prod |
| 6.74 | 26.36 | 1.78 | 501500 | 0.00 | 0.00 | calc_sum |
| 0.09 | 26.39 | 0.03 | 1 | 0.03 | 26.39 | all1 |

The "flat profile" (shown by default, or when using the option -p, suppressed by -P or --no-flat-profile) basically states for each function, listed in the last column (name), how much running time it consumes. The first column (% time) states which percentage of the total running time is spent in the function, while the third column (self seconds) contains the running time measured in seconds consumed by the function. Note that this number does not contain the running times used by calls to other functions from this function, i.e. to *descendants*. The flat profile table is sorted according to the third column. The second column (cumulative seconds) contains the cumulative running times as obtained from the third column. In the fourth column (calls), the number of times each function is invoked is given. In the above example, the number of calls to calc_prof() and calc_sum() are almost the same (calc_sum() exhibits some extra calls in the main program), but the running time spent in calc_prod() (15.08s) is much larger than the time spent in calc_sum() (1.15s). The reason is that calculating the product of two numbers takes more CPU cycles than calculating the sum. The fifth column (self s/call) contains the average number of seconds per call to a function, i.e. the value of the third column divided by the value in the fourth column. The six column (total s/call) states the average time spent in a function *including* its descendants. For this reason, since all1() contains basically all calls to calc_sum() and calc_prod(), the value for all1() contains almost the full running time.

The output of *gprof* contains, after the flat profile, a short explanation of the different columns, equivalent to what was just explained. This can be suppressed by using the options -b or --brief.

Next, the output contains the *call graph* (by default, or when using -q or --graph, suppressed by -Q or --no-graph). This part of the output lists for each function from which functions it was called and which functions it calls itself. For our example, the output looks as follows:

```
                              Call graph

index % time    self  children    called     name
                                              <spontaneous>
[1]     100.0    0.00   26.39                 main [1]
                 0.03   26.36      1/1             all1 [2]
                 0.00    0.00   1000/501500        calc_sum [4]
-----------------------------------------------
                 0.03   26.36      1/1             main [1]
[2]     100.0    0.03   26.36      1          all1 [2]
                24.59    0.00   500500/500500       calc_prod [3]
                 1.78    0.00   500500/501500       calc_sum [4]
-----------------------------------------------
                24.59    0.00   500500/500500       all1 [2]
[3]      93.2   24.59    0.00   500500          calc_prod [3]
-----------------------------------------------
                 0.00    0.00   1000/501500        main [1]
                 1.78    0.00   500500/501500       all1 [2]
[4]       6.7    1.78    0.00   501500          calc_sum [4]
-----------------------------------------------

Index by function name

   [2] all1              [3] calc_prod              [4] calc_sum
```

Each of the *nodes*, separated by horizontal lines, is for one *current* function. The functions are identified by index numbers given in the first column. These index numbers are also written in brackets next to the function names in the sixth column and also summarized at the end of the output. Each node may consist of several lines. First come the functions which call the current function, i.e. its *parents*. For `main()` (index `[1]` here), which is not called by any function, `<spontaneous>` is written in this case. Then, for each node, comes a line for the current function (the name being indented to the left in the `name` column) and finally lines for all functions which are directly called by the current function, i.e. its *children*. This can be translated into a directed graph (see Sec. 6.8) where the functions are represented by nodes and where calls of one function to another are represented by directed edges. The resulting graph for the example program is shown in Fig. 4.3.

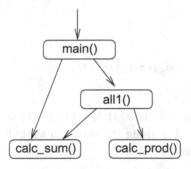

Fig. 4.3 Call graph of the above sample program. Nodes represent functions and each directed edge represents a call of one function to another.

The second column of each entry states the fraction of the total running time which is spent in the current function including all calls to descendants. In the above example, all running time is spent in `main()` and its children, i.e. 100 %. For each function, the third column (`self`) lists the running time spent inside the function, while the fourth column (`children`) states the total running time for calls to all children. Thus, for `main()`, the `self` contribution is zero within the given accuracy and all time is spent for calls to children. In the line corresponding to the current function, the fifth column (`called`) gives the total number $n_{tot}$ of times the function is called during the execution of the program *without* recursive calls (see below). For each parent the format of the fifth column is different. Here, it states how often the current function was called by this parent and, after the slash /, the value $n_{tot}$. For each child, the fifth column shows how often the child is called by the current function, again accompanied by $/n_{tot}$. As example, calc_sum() is called 501500 times, this is 1000 times by main() and 50500 times by all1().

Another output form is the *annotated listing*. It shows the source code of the program. Next to the first line of each function, the number of times the function has been called is printed. One has to compile additionally with the -g option and call *gprof* with the -A or with the --annotated-source option. Other output formats of *gprof* are discussed on its *man* page.

In case the program contains recursive calls, the output of *gprof* will look slightly different. Here, we consider an recursive version of the calculation of the sums, which is again artificially slow compared to dynamic programming approach. The function reads as follows:

```
1  double calc_sum_rec(double *a, int i, int j)
2  {
3    if(i>=j)
4      return(a[i]);
5    else
6      return(a[i]+calc_sum_rec(a,i+1,j));
7  }
```

When replacing the calls in main() and all1() to calc_sum() by calls to calc_sum_rec() and running the program again for n=1000, the resulting flat profile will look as follows:

| %<br>time | cumulative<br>seconds | self<br>seconds | calls | self<br>s/call | total<br>s/call | name |
|---|---|---|---|---|---|---|
| 84.87 | 24.24 | 24.24 | 500500 | 0.00 | 0.00 | calc_prod |
| 14.99 | 28.52 | 4.28 | 501500 | 0.00 | 0.00 | calc_sum_rec |
| 0.14 | 28.56 | 0.04 | 1 | 0.04 | 28.55 | all1 |

The running time spent in calc_prod() has basically not changed. Note that small fluctuations are always observed, even between different runs of identical programs. The reason is that the running time of a process depends on the environment. Of particular importance is how often a memory location can be found in the fast but small on-chip cache memory and how often the comparable slower main memory has to be accessed.

The running time of calc_sum_rec() is much larger than that of calc_sum() due to the much higher number of calls generated by the recursion. Note that in column four (calls) only non-recursive calls are listed.

The effect of the recursion can be seen much better from the call graph, where the entry for `calc_sum_rec()` reads as follows:

```
-------------------------------------------------
                        167166000              calc_sum_rec [4]
            0.01    0.00    1000/501500     main [1]
            4.27    0.00  500500/501500     all1 [2]
[4]    15.0    4.28    0.00  501500+167166000 calc_sum_rec [4]
                        167166000                  calc_sum_rec [4]
-------------------------------------------------
```

The number of recursive calls appears here three times: in the first line, where `calc_sum_rec()` is a parent of itself,; in the fourth line where both the number of non-recursive and of recursive calls (the latter behind the + symbol) are shown; and in the fifth line, where `calc_sum_rec()` is shown as child. For more complex structures of recursive calls like `fA()` calls `fB()` and `fB()` calls `fA()`, these numbers might be different for the different entries.

In some cases, only parts of the program are (fully) profiled. This can happen, if the final program is linked from several object code files, where some had not been compiled with the `-pg` option. Also, one can choose some functions explicitly to be analyzed. Using (multiple times) the option `-p` and/or `-q` followed by a function name will show only the entries which involve the listed functions.

Finally, note that you can build in your own timing measurements into your program. This is useful to measure the timing of some critical part of your program during all simulations, and including the timing results in your standard log files. For this purpose, the C standard library function `times` can be used, declared in `sys/times.h`. It measures the time in units of clock ticks, hence with the highest possible resolution. More information on the usage of the function can be found in the UNIX man page via typing "`man 2 times`" in a shell.

# Chapter 5

# Object-oriented Software Development

In recent years, *object-oriented* programming languages like C++, Smalltalk or Eiffel became very popular. These programming languages are usually used to implement programs which are designed in an object-oriented way. But, *using* an object-oriented language and *designing* a simulation program in an object-oriented style are not necessarily the same. For example, you can set up your whole project by applying object-oriented methods even when using a traditional *procedural* programming language like C, Pascal or Fortran. In general, taking an object-oriented viewpoint facilitates the analysis of problems and the development of suitable programs. An introduction to object-oriented software development can be found for example in Refs. [Rumbaugh et al. (1991); Johnsonbaugh and Kalin (1994); Skansholm (1997)]. In the first section of this chapter, an introduction to object-oriented concepts is given.

As already mentioned, the implementation of an object-oriented design is still possible with a procedural language. This is shown in Sec. 5.2, where *histograms* (see Sec. 8.3.3) are implemented in C.

In the final section of this chapter, the histograms are implemented again, but now using C++. This section also serves as an introduction to C++, which is basically an object-oriented extension of C. Furthermore, C++ offers *templates*, which allow to implement algorithms for unspecified, i.e. arbitrary, data types. This will be covered in another chapter (see Sec. 7.2). In general, the advantage of C++ is that is helps you to organize your programs in terms of objects, but still you have the flexibility to do it in a non-object-oriented way as well.

## 5.1   Object-orientation principles

Here, we start by explaining the main principles of object-oriented design and programming.

- **Objects and methods**

  The real world is made of *objects* such as traffic-lights, books or computers. According to some criteria, you can classify different objects into *classes*. This means, different types of chairs belong to the class "chair". The objects of many classes can have internal *states*. For instance, a traffic-light can be red, yellow or green. This is very simple. But there are much more complex objects; for example the state of a living cell is much more difficult to describe. To cope with this complexity, one often uses a hierarchical description, which is explained below under the item *inheritance*.

  Furthermore, objects are not existing in isolation. Other objects *interact* via *operations* with an object, i.e. it is possible to *access* the objects. For example, you (belonging to the class "human") can read the state of a traffic light, some central computer may set the state of a traffic light or even switch it off, or a car may hit the traffic light such that it is "deleted".

  Similar to the real world, you can have objects in simulation programs as well. The internal state of an object is given by the values of the variables describing the object, called *data members* in C++. An example is shown in Fig. 5.1. Also it is possible to access the objects or to interact with them by calling subroutines (called *methods* in general, in C++ also called *member functions*) associated with the objects.

  Objects and the related methods are seen as coherent units. This means you define, when using an object-oriented language, within one *class definition* the way the objects look, i.e. the data structures, together with the methods which access/alter the content of the objects. The syntax of the class definition depends on the programming language you use. Examples are shown in Sec. 5.3.

  When you take the viewpoint of a pure object-oriented programmer, then all programs can be organized as collections of objects which call methods of each other. This is derived from the structure the real world has: It is a large set of interacting objects. But for writing good programs it is as in real life, taking an orthodox position imposes too many restrictions. You should take the best of both worlds,

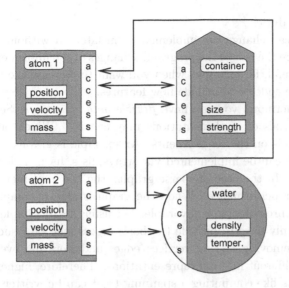

Fig. 5.1 An example for objects. There are objects of different type, here "atoms", "containers" and solvent "water". Each object has internal states or variables and interacts with the other objects via defined access interfaces. For example, in a simulation of proteins dissolved in water, an atom has a position and a velocity, which are influenced via forces from other atoms, the solvent and the container.

the object-oriented and the procedural world, depending on the actual problem. In modern programming, usually only the basic functionality is implemented as methods, while many algorithms are implemented via standard subroutines.

- **Data capsuling**

  When using a computer, you do not care about the implementation. When you press a key on the keyboard, you would like to see the result on the screen. You are not interested in how the key converts your keystroke into an electrical signal, how this signal is sent to the input ports of the chips, how the algorithm treats the signal and so on.

  Similarly, a main principle of object-oriented programming is to hide the actual implementation of the objects. Access to them is only allowed via given interfaces, i.e. via methods. The internal data structures are hidden, this is called `private` in C++. The data capsuling has several advantages:

  - You do not have to remember the implementation of your objects. When using them later on, they just appear as a black box fulfilling some duties.

- You can change the implementation later on without the need to change the rest of the program. Changes of the implementation may be useful, for example, when you want to increase the performance of the code or to include new features.

- Furthermore, you can have *flexible data structures*: Several different types of implementations may coexist. Which one is chosen depends on the requirements. An example is graphs (see Sec. 6.8) which can be implemented via arrays, lists, hash tables or in other ways. In the case of sparse graphs, the list implementation has a better performance. When the graph is almost complete, the array representation is favorable. For using flexible implementations, you only have to provide the basic access methods, such as inserting/removing/testing vertices/edges and iterating over them, for the different internal representations. Therefore, higher-level algorithms like computing a spanning tree[1] can be written in a simple way to work with all internal implementations. When using such a class, the user just has to specify the representation he wants. The rest of the program is independent of this choice.

- Last but not least, software debugging is made easier. Since you have only defined ways, the data can be changed and undesired side-effects become less common. Also the memory management is easier to control.

Nevertheless, for the sake of flexibility, convenience or slight decrease of running time, it is possible to declare internal variables as `public`. In this case they can be accessed directly from outside. Thus, there is always a workaround for everything, but this should be used with extreme caution. In general, the use of public variables is not recommended.

- **Inheritance**
This means lower level objects can be specializations of higher level objects. For example, the class of (German) "ICE trains" is a child class of "trains" which itself is a child class of "vehicles".
In atomistic computer simulations, you may have a basic class of "atoms" containing mass, position and velocity, and built upon this a class of "charged atoms" by including the value of the charge. Then you can use the subroutines you have written for the uncharged atoms, like moving the particles or calculating correlation functions, for the

---

[1]This is a subgraph of a graph, which exhibits no loops but contains all nodes.

charged atoms as well.

A similar form of hierarchical organization of objects works the other way round: Higher level objects can be defined in terms of lower level objects. This is called a *composition* [Meyers (2005)]. For example, a book is composed of many objects belonging to the class "page". Each page can be regarded as a collection of many "letter" objects.

For the physical example above, when modeling chemical systems, you can have "atoms" as basic objects and use them to define "molecules". Another level up would be the "system" object, which is a collection of molecules.

- **Function/operator overloading**

  The inheritance of methods (i.e. functions in C++) to lower level classes is an example of *operator overloading*. It just means that you can have methods for different classes having the same name, sometimes the same code applies to several classes. This can happen also for classes, which are not connected by inheritance. For example, you can define how to add integers, real numbers, complex numbers or larger objects like lists, graphs or documents. In languages like C or Pascal, you can define subroutines to add numbers and subroutines to add graphs as well, but they must have different names. In C++, you can define the operator "+" for many different classes (see one example in Sec. 5.3). Hence, the operator-overloading mechanism of object-oriented languages is just a tool to make the code more readable and clearly structured.

- **Software reuse**

  Once you have an idea of how to build a chair, you can do it several times. Because you have a blueprint, the tools and the experience, building another chair is an easy task.

  This is true for building programs as well: Both data capsuling and inheritance facilitate the reuse of software. For instance, once you have written your class for treating lists, you can include the lists in other programs as well. This is easy, because later on you do not have to care about the implementation. With a class designed in a flexible way, much time can be saved when realizing new software projects.

## 5.2 A sample using C

As mentioned before, when programming in an object-oriented way, you do not necessarily have to use an object-oriented language. It is true that they are helpful for the implementation and the resulting programs will look slightly more elegant and clearer, but you can program everything with a language like C as well. In C, an object-oriented style can be achieved very easily. As an example, a "class" `histo` implementing histograms is outlined next. Histograms are needed for almost all types of computer simulations as evaluation and analysis tools. Formally, histograms are introduced in Sec. 8.3.3.

Initially, you have to think about the data you would like to store. The histogram itself is an array `table` of bins. Each bin just counts the number of events which fall into a small interval

| GET SOURCE CODE |
| --- |
| DIR: oop |
| FILE(S): `histo.h` |

of the corresponding histogram range. To achieve a high degree of flexibility, the range and the number of bins must be variable. It is assumed here for simplicity that all bins have the same width. Thus, the width `delta` of each bin can be calculated easily. For convenience `delta` is stored as well. To count the number of events which are outside the range of the table, the entries `low` and `high` are introduced. Furthermore, statistical quantities like mean and variance should be available quickly and with high accuracy. Thus, several accumulated moments `sum[]` of the distribution are stored separately as well. Here, the number of moments `_HISTO_NOM_` is defined as a macro, converting this macro to variable would be straightforward. Altogether, this leads to the following C data structure:

```
#define _HISTO_NOM_        9          /* No. of (statistical) moments */

typedef struct
{
    double          from, to;   /* range of histogram            */
    double            delta;    /* width of bins                 */
    int             n_bask;     /* number of bins                */
    double          *table;     /* bins                          */
    int             low, high;  /* No. of data out of range      */
    double    sum[_HISTO_NOM_]; /* sum of 1s, numbers, numbers^2 ...*/
} histo_t;
```

Here, the postfix `_t` is used to stress the fact that the name `histo_t` denotes a type. The bins are variables of type `double`, which allows for more general applications. Please note that it is still possible to access the internal structures from outside, but it is not necessary and not recommended. In C++, you could prevent this by declaring the internal variables as `private`, see below. Nevertheless, everything will be done here via special subroutines without the need for a direct access.

First of all, one must be able to create and delete histograms. The function `histo_new()` receives the range [`from`, `to`) of the interval which is to be covered and the number of bins

GET SOURCE CODE

DIR: oop
FILE(S): `histo.c`

as arguments. It returns a pointer to the newly created histogram:

```
1  histo_t *histo_new(double from, double to, int n_bins)
2  {
3    histo_t *his;                              /* histogram pointer */
4    int t;                                     /* loop counter */
5
6    his = (histo_t *) malloc(sizeof(histo_t));
7    if(his == NULL)                            /* enough memory ? */
8    {
9      fprintf(stderr, "out of memory in histo_new()");
10     exit(1);
11   }
12   if(to < from)                              /* boundaries in wrong order ? */
13   {
14     double tmp;
15     tmp = to; to = from; from = tmp;
16     fprintf(stderr, "WARNING: exchanging from, to in histo_new()\n");
17   }
18   his->from = from;
19   his->to = to;
20   if( n_bins <= 0)            /* number of bins should be positive */
21   {
22     n_bins = 10;
23     fprintf(stderr, "WARNING: setting n_bins=10 in histo_new()\n");
24   }
25   his->delta = (to-from)/(double) n_bins;                 /* setup */
26   his->n_bins = n_bins;
27   his->low = 0;
28   his->high = 0;
29   for(t=0; t< _HISTO_NOM_ ; t++)       /* init. accumulated moments */
30     his->sum[t] = 0.0;
31   his->table = (double *) malloc(n_bins*sizeof(double));
```

```
32    if(his->table == NULL)                         /* enough memory ? */
33    {
34      fprintf(stderr, "out of memory in histo_new()");
35      exit(1);
36    }
37    else
38      for(t=0; t<n_bins; t++)                       /* initalize bins */
39        his->table[t] = 0;
40    return(his);
41  }
```

The main data structure is allocated in line 6, while the array for the moments is allocated in line 31. Some simple error-checking is included: It is verified that enough memory is available for the main data structure (lines 7–11) and for the moment array (line 32–36). Furthermore, it is verified that `from` is smaller than `to` (line 12–17) and that the number of bins is positive (lines 20–24). The histogram is initialized in lines 18–19, 25–28 and 37–39.

The function `histo_delete()` receives a pointer to a histogram, frees the memory associated with the histogram and returns nothing:

```
1  void histo_delete(histo_t *his)
2  {
3    free(his->table);
4    free(his);
5  }
```

All histogram objects are created dynamically by calling `histo_new()`, which corresponds to a call of the *constructor* or `new` in C++, as explained in Sec. 5.3 The objects are addressed via pointers. Whenever a function concerning an object of the `histo` class is called, the first argument will always be a pointer to the corresponding histogram. When avoiding direct access to the structure, the realization using C is perfectly equivalent to C++ or other object-oriented languages. Inheritance can be implemented, by including pointers to `histo_t` objects in other type definitions. When these higher level objects are created, a call to `histo_new()` must be included, while a call to `histo_delete()`, corresponding to the *destructor* in C++, is necessary, to implement a correct deletion of the more complex objects.

As a final example, the functions for inserting an element into the table and calculating the mean are presented. It is easy to figure out how other functions can be realized, such as for calculating the variance/higher mo-

ments or printing a histogram as probability density function (see exercise
(1)).

```
void histo_insert(histo_t *his, double number)
{
  int t;                              /* counter to access moments */
  double value;                       /* auxiliary variable */
  value = 1.0;
  for(t=0; t< _HISTO_NOM_; t++)
  {
    his->sum[t]+= value;                      /* raw statistics */
    value *= number;
  }
  if(number < his->from)              /* insert into histogram */
    his->low++;
  else if(number >= his->to)
    his->high++;
  else
    his->table[(int) floor( (number - his->from) / his->delta)]++;
}
```

```
double histo_mean(histo_t *his)
{
  if(his->sum[0] == 0)
    return(0.0);
  else
    return(his->sum[1] / his->sum[0]);
}
```

## 5.3 Introduction to C++ and an example

The C++ language is, technically seen, an object-oriented extension of C.
It allows for complex object-oriented implementations, but you can write
pure C programs as well, if you like. For example, each executable must
contain a main() function, as in C. This section again presents an implementation of histograms, but now in C++. This serves also as introduction
to the programming language. For details, please consult the literature, for
example *the* standard reference [Stroustrup (2000)].

The most fundamental concept of C++ is
the *class*. A class is a description of the data
elements belonging to objects of this class, together with the functions which can be used to

| GET SOURCE CODE |
| --- |
| DIR: oop<br>FILE(S): histo++.h |

access and manipulate the objects. The functions are called *member functions*. Header files (.h or .hpp) in C++ usually[2] contain the class declarations, while the code files (.cpp) usually contain the code of the member functions. A possible class declaration for histograms reads as follows:

```
1  #include <iostream>
2
3  class Histo
4  {
5  private:
6    double          from, to;   // range of histogram
7    double             delta;   // width of bins
8    int            low, high;   // out of range numbers
9    int               n_bins;   // number of bins
10   double            *table;   // bins
11
12  public:
13   Histo(double from, double to, int n_bins);  // constructor
14   Histo();                                     // simple constructor
15   Histo(const Histo& h);                       // copy constructor
16   virtual ~Histo();                            // destructor
17   void insert(double number);                  // insert data point
18   Histo &operator=(const Histo &h);            // assignment operator
19
20   friend std::ostream& operator<< (std::ostream& os, const Histo& h);
21  };
```

Similar to the C header file stdio.h, the header file iostream contains standard types for in-/output. In line 1, this header file is included. Note that in modern C++, the .h suffix is not used any more for standard system header files.

The main part is the declaration of the class Histo. All the *data members* of Histo objects are listed in lines 6–10. This is similar to the members of a structure in C.[3] Here, only members for the bins and for the data outside the histogram range are included. The data members for the moments, which are present in the histo_t type in Sec. 5.2, are put into an inherited class, see below. Note that in addition to /* ... */ comments, which may extend over several lines, in C++ there are comments of style // ..., which extend till the end of the current line. The lines 6–10 are headed

---

[2]Within the new C++ standard, the class declaration files and the C header files carry no appendix .h, while the C header files are identified by a prefix letter 'c'.

[3]In fact, a structure is class in C++ as well, where all members are declared by default as public.

by the keyword `private`. This means that these members are not visible from outside the class. If `h` is a `Histo` object, it is *not* allowed to change for example `h.low`, except `h` is passed to a function which is granted access to objects of this class. This is the principle of data capsuling (see page 139). Mainly *member functions* are allowed to access these data members. Member functions are functions which are also declared inside the class declaration. The member functions of `Histo` are declared in lines 13–18. These lines are preceded by the keyword `public`, which means that they can be used from outside a class. Note that in principle, one can define member functions `private` as well. In line 16, the keyword `virtual` is important for the case that a class is derived by inheritance from `Histo` (which we will do below).[4] In line 20, a function is declared, which is *not* a member function. The keyword `friend` means that the function is also allowed to access `private` members of the class `Histo`. This should be used with caution, because it breaks encapsulation. There exists a third type of access control for members, which is called `protected`. This means that these members in general cannot be accessed from outside a class, but from *child* classes which are inherited from the class.

Whenever a `Histo` object is created, the data elements should be initialized. This is done by a *constructor*. For the C example in the previous section, this was done by the function `histo_new()`. A constructor has always the same name as the class and is always called automatically, if an object of the class is created.

In the class declaration, there a *three* different constructors declared in lines 13–15. This multiple declaration is no problem in C++, since the compiler can distinguish the three

| GET SOURCE CODE |
| --- |
| DIR: oop |
| FILE(S): `histo.cpp` |

variants, because they have different argument lists. This is called *overloading*. The first constructor requires as arguments the range [`from`, `to`) of the histogram and the number of bins `n_bins`. Nothing (not even `void`) is returned by a constructor. The implementation reads as follows:

---

[4]If a pointer points to objects both of the parent class and of the child class, automatically the corresponding virtual function of the suitable class will be used, depending on the type of the object. This is important when deallocating memory, to avoid memory leaks.

```
1   Histo::Histo(double h_from, double h_to, int h_n_bins)
2   {
3     from = h_from; to = h_to;        // store parameters
4     if(to < from)                    // boundaries in wrong order ?
5     {
6       double tmp;
7       tmp = to; to = from; from = tmp;
8       fprintf(stderr, "WARNING: exchanging from, to in Histo()\n");
9     }
10    low = 0;
11    high = 0;
12    n_bins = h_n_bins;
13    delta = (to-from)/n_bins;        // calculate bin width
14    table = new double[n_bins];      // get memory for bins;
15
16    for(int t=0; t<n_bins; t++)      // Initialize bins
17      table[t] = 0;
18  }
```

Note that in the function declaration (line 1), the function name is preceded by the class name followed by a double colon. It must occur here, because different classes may have members of the same name, hence the class name must be given to distinguish them.[5] The data members from and to are initialized in line 3. Inside member functions, the data members can be accessed directly, just like local variables in functions. In lines 4–9 it is assured that from < to holds. In lines 10-13, additional data elements are initialized. In line 14, the table of bins is allocated dynamically. In C++ this is done with the command new, which works like malloc() in C.[6] After the new command, the type of the array is written (double here) and the number of requested elements is stated in [ ] brackets. If the brackets are omitted, just one single variable is generated, no array. Finally (lines 16–17), the bins are initialized.

If you want to define an object his of the class Histo, with given values for the parameters, you have to use

```
Histo his(0.0, 10, 100);
```

which creates a histogram with 100 bins $[0, 0.1), [0.1, 0.2), \ldots, [9.9, 10)$. This looks like a variable definition in C, but with the object name followed

---

[5]You could, in principle, use the name of a constructor of one class as name for a data member in another class, without conflict.

[6]You can use malloc() in C++ as well, if you like. More sophisticated is the use of template vectors std::vector<>, see Sec. 7.2.

by the arguments for the constructor in brackets. In case you are lazy and do not want to state arguments, you can use a standard constructor, which looks as follows:

```
Histo::Histo()
{
    Histo(0, 1, 10);    // Use standard values
}
```

Here you can see that, as for data members, member functions can be directly used/called inside (other or the same) member functions, just as local variables. If an object his2 is defined using the standard constructor, it resembles a variable definition in C, for example:[7]

```
Histo his2;
```

A third type of constructor is the so called *copy constructor*. It is used when an object is initialized by copying from another object:

```
Histo his3 = his;
```

In principle, you do not have to define your own copy constructor. If no explicit copy constructor is given, the new object (his3) is just a memberwise copy of the given object (his). But this means that the pointer element table in his3 points exactly to the same memory area as in his. This is usually not what you want. Instead, one wants that his3 has its own table, but filled with the same numbers as his. This is achieved by the following explicit copy constructor:

```
Histo::Histo(const Histo &h)
{
    from = h.from;              // store parameters
    to = h.to;
    low = h.low;
    high = h.high;
    n_bins = h.n_bins;
    delta = h.delta;           // bin width
    table = new double[n_bins];  // get memory for bins;

    for(int t=0; t<n_bins; t++)  // copy bin entries
        table[t] = h.table[t];
}
```

---

[7]You can also include empty brackets, indicating explicitly that no arguments are passed.

It looks pretty similar to the first constructor, but with some differences. The initial values of the data members are taken from the object h passed to the constructor. Formally, it is written as argument, although when using it, the object where the data is taken from is written after an = assignment operator. The variable declaration of h is preceded by the & character, which means that the variable is passed *by reference*. This means that only the address is passed, as if the object would be passed as a pointer. Nevertheless, when passed by reference, the object can be accessed directly, without using the * operator, in contrast to pointers. Note that the data elements of the object h are accessed like elements in structures. This is allowed here, since the copy constructor is a member function of the class, although the object which is accessed is different from the object for which the member function was called.

Objects which are created somewhere, must be deleted after usage. In the C implementation presented in Sec. 5.2 this is done by the function `histo_delete()`. In C++, the deletion of an object is the task of the *destructor*, which is always a member function with the name of the class preceded by a ~ character. Here, the destructor must free the table of bins:

```
1  Histo::~Histo()
2  {
3    delete[] table;
4  }
```

The effect of the `delete` in line 3 is equivalent to the function `free()` in C. If an array is deleted, like here, one must write [] brackets after the `delete` command.

Next, a member function `insert()` for inserting a number is considered. As argument it takes the number to be inserted. Here, just as for ordinary functions, a return type should be stated, `void` in this case. Note that the histogram in which the number is inserted is not passed explicitly as argument, since the object for which a member function is called is accessible inside the member function through "local" variables:

```
1  void Histo::insert(double number)
2  {
3    if(number < from)                // insert into histogram
4      low++;
5    else if(number >= to)
6      high++;
7    else
8      table[(int) floor( (number - from) / delta)]++;
9  }
```

To print a histogram, you could define another member function, which uses for example printf() functions to print the content of the array table. Here, another C++-specific solution is shown. In C++, printing can be done via *streams*. A standard stream for output, similar to stdout in C, is std::cout. You can write the string "number: " to the standard output followed by the content of the variable num followed by a new line, via:

```
std::cout << "number: " << num << std::endl;
```

This is equivalent to printf("number: %d\n", num). Hence, the *operator* << is used for output. Several objects to be printed can be separated by multiple occurrences of <<. Note that std::cout belongs to the *namespace* std. Thus, std:: does *not* refer to a class name. Namespaces are used in C++ for allowing different objects to have the exactly same name, within different namespaces. The standard namespace std contains many frequently used objects. If you are lazy, you can write using namespace std just below all #include commands at the top of the file. Now you can write cout instead of std::cout, but only if there are no other objects, variables or types with the name cout. Finally, std::endl is the end-of-line signature, equivalent to "\n".

If you want to write a histogram in the same way to std::cout like a standard variable, for example by using the << operator. In this case one has to define the operator << accordingly. This works as follows. The operator obtains the stream (os) and the Histo object (h) to be printed as arguments and returns the stream again. Note again that << has access to h, because the operator is declared as friend in the header file.

```
1  std::ostream& operator << (std::ostream& os, const Histo& h)
2  {
3
4    int i;
5    os << "# below:" << h.from << std::endl;
6    os << "# above: " << h.to << std::endl;
7    os << "# from: " << h.from << " delta: " << h.delta;
8    os << " nbins: " << h.n_bins << std::endl;
9    for(i=0; i<h.n_bins; i++)
10     os << i << " " << h.table[i] << std::endl;
11
12   return(os);
13
14 }
```

Also the assignment operator = is often implemented explicitly for C++ classes. If no implementation is given, objects are copied bitwise, which may not be desirable, similar to the case of the copy constructor. For Histo, the assignment operator looks as follows:

```
1  Histo& Histo::operator=(const Histo &h)
2  {
3    if( this != &h )                  // no self assignment
4    {
5      from = h.from;                  // store parameters
6      to = h.to;
7      n_bins = h.n_bins;
8      delta = h.delta;                // bin width
9      delete[] table;                 // delete old table
10     table = new double[n_bins];     // get memory for bins;
11     low = h.low;
12     high = h.high;
13
14     for(int t=0; t<n_bins; t++)     // copy bin entries
15       table[t] = h.table[t];
16   }
17   return( *this);
18 }
```

The implementation looks very similar to the copy constructor.[8] Note that assignment of an object to "itself" should be prohibited. This is ensured in line 3 via the local variable this, which points inside all member

---

[8]Hence, one could use a private function do_copying(), which does the copying wherever needed.

functions always to the object for which the member function is called. Another difference to the copy constructor is that the member `table` should at first be deleted, and then created again, because it might change size. Furthermore, the operator should return a reference to the object (`Histo &` at the binning of line 1), since the result of an assignment in C/C++ is the value assigned. For this purpose the object for which the member function is called must be returned via `*this` in line 17.

Next, it is shown how the class `Histo` can be augmented via inheritance by including a member array for storing the accumulated moments. For this purpose, the class `HistoM` is declared as follows:

```
1  class HistoM: public Histo
2  {
3    private:
4      int      num_moments;    // how many moments are stored
5      double        *sum;      // sum of 1s, numbers, number^2, ...
6
7    public:
8      HistoM(double from, double to, int n_bins);  // constructor
9      HistoM(const HistoM& h);                      // copy constructor
10     ~HistoM();                                     // destructor
11
12     void insert(double number);                   // insert data point
13     double mean();                                 // return mean
14 };
```

The `: public Histo` in line 1 indicates that the class is inherited from `Histo`. `HistoM` contains all members of `Histo` plus the members stated here.[9] The keyword `public` indicates that all public members of `Histo` are public members of `HistoM` as well, otherwise they would be `private`. Furthermore, one can access all `public` and all `protected` members of `Histo` in member functions of `histoM`. The constructor for `HistoM` looks as follows:

---

[9]Note that multiple inheritance is possible as well, if several parent classes are stated, which are separated by commas.

```
1  HistoM::HistoM(double h_from, double h_to, int h_n_bins) :
2    Histo(h_from, h_to, h_n_bins)
3  {
4    num_moments = 8;                    // allocate moments
5    sum = new double[num_moments];
6
7    for(int m=0; m<num_moments; m++)
8      sum[m] = 0;
9  }
```

Whenever an object of the class `HistoM` is initialized, first the members of the part inherited from `Histo` are initialized, this is indicated by the : `Histo(h_from, h_to, h_n_bins)` in lines 1–2. Nevertheless, all **public** and **protected** members of `Histo` could be initialized explicitly inside this constructor, which is not done here.

The destructor for `HistoM` deals only with the non-inherited members, since the destructor for the parent class is called automatically:

```
1  HistoM::~HistoM()
2  {
3    delete[] sum;
4  }
```

You are asked to implement a copy constructor for `HistoM` in exercise (2).

Finally, it is shown how the member function `insert()` is augmented for `HistoM`. Note that in principle it is not necessary to implement all member functions of the parent class again. In the same way, you could be satisfied with the `<<` operator defined for `Histo` and just print the histogram when printing an `HistoM` object. For `insert()`, we want to update the moments, hence we need a new implementation, based on the implementation for `Histo`:

```
1  void HistoM::insert(double number)
2  {
3    double value = 1.0;          // auxiliary variable
4
5    for(int m=0; m<num_moments; m++)
6    {
7      sum[m]+= value;;           // raw statistics
8      value *= number;
9    }
10   Histo::insert(number);       // perform insert for parent class
11 }
```

Note that the member function `insert()` of `Histo` can be called directly. In this case it will be automatically called for the current object as well, for which `HistoM::insert()` was called.

You are asked in exercise (2) to implement a simple member function `HistoM::mean()` which returns the mean of the numbers stored in the histogram. Within the scope of this book, it is impossible to go beyond the fundamentals of C++. More details can be found in the C++ literature [Stroustrup (2000)], in particular about *templates*, which are functions or classes where the type of the arguments needs not to be specified. This allows objects to be implemented which work with a wide range of types. Of particular importance are *container* classes, which can store "anything". Some container classes are implemented in the *standard template library*, which is introduced in Sec. 7.2.

# Exercises

(solutions: can be downloaded from http://www.worldscientific.com/r/9019-supp)

(1) **Histogram probability density function**

Design, implement and test a function, which
for a histogram **his** writes to a file

| SOLUTION SOURCE CODE |
| --- |
| DIR: oop |
| FILE(S): main_histo.c |
| histo_fprint_pdf.c |

- values of all moments (preceded by #)

- the number of outliers in **his->low**,
  **his->high** (preceded by #)

- the histogram as probability density
  function, see page 293 in Sec. 8.3.3.

The function prototype reads as follows:

```
/************* histo_fprint_pdf() ********************/
/** Prints moments of histogram and bin counts for  **/
/** outliers to file.                                **/
/** Prints also histogram as pdf                     **/
/** Prints also Gaussian standard error bars of      **/
/** bin counts (normalized)                          **/
/** PARAMETERS: (*)= return-parameter                **/
/**       file: file pointer                         **/
/**       histo: histogram                           **/
/** RETURNS:                                          **/
/**     (nothing)                                     **/
/*****************************************************/
void histo_fprint_pdf(FILE *file, histo_t *his)
```

Remark: You can include, for simplicity, the standard "Gaussian" error bar
for the bin entries, see Eq. (8.65).

You may use the **main()** function provided in **main_histo.c** to test your
function, if you compile with **-DSOLUTION**.

(2) **C++**

Design, implement and test a copy constructor for the class **HistoM**. Use the copy constructor for **Histo** (see page 149) and the
constructor for **HistoM** (see page 153) as example.

| SOLUTION SOURCE CODE |
| --- |
| DIR: oop |
| FILE(S): histoSOL.cpp |

The function prototype reads as follows:

```
/*********************** HistoM() ********************/
/** Copy Constructor: Creates a histogram (including **/
/** moments) as copy of argument histogram           **/
/** PARAMETERS: (*)= return-parameter                 **/
/**        h: histogram                               **/
/** RETURNS:                                          **/
/**      ---                                          **/
/*****************************************************/
HistoM::HistoM(const HistoM &h) : Histo(h)
```

Furthermore, write and test a member function for class HistoM, which returns the mean of the numbers stored in the histogram.

The function prototype reads as follows:

```
/*********************** mean() ********************/
/** Calculates mean from  histogram                **/
/** PARAMETERS: (*)= return-parameter              **/
/**                                                **/
/** RETURNS:                                        **/
/**     mean                                        **/
/*************************************************/
double HistoM::mean()
```

# Chapter 6

# Algorithms and data structures

If you ever want to change, extend, or even develop a full simulation program, you could, in principle, just use the techniques you have learnt in the previous chapters, and directly implement the simulation methods using arrays and C `structs` as main data structures. Nevertheless, the efficiency of such a simulation program can be greatly enhanced, if you know about fundamental algorithms and data structures in computer science. There are many cases in science, where researchers first used straightforward implementations for their problem and even got some results which were published in scientific papers. Later on, it was found out in these cases that by using simple techniques as described in this chapter, the problem under investigation could be tackled much better, i.e. larger system sizes could be studied, and more data could be gathered, hence reducing the statistical error bar of the data (c.f. Chap. 8). Then, using the better approach, it turned out that the conclusions, which were previously drawn using the straightforward program, were wrong. Hence, if you want to be among those who write efficient programs which can simulate large systems, then you should read this chapter carefully.

We begin by introducing the $\mathcal{O}$ *notation*, which is used to describe how fast a program runs. In the next couple of following sections, five basic programming techniques are introduced: *iteration, recursion, divide-and-conquer, dynamic programming*, and *backtracking*. Although in these chapters, the examples are given using the C programming languages, it is important that you understand the basic principles behind the C code. The basic recipe describing how a problem is solved, independent of an implementation in a programming language, is called the *algorithm*. Almost all clever algorithms known in computer science are based on one or more of these five fundamental algorithmic techniques.

In the final set of sections of this chapter, some sophisticated data structures are introduced. They allow in most cases to organize the simulation data in a way that suits the given problem better and allows for a quicker processing. Here, we present *lists*, *trees*, and *graphs*, together with some sample algorithms to create, modify and use them.

Note, as always in this book, that only few introductory samples (although covering a large fraction of the basic applications) can be given. The important message is that you have to learn to think in terms of data structures, which allows you to aim at looking for efficient implementations. Once you have a little bit of experience, a kind of library of samples in your head, this will become increasingly easier for you. In fact, many more data structures, usually based on the basic data structures presented here, and an overwhelming amount of algorithms acting on these data structures exist. If you want to learn more, and later on you want to do this, you should consult more specialized text books like Refs. [Aho et al. (1974); Cormen et al. (2001); Sedgewick (1990)].

## 6.1 $\mathcal{O}$ notation

For an arbitrary algorithm, to describe the dependence between a suitably chosen measure $n$ of the problem size and the running time $T$, the $\mathcal{O}$ notation is used.

**Definition 6.1** Let $T, g : \mathbb{N} \to \mathbb{R}$ be two real-valued functions.
We write $T(n) = \mathcal{O}(g(n))$, if there exist two positive numbers $c_1, c_2 > 0$ and an integer $n_0$, such that $c_1 g(n) \leq T(n) \leq c_2 g(n)$ is valid for all $n > n_0$. We say, $T(n)$ is *of order* of $g(n)$.

Since constants are ignored when using the $\mathcal{O}$ notation, one speaks of the *asymptotic* running time or *time complexity*. In theoretical computer science, usually one states an upper bound over all possible inputs of size $n$, i.e., the *worst-case running time*.

As an example, we look at a function which adds up $n$ numbers:

```
 1  double compute_sum(int n, double *number)
 2  {
 3    double sum=0.0;                              /* for summing up */
 4    int t;                                       /* loop counter   */
 5
 6    for(t=0; t<n; t++)                                /* main loop */
 7      sum += number[t];
 8
 9    return(sum);
10  }
```

The loop lines 6–7 is executed $n$ times. For each loop iteration, the following operations are performed (details might depend on the compiler, microprocessor used, etc.):

> **GET SOURCE CODE**
>
> DIR: `algorithms`
> FILE(S): `sum.c`

- The value of the variable t is read. Note that usually the compiler produces code such that this value is stored in a register of the microprocessor, i.e. the access is fast.
- The position in memory of the t'th element of the array **number** is calculated (this is `number+t*sizeof(real)`)
- The t'th element of the array **nummber** is fetched from memory. This is a bit slower than the access to a register.[1]
- The value of **sum** is read.
- Then the sum of **number[t]** and **sum** is calculated.
- The result is stored again in **sum**.
- The value of t is increased (in the register) by one.
- It is checked whether t<n (n also assumed to be in a register, otherwise it has to be fetched from memory as well).

On modern computers, all these operations take a running time which is independent of $n$, hence, we can assume that the total running time of one loop iteration is a sum of constant values, i.e. a constant $C$ itself.[2] Hence, the full running time of the function, i.e. the running time of the

---

[1] In modern computers, there are different levels of memory like 1st and 2nd level cache, main memory and swap space. Cache memories are much faster than main memory, but also much smaller. Swap memory is located on hard disk and only used if the main memory is too small for the program. Nevertheless, swapping is very slow and should be avoided.

[2] The actual CPU time needed for one iteration also depends on external factors, like whether the cache can be used or not, what other programs are running in multitasking mode etc. For assessing the running time of an algorithm, this is not taken into account, because it does not depend on the algorithm itself.

"summation algorithm", is $C \times n$, i.e. $\mathcal{O}(n)$. Note that in this case the running time basically does not depend on the actual numbers which are summed up, i.e. the worst case running time equals the typical running time. Nevertheless, there are examples where the running time might depend drastically on the actual problem instance given.

In Table 6.1, orders of running times, which occur typically in the context of algorithms, are presented, accompanied by the resulting values for problem sizes 10, 100, and 1000.

Table 6.1　Growth of functions as a function of input size $n$.

| $T(n)$ | $T(10)$ | $T(100)$ | $T(1000)$ |
|--------|---------|----------|-----------|
| $n$ | 10 | 100 | 1000 |
| $n \log n$ | 10 | 200 | 3000 |
| $n^2$ | $10^2$ | $10^4$ | $10^6$ |
| $n^3$ | $10^3$ | $10^6$ | $10^9$ |
| $n^{\log n}$ | 10 | $10^4$ | $10^9$ |
| $2^n$ | 1024 | $1.3 \times 10^{30}$ | $1.1 \times 10^{301}$ |
| $n!$ | $3.6 \times 10^6$ | $10^{158}$ | $4 \times 10^{2567}$ |

Usually one considers problems *easy,* if the running time is bounded by a polynomial, all others are considered *hard*. The reason can be understood from the table: Even if the polynomial functions may take higher values for small $n$, asymptotically non-polynomial functions diverge much faster. Let us consider, e. g. the relative performance of two computers, one being twice as fast as the other one. In a linear-time problem, the faster computer is able to solve a problem which is twice as large as the problem solvable in the same time on the slower computer. If the running time grows, however, as $2^n$, the faster computer is just able to go from $n$ to $n + 1$ compared with the slower one. We see that for such hard problems, the utility of higher-speed computers is very limited – a substantial increase in the size of solvable problems can only be achieved via the use of better algorithms.

## 6.2　Iteration and recursion

If a program has to perform many similar tasks, this can be expressed as an iteration, also called a loop, e. g. with the `for`-statement from C/C++. Sometimes it is more convenient to use the concept of *recursion*, especially if the quantity to be calculated has a recursive definition. One speaks of recursion if an algorithm/function calls itself (maybe indirectly through other algorithms/functions). As a simple example we present a C function for the calculation of the factorial $n!$ of a natural number $n > 0$. Its recursive definition is given by:

$$n! = \begin{cases} 1 & \text{if } n = 0 \text{ or } n = 1 \\ n \times (n-1)! & \text{else} \end{cases} \tag{6.1}$$

This definition can be translated directly into a C function:

```
1  int factorial(int n)
2  {
3    if(n <= 1)
4      return(1);
5    else
6      return(n*factorial(n-1));
7  }
```

In line 3 the test is whether $n \leq 1$ instead of testing for $n = 0$ or $n = 1$. Therefore, it is guaranteed that the algorithm returns something on all inputs.

| GET SOURCE CODE |
|---|
| DIR: **algorithms** |
| FILE(S): **fac_rec.c** |

For $n > 1$, during the execution of **factorial(n)**, a sequence of nested calls of the function is created up to the point where the function is called with argument 1. The call to **factorial(n)** begins before and is finished after all other calls to **factorial(i)** with $i < n$. The hierarchy in Fig. 6.1 shows the calls for the calculation of **factorial(4)**.

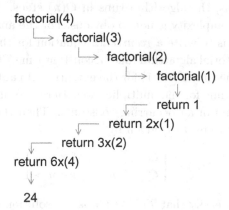

Fig. 6.1   Hierarchy of recursive calls for calculation of factorial(4).

Note that the present implementation of the algorithm allows only for small factorials to be calculated, due to the finite number of bits used to store integers. For 32 bit (signed), the largest value for which the program is correct is $n = 13$. In case you want to calculate the factorial of larger numbers, you can e.g. use libraries which allow for arbitrary precision, see Chap. 7.

Every recursive algorithm can be rewritten as a sequential algorithm, containing no calls to itself. Instead loops are used. Usually, these sequential versions of recursive algorithms are

| GET SOURCE CODE |
| --- |
| DIR: algorithms<br>FILE(S): fac_seq.c |

faster by some constant factor but harder to understand, at least when the algorithm is more complicated than in the present example. The sequential version for the calculation of the factorial reads as follows:

```
1  int factorial(int n)
2  {
3    int t;                          /* loop counter */
4    int fac;                        /* result */
5
6    fac = 1;
7    for(t=2; t<=n; t++)             /* main loop */
8      fac *= t;
9
10   return(fac);
11 }
```

The sequential factorial algorithm contains one loop which is executed $n - 1$ times. Thus, the algorithm runs in $\mathcal{O}(n)$ steps. For the recursive variant the time complexity is not so obvious. For the analysis of recursive algorithms, one has to write a *recurrence* equation for the execution time. For $n = 1$, the factorial algorithm takes a constant time $T(1)$. For $n > 1$ the algorithm takes the time $T(n-1)$ for the execution of `factorial(n-1)` plus another constant time for the multiplication. Here and in the following, let $C$ be the maximum of all occurring constants. Then the running time is bounded from above by $\tilde{T}(n)$ given by

$$\tilde{T}(n) = \begin{cases} C & \text{for } n = 1 \\ C + \tilde{T}(n - 1) & \text{for } n > 1. \end{cases} \tag{6.2}$$

One can verify easily that $\tilde{T}(n) = Cn$ is the solution of the recurrence, i.e., both recursive and sequential algorithms have the same asymptotic time complexities. There are many examples where a recursive algorithm is asymptotically faster than a *straightforward* sequential solution (i.e. sequential implementations of basically recursive algorithms are not meant here). An example will be given in the following section, see also [Aho et al. (1974)]. Exercise (1) at the end of the chapter provides another example for a problem which can be solved very elegantly by recursion.

## 6.3 Divide-and-conquer approach

The basic idea of *divide-and-conquer* is to divide a problem into smaller subproblems, solve the subproblems separately, and then combine their solutions to form the final solution. Recursive calls of the algorithm are usually applied here as well.

As an example we consider sorting problems. Given are $n$ data sets $A_i$ ($i = 1, 2, \ldots, n$). These can be, in the simplest case, natural numbers, or strings or complex data structures having a *key* exhibiting a natural ordering "$<$". We want to find a permutation $B_i$ of them such that they are sorted in (say) increasing order: $B_i < B_{i+1}$ for all $i < n$. First, we quickly explain a simple recursive algorithm for sorting elements, which does not use the divide-and-conquer principle. This simple approach starts by scanning through the array $A$ and looks for the smallest element. It is deleted from $A$ and stored in $B_1$. Then one again looks for the smallest element in the remaining array $A$, removes it from $A$ and stores it in $B_2$. This iteration is repeated, until all elements are treated. Since the iteration is performed $n$ times, and because looking for the smallest element each time takes $\mathcal{O}(n)$ accesses to the array $A$, the full algorithm has a complexity of $\mathcal{O}(n^2)$.

Next, an approach based on the divide-and-conquer principle is presented, called *mergesort*. As it will be shown below, the algorithm will require only $\mathcal{O}(n \log n)$ steps. Note that no proof will be given that the algorithm actually works, since this is beyond the scope of this book. Nevertheless, the reason that the algorithm is correct should be obvious from the following discussion: The basic idea of mergesort is to part the set which is to be sorted into two subsets of roughly equal size, sort them recursively and finally *merge* the two sorted sequences into one sorted sequence. The merging is performed by iteratively removing the smallest element among both sequences.

As an example, the approach is applied to an array consisting of animal data, each animal is described by an ID and by its weight. The animals should be sorted in ascending weights.

| GET SOURCE CODE |
| --- |
| DIR: algorithms |
| FILE(S): mergesort.c |

The data is an array of type `animal_t`, which is defined via a structure:

```
typedef struct
{
  int weight;
  int animal_id;
} animal_t;
```

The function performing the actual sorting reads as follows:

```
1   animal_t *mergesort(int num, animal_t *data)
2   {
3     animal_t *result, *result1, *result2;        /* sorted elements */
4     int size1, size2;                  /* sizes of two subsets of data */
5     int t;        /* counter for putting elements into result array */
6     int t1, t2;        /* for getting elements from result1/2 arrays */
7
8     result = (animal_t *) malloc(num*sizeof(animal_t));
9
10    if(num==1)                               /* solve trivial case */
11      result[0] = data[0];
12    else                                         /* main work */
13    {
14      size1 = num/2; size2 = num - size1;
15      result1 = mergesort(size1, data);          /* sort 1st half */
16      result2 = mergesort(size2, data+size1);    /* sort 2nd half */
17
18      t=0, t1=0; t2=0;
19      while(t<num)                 /* merge result1,result2 into result */
20        if( (t2==size2) ||
21            ((t1 < size1) && (result1[t1].weight<result2[t2].weight)))
22          result[t++] = result1[t1++];
23        else
24          result[t++] = result2[t2++];
25
26      free(result1); free(result2);
27    }
28    return(result);
29  }
```

In line 11, the trivial case is treated, where only one element is to be sorted. The main part of the algorithm is in lines 14–24. In lines 14–16 the set of data is parted into two (almost) equal-sized subsets (the second subset is by one element larger if the number of elements is odd).[3] Then, in lines 18–24, the two sorted subsets `result1`, `result2` are merged into `result`. The counter t indicates the position where the next element is to be put in `result`, while t1 and t2 indicate always the smallest element not yet treated in `result1` and `result2`, respectively.

Please note that this implementation is just for the data type `animal_t`. If you want to write a function which sorts general data, the function would need also to be provided another function which allows to compare two elements of the given set of data. In this way the C library function `qsort()` is implemented, which is explained in Sec. 7.1.

As an example, in the upper part of Fig. 6.2 the hierarchy of recursive calls of `mergesort`$(4, \{5, 2, 3, 1\})$ is displayed. In the lower part the merging of the sorted subset is shown. For $n = 2^k$ one obtains $k + 1$ layers in the hierarchy of calls.

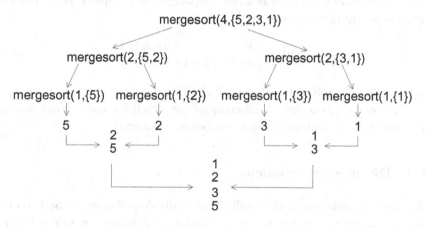

Fig. 6.2   Call of mergesort$(4, \{5, 2, 3, 1\})$.

The division of the sets and the merge-operation takes $\mathcal{O}(n)$ time, while each recursive call takes $T(n/2)$. Hence, the recurrence for this algorithms

---

[3]Note that in line 16, when calculating `data+size1`, the address of the element `data[size1]` is calculated. Hence, it does *not* mean that `size1` bytes are added to `data`, instead `sizeof(animal_t)*size1` bytes are added to `data`.

reads:

$$T(n) = \begin{cases} C & (n = 1) \\ Cn + 2T(n/2) & (n > 1) \end{cases} \tag{6.3}$$

If $n$ is large enough, this recurrence can be solved by $T(n) = \frac{C}{\log 2} n \log n$. This can be seen by inserting the solution into the second line of the equation

$$\begin{aligned} T(2n) &= C2n + 2T(n) \\ &= C2n + 2\frac{C}{\log 2} n \log n \\ &= \frac{C}{\log 2} 2n \log 2 + \frac{C}{\log 2} 2n \log n \\ &= \frac{C}{\log 2} 2n \log(2n) \end{aligned} \tag{6.4}$$

Consequently, the divide-and-conquer realization of sorting is asymptotically faster than the simple recursive sort-algorithm. Finally, note that the general recurrence equation

$$T(n) = \begin{cases} k & \text{for } n = 1 \\ aT(n/c) + kn & \text{for } n > 1 \end{cases} \tag{6.5}$$

is, for $a > c$ (!), solved by $T(n) = \mathcal{O}(n^{\log_c(a)})$.

In exercise (2) a divide-and-conquer algorithm for quickly calculating the power $a^n$ ($n$ an integer) of a number is considered.

## 6.4   Dynamic programming

Another problem where the application of divide-and-conquer and  recursion seems quite natural is the calculation of *Fibonacci numbers* fib($n$). Their definition is as follows:

$$\text{fib}(n) = \begin{cases} 1 & (n = 1) \\ 1 & (n = 2) \\ \text{fib}(n-1) + \text{fib}(n-2) & (n > 2). \end{cases} \tag{6.6}$$

Thus, for example, fib(4) = fib(3)+fib(2) = (fib(2)+fib(1))+fib(2) = 3, fib(5) = fib(4) + fib(3) = 3 + 2 = 5. The functions grow very rapidly: fib(10) = 55, fib(20) = 6765, fib(30) = 83204, fib(40) > $10^8$. Let us assume that this definition is translated directly into a recursive algorithm. Then a call to fib($n$) would call fib($n-1$) and fib($n-2$). The recursive call of fib($n-1$) would call *again* fib($n-2$) and fib($n-3$) [which is also called from the two calls of fib($n-2$), etc.]. The total number of calls increases rapidly with $n$, even more than fib($n$) itself increases with $n$. In Fig. 6.3, the top of a hierarchy of calls is shown. Obviously, every call to fib with a specific argument is performed frequently, which is definitely a waste of time.

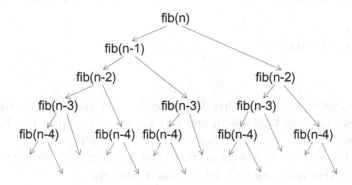

Fig. 6.3   Hierarchy of calls for fib($n$).

Instead, one can apply the principle of *dynamic programming*. The basic idea is to start with small problems, solve them and store the solutions for later use. Then one proceeds with

| GET SOURCE CODE |
| --- |
| DIR: `algorithms` |
| FILE(S): `fibonacci.c` |

larger problems by using divide-and-conquer. If, for the solution of a larger problem, a smaller one is necessary, it is already available. Therefore, no direct recursive calls are needed. As a consequence, the performance increases drastically. The divide-and-conquer algorithm for the Fibonacci numbers reads as follows, the array f [] is used to store the results:

```
1  double fib(int n)
2  {
3     double *f;              /* stores Fibonacci numbers: f[k]=fib(k+1) */
4     int t;                                         /* loop counter */
5     double result;
6
```

```
 7    if(n<3)
 8      return(1.0);
 9    else
10    {
11      f = (double *) malloc(n*sizeof(double));      /* allocate memory */
12
13      f[0] = 1; f[1] = 1;  a /* start values for dynamic programming */
14      for(t=2; t<n; t++)              /* main loop: dynamic programming */
15        f[t] = f[t-1] + f[t-2];
16      result = f[n-1];                          /* save final result */
17
18      free(f);
19      return(result);
20    }
21  }
```

Note, since we have to free the array f[] before we return the final result, we have to buffer it in the variable result. Since the function contains just one loop in the lines 14–15, it runs in $\mathcal{O}(n)$ time.

Finally, we should point out that there is an explicit formula which allows a direct calculation of the Fibonacci numbers:

$$\mathrm{fib}(n) = \frac{1}{\sqrt{5}} \left( \left( \frac{1+\sqrt{5}}{2} \right)^n - \left( \frac{1-\sqrt{5}}{2} \right)^n \right). \tag{6.7}$$

An example for a more sophisticated application of the dynamic programming principle can be found in exercise (3).

## 6.5  Backtracking

The last basic programming principle which is presented here is *backtracking*. This method is applied when there is no direct way of computing a solution. This is typical of many combinatorial problems, like optimization of functions over discrete variables. The basic idea is that one has to try some (sub-)solutions, discard them if they turn out not to be good enough and try some other (sub-)solutions. Hence, all the time variables are assigned and later reassigned. This is done in a controlled way, such that all interesting possible assignments are tried, until a solution is found. This is the basic principle of backtracking.

As an example, in the following we will present a backtracking algorithm for the solution of the $N$-queens problem.

## N-queens problem

$N$ queens are to be placed on an $N \times N$ chess board in such a way that no queen checks against any other queen. This means that, in each row, in each column and in each diagonal at most one queen is placed, see Fig. 6.4.

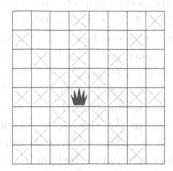

Fig. 6.4   A standard $8 \times 8$ chess board with a queen placed near the center. The crosses indicate the squares, where no other queen can be placed for the 8-queens problem.

A naive solution of the problem works by enumerating all possible configurations of $N$ queens and checking, for each configuration, whether any queen checks against another queen. By restricting the algorithm to place at most one queen per column, there are $N^N$ possible configurations. This is a very strongly increasing running time, which can be decreased by backtracking.

The idea of backtracking is to place one queen after the other. One stops placing further queens if a non-valid configuration is already obtained at an intermediate stage. Then one goes one step back, removes the queen which was placed at the step before, places it elsewhere, if possible, and continues again. Note that it also will occur that several queens are removed again, i.e., several steps are taken back, see the example below.

The algorithm starts in the last column and places a queen. Next a queen is placed in the second last column by a recursive call and so forth. If all columns are filled, a valid configuration has been found. If at any stage it is not possible to place any further queen in a given column, then the backtracking step is performed: the recursive call finishes, the queen which was set in the recursion step before is removed. Then it is placed elsewhere and the algorithm proceeds again. The argument n denotes the column where the next queen is to be placed. To match the C standard of starting arrays at index 0, we number the rows and columns from 0 to $N - 1$. Hence, initially the algorithm is called with n= $N - 1$.

We use an array pos[], where pos[c] stores the position of the queen in column c. If pos[c] = 0, no queen has been placed in that column. Hence, initially all pos[c] values are 0.

The array pos[] contains the complete information, hence it would suffice to use this array. Then, if one wants to test whether a queen in column $c$ checks again any other queen, one has to run through the fields pos[c+1] to pos[N-1]. Hence, any such test would require $\mathcal{O}(N)$ operations. The test can be performed in $\mathcal{O}(1)$ steps, if additional arrays are used for all rows and all diagonals of the chess board, which indicate whether a queen is present or not in the row or diagonal. Hence, this is an example of how one can *save running time at the expense of memory and by using more sophisticated data structures*. This is a general balancing principle, which you should always keep in mind when designing your simulations.

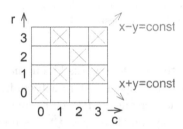

Fig. 6.5 "Up" and "down" diagonals $\{(c, r)\}$ on the chessboard are characterized by $c - r$ =const and $c + r$ =const, respectively.

Coming back to the $N$-queens problem, note that diagonals come in two types. If we denote the coordinates by $(c, r)$, "up diagonals" are characterized by $c - r$ =const, while "down

| GET SOURCE CODE |
| --- |
| DIR: **algorithms** |
| FILE(S): **queens.c** |

diagonals" are characterized by $c + r$ =const, see Fig. 6.5. Here we use row, diag_up and diag_down, a value 0 always indicates that no queen is present, while a value 1 means that a queen has been placed. Note that there are $2N - 1$ diagonals of each kind, hence these arrays must be larger (allocated in the main program, which is not shown here). Also note that for the up diagonals, $c - r$ can be as small as $-(N - 1)$, hence we shift the index of this array by this amount to start as usually at index 0. Using these additional arrays, the backtracking function for the queens problem reads as follows:

```
1   void queens(int c, int N, int *pos, int *row,
2                 int *diag_up, int *diag_down)
3   {
4     int r, c2;                              /* loop counters */
5     if(c == -1)                             /* solution found ? */
6     {
7       /* omitted here */                    /* print solution */
8     }
9     for(r=N-1; r>=0; r--)      /* place queen in all rows of column c */
10    {
11      if(!row[r]&&!diag_up[c-r+(N-1)]&&!diag_down[c+r])   /*  place ? */
12      {
13        row[r] = 1; diag_up[c-r+(N-1)] = 1; diag_down[c+r] = 1;
14        pos[c] = r;
15        queens(c-1, N, pos, row, diag_up, diag_down);
16        row[r] = 0; diag_up[c-r+(N-1)] = 0; diag_down[c+r] = 0;
17      }
18    }
19    pos[c] = 0;
20  }
```

In Fig. 6.6 the way in which the algorithm solves the problem for $N = 4$ is shown. It starts by calling queens(3,...) where a queen is placed in column 3 and row 3. Then queens(2,...) is called. The positions where no queen is allowed are marked with a cross. For column 2 no queens in row 3 and row 2 are allowed. Thus, a queen is placed in row 1 and queens(1,...) is called. In column 1 it is now impossible to place any queen. Hence, the call to queens(1,...) finishes. The queen in column 2 is placed one row below, i.e., row 0 (second line in Fig. 6.6). Then, by calling queens(1,...), a queen is placed in row 2 and queens(0,...) is called. Now, no queen can be placed in the first column, hence queens(0,...) returns. Since there was only one possible position in column 1, the queen is removed and also the call queens(1,...) finishes. Now, both possible positions for the queen in column 2 have been tried. Therefore, the call for queens(2,...) finishes as well and we are back at queens(3,...). Now, the queen in the last column is placed in the row 2 (third line in Fig. 6.6). From here it is straightforward to place queens in all columns and the algorithm succeeds.

Although this algorithm avoids many "dead ends", it still has an exponential running time as a function of $N$. Nevertheless, there are better but very specialized algorithms, where the running time increases only lin-

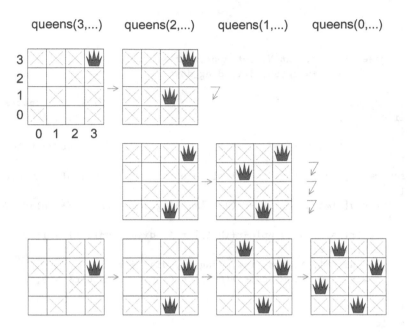

Fig. 6.6   How the algorithm solves the 4-queens problem.

early with $N$ [Abramson and Yung (1986); Abramson and Yung (1989); Sosic and Gu (1991)].

## 6.6   Lists

In the previous sections, elementary algorithms were presented. Now we turn to data structures, which are special ways to arrange data in memory such that algorithms can be implemented as efficiently as possible. Note that for different types of tasks, different data structures are most suitable, hence one always has to have the target application in mind. This also explains why a careful planning of the simulation program before actually starting programming is very recommendable, see Chap. 3. Furthermore, note that usually data structures which allow for very fast operations sometimes require some degree of redundancy. Very often, the quicker the operations are, the more memory you have to consume. Hence, efficiency in terms of running time comes at the expenses of efficiency in terms of memory, and vice versa.

Assume that you are writing a simulation package for describing the social interactions of persons. For each person, there is always a (sometimes empty) set of other people who are currently interacting with the person. This set may change in time. Hence, we would like to have a data structure to store all current acquaintances in a flexible way, i.e., persons can be easily added and removed from the set. Here, we present a data structure called a *list*, which allows to solve this task. Lists are, for example, used in atomistic simulation packages for implementing short-range interactions. In this case, each atom interacts only with atoms not too far away. In this case, for each atom a list is used (a so-called "Verlet table") to store the atoms which are currently in the neighborhood.[4] Another example for lists are waiting queues, which consist of tasks given in a linear order, where one task has to be performed after the other. Here, also the order of the elements is important, as for many applications, in contrast to acquaintance lists or the Verlet tables. Note that the example applications and implementations given below are rather simple, in contrast to the real applications just mentioned, to keep the subject as simple as possible.

Lists are generalizations of arrays of elements (of arbitrary type). Lists also exhibit a linear order like arrays, i.e., there is always a first element, for each element there is a successor, except for the last element of the list. For an array, the order of the elements is fixed. If you want to remove, say, the 5th element in an array of 100 elements, and you want to keep the order of the remaining elements, then you have to copy the 6th element to the 5th position, then the 7th element to the 6th, and so on, see Fig. 6.7. Hence, for a list of $N$ elements, $O(N)$ operations are needed to remove an element from the list while keeping the order. Also $O(N)$ operations are required when an element is inserted into an array. Furthermore, the size of an array is fixed, hence usually one stores the number of the last element which is used and the maximum number of allowed elements. Now, if the number of elements to be stored grows larger than the number of allowed elements, one has to use `realloc` to obtain more memory, which might lead to a considerable amount of memory copy operations, if the free amount of memory right behind the array does not provide enough available memory to satisfy the request.

---

[4]Usually, for better performance, the simulation volume is divided into boxes, which are of the size of the interaction range. For each box, a list of the atoms currently in this box is stored, which reduced the memory usage compared to Verlet tables [Allen and Tildesley (1989)].

For lists, on the other hand, the size and the order are not fixed. Hence, they are designed in such a way that insert and delete operations can always be performed in constant running time.

| GET SOURCE CODE |
| --- |
| DIR: `algorithms` |
| FILE(S): `list.h` |

The most natural implementation for this purpose is using *pointers*, see Sec. 1.4. The basic idea is that each element carries a pointer, which points to its successor. Here, we show a sample C structure, which implements lists, and, for simplicity, just stores integer numbers.

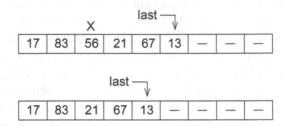

Fig. 6.7   Removing an element from an array, such that the order of the remaining elements is conserved. Unused array elements are indicated by a line —. The last element used in the array is stored in the variable `last`. Top: array before removing the third element, which contains the number 56 (marked by an X). Bottom: final situation, all elements to the right of the third position have moved one position to the left.

```
1  struct elem_struct
2  {
3    int              info;           /* holds "information" */
4    struct elem_struct *next;  /* pointer to successor (last: NULL) */
5  };
6
7  typedef struct elem_struct elem_t;          /* new type for nodes */
```

Each element consists of two variables. The actual data is stored in `info`, while the successor in the list is stored in `next`. We have used also a `typedef` command, which allows, for convenience, to refer to the new list data type in the same way as for a predefined data type.

A list is a collection of elements, which are linked in a way such that a linear order is represented, e.g. see Fig. 6.8. Such a list is denoted as *single-linked*, because each element stores one pointer, i.e. link. Lists where each element stores two pointers, one to its successor and one to its predecessor, are also widespread and called *double-linked lists*. Double-linked lists require more memory, due to the additional elements, but some

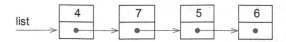

Fig. 6.8   A list consisting of four elements containing the integer numbers 4, 7, 5 and 6. The list is represented by a pointer `list` which points to the first element. Each element contains a pointer to its successor (represented by arrows in the figure), except the last element, which contains a NULL pointer, represented by a filled circle.

list operations can be performed faster, see below the `remove_element()` operation. Here, for brevity, we only consider single-linked lists.

Next, we present some operations needed to work with lists. Following an object-oriented approach, we first consider functions which create and delete elements. The function

| GET SOURCE CODE |
| --- |
| DIR: algorithms<br>FILE(S): list.c |

`create_element()` receives the integer number the new element will store as parameter and returns a pointer to the new element. The function allocates the memory (line 5), stores the number (line 6) and initializes the `next` pointer to NULL (line 7).

```
1  elem_t *create_element(int value)
2  {
3     elem_t *elem;
4
5     elem = (elem_t *) malloc (sizeof(elem_t));
6     elem->info = value;
7     elem->next = NULL;
8     return(elem);
9  }
```

The `delete_element()` function receives a pointer to the element to be deleted and returns an integer number indicating whether the operations were performed successfully (0) or not (1).

```
1   int delete_element(elem_t *elem)
2   {
3     if(elem == NULL)
4     {
5       fprintf(stderr, "attempt to delete 'nothing'\n");
6       return(1);
7     }
8     else if(elem->next != NULL)
9     {
10      fprintf(stderr, "attempt to delete linked element!\n");
11      return(1);
12    }
13    else
14    {
15      free(elem);
16      return(0);
17    }
18  }
```

The function contains two consistency checks. First, it is not allowed to delete "nothing" (lines 3–7). Second, it is only allowed to delete isolated elements (lines 8–12), which is indicated by a successor given by a NULL pointer. The actual deletion consists just of the **free()** command (line 15).

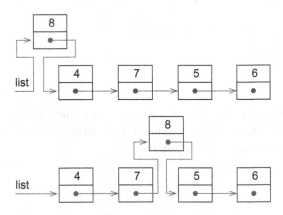

Fig. 6.9  Inserting an element containing an 8 into the list from Fig. 6.8. Either the element is inserted at the beginning (top) or elsewhere into the list (bottom). In both cases the final situation after the insertion is shown.

To assemble lists from elements, we use the **insert_element()** function, which inserts an element (represented by a pointer **elem**) into a list (represented by a pointer `list` to the first element), *after* a given element `where`, which is already member of the list. There are two cases to be considered. Either the new element is inserted at the beginning of the list, in this case `where` contains the NULL pointer. Note that this case applies in particular when the list is empty, hence the new element is used to create a new list. The second case is the general case that `where` points to an existing member of the list. The two cases are illustrated in Fig. 6.9.

The function `insert_element()` receives exactly the three pointers `list`, `elem`, and `where` as arguments. Note that in the case the new element is inserted at the beginning, the pointer to the first element of the list, which represents the list itself and is defined outside `insert_element()`, has to be changed. To allow for this update, the function returns always a pointer to the current first element.

```
1   elem_t *insert_element(elem_t *list, elem_t *elem, elem_t *where)
2   {
3     if(where==NULL)                        /* insert at beginning ? */
4     {
5       elem->next = list;
6       list = elem;
7     }
8     else                                   /* insert elsewhere */
9     {
10      elem->next = where->next;
11      where->next = elem;
12    }
13    return(list);
14  }
```

In the case where the element is inserted at the beginning (lines 3–7), one has to change the pointer to the first element and to assign the successor `next` of the inserted element `elem`, as indicated in Fig. 6.9. In the other case (lines 8–12), also the successor of the inserted element is assigned, but also the predecessor is assigned to point to `elem`. Note that many other strategies for building lists are possible. One could, for example, insert elements such that they are ordered according to some criterion. Also, there are some special list types where inserting elements happens always at the beginning (for so-called *stacks*) or at the end (for *waiting queues*). For stacks, elements which arrived last are taken out first, this is called *last-in first-out* (LIFO). For waiting queues, elements which arrived first are taken out first, this is called *first-in first-out* (FIFO). As usual, we refer the reader to specialized textbooks for other types of inserting operations.

The use of the functions is illustrated by the following piece of code, which exactly creates the list (4, 7, 5, 6) by inserting the elements in reverse order all at the beginning of the list, one after the other.

```
1    elem_t *list, *elem;
2
3    list = NULL;
4    elem = create_element(6);
5    list = insert_element(list, elem, NULL);
6    elem = create_element(5);
7    list = insert_element(list, elem, NULL);
8    elem = create_element(7);
9    list = insert_element(list, elem, NULL);
10   elem = create_element(4);
11   list = insert_element(list, elem, NULL);
```

Once a list is created, one would like to do something with it. For example one could just print all members in the given order from the first to the last element. This is performed by the function print_list(), which receives a pointer list to the first element and returns nothing.

```
1  void print_list(elem_t *list)
2  {
3    while(list != NULL)                /* run through list */
4    {
5      printf("%d ", list->info);
6      list = list->next;
7    }
8    printf("\n");
9  }
```

The function just prints the list to stdout. The functions illustrated nicely the main operation for lists, the iteration over all members of the list. This works (lines 3–7) by starting at the beginning of the list, and moving the pointer along the list by iteratively assigning it to its successor (line 6). Note that the pointer list is passed by value, hence its value outside the function print_list() remains unaffected by changing list inside the function.

Finally, we address the removal of an element from the list. This is illustrated in Fig. 6.10. We assume that the pointer elem to the element to be removed is given, as well as a pointer list to the first element of the list. Now we face the difficulty that we have to know the predecessor of the

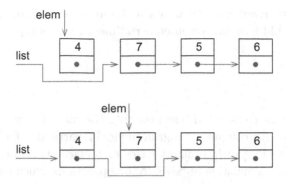

Fig. 6.10   Removal of an element from a list. Either the element to be removed is the first (top) or elsewhere inside the list (bottom). In both cases, the situation after the removal has been completed is shown.

element to be removed: In case there is no predecessor, i.e., if element is the first element of the list, list has to be changed (top of Fig. 6.10). If elem is not the first element, we have to reconnect the predecessor of elem to its successor (bottom of Fig. 6.10). In both cases, we have to run through the list, starting at the first element, until we reach the predecessor of the elem. This requires $\mathcal{O}(N)$ steps if the list contains $N$ elements, i.e., does not happen in constant $\mathcal{O}(1)$ time. This shows a drawback of single-linked lists, as mentioned above. This can be cured either by using double-linked lists, or by designing the remove_element() operation such that always a pointer to the predecessor of the element to be removed is given (NULL if the first element is to be removed). Note also that in the special case of stacks and waiting queues, removal of elements happens always at the beginning, hence can be performed in $\mathcal{O}(1)$ time as well. Instead of presenting a C implementation for remove_element(), we leave this to you as exercise 5, see end of the chapter.

Although we have presented in some detail how lists are implemented, note that usually you do not have to program a complete and most flexible lists package yourself. There are many freely available libraries, which contain implementations of lists, e.g. see Sec. 7.2, where a very flexible implementation using the *Standard Template Library* is explained. Still, although these libraries are available, it is still very instructive to implement one basic set of list operations at least once yourself. In this way you learn how to think in terms of data structures and you get used to work with pointers. Everyone who wants to be an expert simulation program developer should have implemented a rudimentary lists type at least once.

## 6.7 Trees

Now, we present trees, which are probably the most important advanced data structure for computer programming, in particular for many types of simulations. They help in many cases to replace operations which take time $\mathcal{O}(N)$ for a simulation of "size" $N$ by operations which only take time $\mathcal{O}(\log N)$. Hence, trees and other data structures which are derived from trees help a great deal in speeding up simulations, thereby allowing for larger systems to be treated.

For the data structures presented in the previous section, arrays and lists, always a linear order was used or assumed. Very often, additionally a hierarchical order is present or can be imposed. A very simple example of a system exhibiting a hierarchy is a company or a governmental organization. To consider a toy example, let us assume that a railway company consists of a planning division, an accounting division and a division operating the transportations. The planning division may be subdivided into a department for planning the network, another one for planning the schedules, a third for developing new pricing schemes, and a fourth responsible for buying new locomotives and wagons. The account division may be subdivided into four departments responsible for personal, maintenance, purchasing, and revenue, respectively. The operation division might be subdivided into departments for the freight trains, for passenger trains, for night trains, for busses, and one for actually performing the maintenance. The natural way to represent the structure of the railway company in the computer is a tree, as shown in Fig. 6.11.

Now we introduce some definitions. The elements of the tree are denoted as *nodes* or as *vertices*, the latter one is more generally used for graphs, see Sec. 6.8. For the railway company, e.g. "railway company", "accounting" and "night trains" are nodes. When two nodes A,B are connected by an arrow A→B, usually called a *link* or an *edge*, node A is called *predecessor* or *parent* of B, while B is called *successor* or *child* of A. In Fig. 6.11, "personal" is a child of "accounting". The children of a node A are also defined as *descendents* of A. Furthermore, all children of descendents of A are also defined as descendents of A. The single node without parent is called the *root* of the tree. Hence, all nodes, where A is located on a path from the root

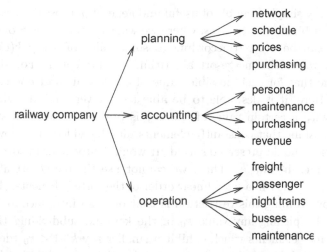

Fig. 6.11 Example of a hierarchical structure, represented as a tree: a railway company. Note that a tree is often drawn with the root node (here "railway company") not located at the bottom.

to the node, are descendents of A. In the example tree, "railway company" is the root of the tree. On the other hand, each node A together with all its descendents also forms a tree, called a *subtree*, with A being the root of the subtree. Nodes without children are called *leaves*, e.g. "network" and "busses". In principle, the number of children per node, also for the non-leaves, may vary within the tree. In case each node has at most two children, the tree is called a *binary tree*. To reach a node A, one can always start at the root of a tree and follow some links in the direction of the arrow, until the node A is reached. The sequence of visited nodes is called a *path*.[5] For the example, "schedule" can be reached via the path "railway company" → "planning" → "schedule". Note that for each node, the path to reach it from the root is unique. The number of traversed links between the root and node A is called the *height* or *level* of node A. Hence, the root has always height zero. In the example tree Fig. 6.11, node "planning" has height one, while node "prices" has height two. For this example, all leaves have the same height. In general, the leaves of a tree can have different heights, for example, if the railway company has a division "public relations" which is not subdivided into departments. Finally, the *height of the tree* itself is the largest height of any leaf.

---

[5]Equivalently, one can also call the sequence of traversed links a path.

As a very simple example of useful and general-purpose data structures, we consider binary *search trees*, where we want to store a set $S$ of elements $e$, which can be ordered according to some value of a key $k(e)$, like in Sec. 6.3, where the mergesort algorithm was presented. For simplicity, we assume that for each possible value of the key at most one element is present. Here, it is desirable to be able to test very quickly whether an element with a certain value of the key is present in the data structure. Also, we assume that constantly elements may be added or removed while the elements should preserved sorted. It would be too slow to sort the data again after each change. Thus, we cannot use the mergesort algorithm, since it results only in a fixed linear order of the sorted elements. The basic idea for overcoming this restriction is that one can introduce an artificial hierarchy by picking any value $k_0$ of the key and subdividing the set of elements into elements which exhibit a smaller key $k(e) < k_0$ (denoted as subset $S(-\infty, k_0)$), another subset containing elements exhibiting $k(e) > k_0$ $(S(k_0, \infty))$ and possibly one remaining element with $k(e) = k_0$. The subsets $S(-\infty, k_0)$ and $S(k_0, \infty)$ can be subdivided in the same way. For example, using a key value $k_{1,0} < k_0$, $S(-\infty, k_0)$ can be subdivied into a subset $S(-\infty, k_{1,0})$ containing elements smaller than $k_{1,0}$, into a subset $S(k_{1,0}, k_0)$ containing elements between $k_{1,0}$ and $k_0$, and finally possibly the element with exactly the key value $k_{1,0}$. In the same way, the other subset $S(k_0, \infty)$ can be subdivided using a key $k_{2,0} > k_0$. By this hierarchical splitting of the set of elements a tree structure is generated. A binary tree can be used to store the elements via representing this hierarchy. In this case, for convenience, only the key values of existing elements are used for the hierarchical subdivision of the set. Hence, at each node an element $e$ is stored. For this node, one subtree (usually denoted as *left* subtree) contains all elements $e'$ with key values $k(e') < k(e)$ the other (*right*) subtree the elements exhibiting $k(e') > k(e)$. A possible binary search tree containing elements with key values 13, 15, 20, 23, 24, 25, 27 is shown in Fig. 6.12.

A very natural way to represent trees in memory is to use a C structure for each element and pointers to represent the links. This is similar to the way we have represented lists

| GET SOURCE CODE |
|---|
| DIR: `algorithms` |
| FILE(S): `tree.h` |

in Sec. 6.6. The only difference is, since we are dealing with binary trees where each node has up to two children, that we store two pointers for each element. For leaves, both pointers have the value NULL. The following data structure can be used:

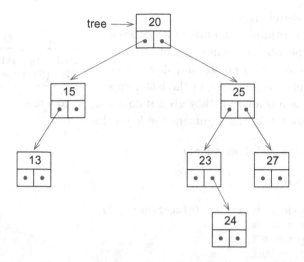

Fig. 6.12  A binary search tree containing elements with key values 13, 15, 20, 23, 24, 25, 27 (only key values shown). Each element has up to two children, i.e. subtrees. At the root, the element with key value $k_0 = 20$ is stored. The left subtree contains the elements $S(-\infty, 20)$ with key values smaller than 20, while the right subtree contains the elements $S(20, \infty)$.

```
1  struct node_struct
2  {
3      int                  key;                        /* holds key */
4      struct node_struct  *left;     /* to left subtree (NULL: none) */
5      struct node_struct  *right;    /* to right subtree (NULL: none) */
6  };
7
8  typedef struct node_struct node_t;   /* define new type for nodes */
```

We have defined a new data type **node_t** for convenience. For simplicity, only the key values are stored here. Other data belonging to an element, usually necessary for a real application, could be stored in other fields added to the structure. Note that if one wants to have trees of several types, e.g. one to store data of atoms and one to store data of full molecules (e.g. each containing a set of atoms), one would need to define two different tree node types. As a more flexible alternative, one could add one field with a **void *data** pointer, which points to an arbitrary memory area, where the actual data, independent of the tree structure, is stored. This would allow for a tree containing arbitrary objects. Since there are existing excellent libraries having data structures for this purpose, e.g. see Sec. 7.2, we do not go into details here.

Instead, we continue with our simple example. Following the object-oriented spirit, first we need functions which create and destroy isolated elements, respectively, in the latter case

| GET SOURCE CODE |
| --- |
| DIR: algorithms<br>FILE(S): tree.c |

under the assumption that they are not contained in any tree. The function creating a node takes as argument the key value:

```
1  node_t *create_node(int value)
2  {
3    node_t *node;
4
5    node = (node_t *) malloc (sizeof(node_t));
6    node->key = value;
7    node->right = NULL;
8    node->left = NULL;
9    return(node);
10 }
```

For creating a node, first one has to allocate the memory (line 5), then the key value is assigned (line 6) and the pointers to the children are initialized as NULL pointers (line 7–8). The function for deleting a given node (passed as pointer) is also straightforward:

```
1  int delete_node(node_t *node)
2  {
3    if(node == NULL)
4    {
5      fprintf(stderr, "attempt to delete 'nothing'\n");
6      return(1);
7    }
8    else if( (node->left != NULL)||(node->right != NULL))
9    {
10     fprintf(stderr, "attempt to delete linked node!\n");
11     return(1);
12   }
13   else
14   {
15     free(node);
16     return(0);
17   }
18 }
```

For deleting a node, it is first tested whether actually something has been passed to the function (lines 3–7) and whether the node has no children (lines 8–12). If these tests are passed successfully, the node is freed (line 15).

To actually create a sorted tree, we insert nodes (exhibiting keys) into a sorted tree such that the order is always preserved. Note that the tree can be empty as well, which will be the case in particular when a new tree is created.

The basic idea is that the algorithm starts at the root of the tree and searches for the occurrence of the key. Since the tree is sorted the search can be performed very quickly: If the key value is stored at the root, it is already present, hence the tree is not changed. If the given key is smaller than the key at the root, the wanted key is for sure in the left subtree, hence the algorithm continues to search there. If the wanted key value is larger than the key at the root, the algorithm continues in the right subtree. The search continues iteratively by branching either into left or right subtrees, until the key is found or an empty subtree is reached. In the latter case, it is clear that the key is not present in the tree, hence it can be attached at the position, where the empty subtree was reached. As an example, in Fig. 6.13 it is shown how a node with key value 22 is inserted into the tree from Fig. 6.12.

The function `insert_node()` does the job of inserting a node in the right order into the tree. It receives a pointer to the tree, i.e. to its root node, and a pointer to the node to be inserted. It returns a pointer to the root of the tree, which changes only if the tree is previously empty. Another return parameter is a flag, which is one if the node is already contained in the tree. A pointer to this flag is passed as last parameter. The C code for the function reads as follows.

```
1   node_t *insert_node(node_t *tree, node_t *node, int *in_p)
2   {
3     node_t *current;
4
5     if(tree==NULL)
6       return(node);
7     current = tree;
8     *in_p = 0;                          /* default: not contained */
9     while( current != NULL)             /* run through tree */
10    {
11      if(current->key==node->key)    /* node already contained ? */
12      {
13        *in_p = 1;
14        return(tree);
15      }
16      if( node->key < current->key)            /* left subtree */
17      {
18        if(current->left == NULL)
19        {
20          current->left = node;                /* add node */
21          return(tree);
22        }
23        else
24          current = current->left;       /* continue searching */
25      }
26      else                                     /* right subtree */
27      {
28        if(current->right == NULL)
29        {
30          current->right = node;               /* add node */
31          return(tree);
32        }
33        else
34          current = current->right;      /* continue searching */
35      }
36    }
37  }
```

To create a new tree, the function is called with `tree=NULL` (equivalently, a given node can be used as the root of a new tree, without calling the function). This case is treated in lines 5-6. The tree is searched for the given key value in lines 7–33, starting at the root (line 7). If the key value is found the function finishes, the tree remains as it is (lines 11–15). Otherwise, the search continues in the left (lines 17–25) or right (lines 27–

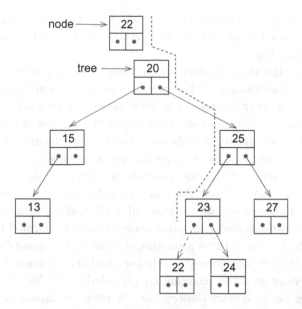

Fig. 6.13  Inserting a new key value 22 into the tree from Fig. 6.12. The dashed line next to the tree indicates the path along which the tree is traversed in the search for key value 22. The search stops at the left child of the node with key 23, hence the new node becomes the new left child (indicated by a dashed arrow).

35) subtree. In case the corresponding subtree does not exist, the new node is inserted into the tree (lines 19–22 and lines 29–33).

Note that another function which just tests whether a given key value is contained in the tree can be written in a similar way. This is achieved by first returning "yes" (1 in C) if the key value is found, second by leaving out the lines which insert the given element into the tree, and third by returning "no" (0 in C) in case the end of the function is reached. The details are left to you as exercise (7).

During the search for a given key, the algorithm moves the `current` pointer one level up in the tree per iteration of the main loop (lines 8–33). Hence for a tree of height $L$ it takes at most $L$ iterations to locate a key in a tree, the algorithm runs in $\mathcal{O}(L)$ time.

In case a tree with $L+1$ levels contains the maximum number $2^l$ of nodes per level $l = 0, \ldots, L$, the tree can be considered as "full". In this case the tree contains a total of $N = 2^{L+1} - 1$ nodes, hence the number of levels depends basically logarithmically on the number of nodes $L = \mathcal{O}(\log N)$. This is typically also true in case the tree is not "full". Hence, in these cases inserting and locating elements in a binary search tree takes time $\mathcal{O}(\log N)$. For this reason it is usually very efficient to store data, which is annotated with a key, in a tree.

Whether this is the case, depends on the order the key values are inserted into the tree. For example, if the key values are inserted in already descending or ascending order, the resulting "tree" in fact will be a list, see Fig. 6.14, where the number of levels grows linearly with the number of nodes, instead of growing logarithmically. In this case, inserting and locating nodes would take linear time again. A list is a special case of an *unbalanced* tree, which, in general, contains nodes, whose two subtrees differ significantly in height, for example at least by two. There are several extensions of binary search trees, where the trees are always kept balanced, which guarantees that the main tree operations, also removal of nodes, can be performed in $\mathcal{O}(\log N)$ time. The basic idea of these algorithms is that first elements are inserted as explained above (or removed, see below), and the additional reorganizations are performed if the resulting tree is unbalanced. We do not go into details here; please consider specialized literature [Cormen et al. (2001); Sedgewick (1990)].

Instead, we discuss how a binary search tree can be printed using a simple algorithm, such that all nodes are printed in increasing order. This can be achieved by a recursive approach: First the left subtree is printed, then the root, finally the right subtree. This job is done by the following function, which obtains a pointer to the tree as argument.

```
1   void print_tree(node_t *tree)
2   {
3     if(tree != NULL)
4     {
5       print_tree(tree->left);
6       printf("%d ", tree->key);
7       print_tree(tree->right);
8     }
9   }
```

This way to print a tree is also called *inorder* printing. Note that other orders are possible, e.g. *preorder*, where first the root of the current subtree is printed, then the left and finally the right subtree. For the tree shown in Fig. 6.12, the *inorder* output looks as follows:

13 15 20 23 24 25 27

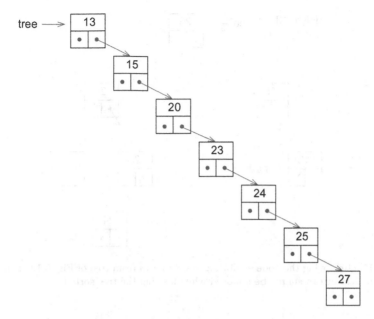

Fig. 6.14 A binary search tree containing elements with key values 13, 15, 20, 23, 24, 25, 27 (only key values shown), if the key values are inserted in ascending order. The resulting tree is in fact a list.

Finally, we consider the case of node removal. Again, this has to be done in a way such that the tree remains sorted. For this, we have to consider three different cases. The simplest case is if the node to be removed is a leaf, i.e. it has no children. In this case, the node can be simply removed by setting the pointer of the parent to the node to NULL. As example, the removal of a leaf (node 13) from the tree shown in Fig. 6.12 is presented in Fig. 6.15.

Also the case when the node to be removed has exactly one child is simple to treat. One just replaces the node by its child, i.e. the pointer from the parent to the node is redirected to the child. As example, the removal of a node 15 from the tree of Fig. 6.12 is presented in Fig. 6.16.

The most complicated case occurs when the node A to be removed has two children. In this case, one searches in the *right* subtree of this node for the node S with the *smallest* key, i.e. the *leftmost* key in the right subtree. This can be achieved by starting at the root of the right subtree (i.e. the right child of node A), moving iteratively to the left child until a node S without left child has been reached. In this way it is guaranteed that S

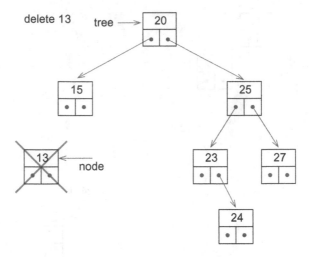

Fig. 6.15   Removing the node exhibiting key value 13 from tree of Fig. 6.12. Since this node is a leaf, it can simply be removed while keeping the tree sorted.

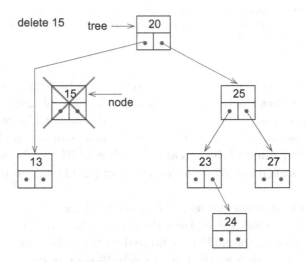

Fig. 6.16   Removing the node exhibiting key value 15 from the tree of Fig. 6.12. Since this node has one child, it can simply be replaced by its child via redirecting the pointer $20 \rightarrow 15$ to $20 \rightarrow 13$.

exhibits the smallest key in the subtree. Note that when printing the tree in *inorder* fashion, node S would be printed right after node A. Next, the keys (and other possibly present data, *except* pointers to children) of the nodes A and S are exchanged. Finally, node S is removed. Since it has no left child, it has in particular at most one child. Thus, it can be removed as described above for case one or two. As example, the removal of a node 20 (the root) from the tree of Fig. 6.12 is presented in Fig. 6.17.

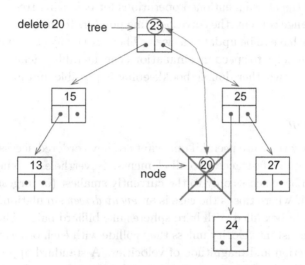

Fig. 6.17 Removing the node exhibiting key value 20 from the tree of Fig. 6.12. Since this node has two children, the node S exhibiting the smallest key within the right subtree is located, i.e. the node with key 23. The contents of these two nodes are exchanged. Finally, node S, which now contains the data of the target node, is removed.

In all three cases, the function `remove_node()` removing the node should return a pointer to this node, since after the removal it will not have been deleted from memory. For this purpose, the function `delete_node()` should be used. We do not present the C implementation of `remove_node()`, but leave this as exercise (8) to the reader. Note that there could be several versions of `remove_node()`, e.g. depending on whether a pointer to the node is already given, or whether only the key value is known. In the latter case, the tree has to be searched for the key, which takes $\mathcal{O}(\log N)$ time, as discussed above. On the other hand, consider again the example of the animals exhibiting weights (= keys) and IDs (Sec. 6.3). In this case, one could, for example, store the animals additionally in an array indexed by the animal ID. Now, it would help to store in the array along with the animal data also pointers to the corresponding nodes in the tree. Hence, if an animal with a given ID is to be deleted, one could access the tree node directly through the pointer stored in the array without searching for the key, i.e. in time $\mathcal{O}(1)$. This way of storing data, a kind of *double bookkeeping*, is an example of where faster implementation at the expense of memory consumption can be achieved. Note that quite a few lines of additional code are necessary in this case: since for deleting elements (or when using the above mentioned operations for balancing trees) sometimes nodes exchange content, the corresponding pointers from the array elements to the nodes have to be updated as well. There are many cases where double bookkeeping helps to speed up simulations considerably. Hence, you should always ponder whether double bookkeeping is possible and meaningful.

### 6.7.1 *Heaps*

Sorted binary trees are useful if one wants to have ordered access to a set of elements, e.g. for fast access to all elements. Nevertheless, sometimes it is sufficient to have access only to the currently smallest (or largest) element. One example where this is the case is an *event-driven* simulation. Consider, for instance, a box filled with hard spheres like billiard balls. These spheres move with constant velocity unless they collide with each other, where they change direction and magnitude of velocities. A standard approach would be to actually simulate the trajectories of spheres by moving the spheres in little steps, i.e. by performing a *Molecular dynamics simulation* [Allen and Tildesley (1989); Haile (1992); Rapaport (1995)]. This is very time-consuming and indeed not necessary since the constant-velocity movement between the collisions is trivial and can be calculated exactly. Hence, one needs only to consider the *events*, given by pairs of spheres $(i, j)$ which collide and the corresponding times $t_{ij}$ where this happens. An event-driven simulation in this case consists of treating all collision events one after the other in the order of increasing times $t_{i,j}$. This order is important, because after a collision $(i, j)$ the participating spheres $i$ and $j$ move in new directions. Hence, new collisions will usually be obtained, say $i$ will collide with particle $k$ at $t_{ik} > t_{ij}$, see Fig. 6.18. Now, it might be that previously a collision of sphere $k$ with another sphere $l$ had been calculated with $t_{lk} > t_{ik}$, hence now the collision $(l, k)$ might actually not take place since $(i, k)$ will happen earlier.

Thus, at each stage of the event-driven simulation one has to determine the event with the smallest even time. For the simplest approach, one would run through the list of all $N$ events and pick the smallest one, but this requires $\mathcal{O}(N)$ steps each time. One better uses a binary search

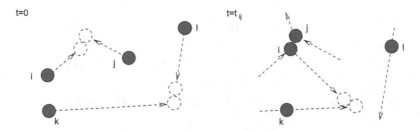

Fig. 6.18   Collisions of hard spheres. Left: In the beginning two collisions are calculated: $(i, j)$ and $(l, k)$ at times $t_{ij} < t_{lk}$. Right: after the collision $(i, j)$ has been evaluated, a new collision $(i, k)$ has been obtained, with $t_{ik} < t_{lk}$, hence $(k, l)$ does not take place currently (but later it might turn out that sphere $k$ might collide with sphere $l$ anyway. This can even happen at a collision time which is smaller than the initially calculated time $t_{lk}$, if the collision $(i, k)$ accelerates sphere $k$ towards the trajectory of $l$.

tree to store the events, where all operations can be performed in $\mathcal{O}(\log N)$ time. The next event to be considered will always be located at the leftmost entry of the tree. Nevertheless, since new events will be generated typically in increasing times, a simple tree implementation would lead to an unbalanced tree, basically a list, as discussed on page 190. Therefore, additional algorithms for balancing the tree would be necessary.

Nevertheless, for the case one has to access always only the smallest element of a set, there is a simpler way to achieve $\mathcal{O}(\log N)$ running time for each basic operation. This is achieved by so-called *priority queues*. There are many different implementations of priority queues. Here, we consider one which is based on a *heap*. This is also a binary tree, but it is *partially sorted* such that for each subtree, the smallest (or largest, depending on the application) element is located at the root of the subtree (*heap property*). Hence, the root of the entire tree will always contain the smallest element of the tree, which thus can be accessed in even $\mathcal{O}(1)$ time. In Fig. 6.19 a heap containing some natural numbers is shown. Note that for each given set of elements, many different valid heaps are possible.

We will consider a heap implementation where all levels of the heap are completely filled except the highest level. The highest level will be filled "from the left", which is also the case in Fig. 6.19. This allows for a very simple storage of a heap using an array: The elements are stored consecutively level by level, starting at index 0 (C convention), where the root is stored. Going from left to right within a level of a tree, the elements are stored in the array one after the other.

Accessing the elements stored in the array is very simple: For the node

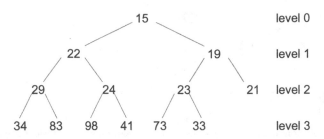

Fig. 6.19 A heap containing the elements 15, 19, 21, 22, 23, 24, 29, 33, 34, 41, 73, 83, 98. For each subtree, the smallest element is stored at the root of the subtree. For instance, number 22 is smallest among the numbers 22, 24, 29, 34, 41, 89, 98.

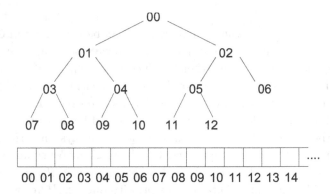

Fig. 6.20 How a heap (top) is stored in an array (bottom). For each node, the index of the array entry is given, where the element is stored. The numbers shown in Fig. 6.19 would be stored in the order 15, 22, 19, 29, 24, 23, 21, 34, 83, 98, 41, 73, 33 in the array.

stored at array index $i$, the parent is located at index $(i-1)/2$, where the division is treated here as integer operation. Hence, for both nodes stored at indices 3 and 4, the parent is stored at index $2/2 = 3/2 = 1$. Furthermore, the left and right children can be found at array indices $2i + 1$ and $2i + 2$, respectively.

Next, a small C implementation of a heap is shown. As an example, the heap stores events, characterized by some ID (which could be used, for example to locate the events in an array) and by the time of the event. Each element of the heap is of the following

| GET SOURCE CODE |
| --- |
| DIR: **algorithms** |
| FILE(S): **heap.h** |

type

```
1  typedef struct
2  {
3    double    t;              /* time of event */
4    int      event;           /* ID of event */
5  } heap_elem_t;
```

The heap consists here of an array **heap** of type **heap_elem_t** and of a variable **heap_num**, which counts the current number of elements in the heap. One could pack these two variables together with (say) a third variable holding the maximum allowed number of elements, into another C **struct**, which is not shown here for conciseness.

The most fundamental operation is to insert a new element into the heap. This works in the following way: First, the element is placed in the first empty entry of the array, i.e., at index

| GET SOURCE CODE |
| --- |
| DIR: **algorithms**<br>FILE(S): **heap.c** |

**heap_num** which is then increased by one. This guarantees that the level-by-level organization is kept. Now, the heap property might be violated, i.e., there might be subtrees where the new element is the smallest one, but not located at the root of the subtree. To restore the heap property, one iteratively compares the new element with the element stored at its current parent. If the new element is smaller, it is exchanged with the parent element. In this way, the new element moves toward the root of the tree. This step is repeated, until the element at the current parent is smaller than the new element. Now the heap property is restored. In Fig. 6.21 an example is given of how to insert a number into the heap from Fig. 6.19.

The function **heap_insert()** has the following arguments: the heap, a pointer to the current number of elements, the time and the ID of the event to be inserted. The C source looks as follows:

```
1   void heap_insert(heap_elem_t *heap, int *num_p, double time, int ev)
2   {
3     int pos, parent;                     /* heap positions */
4     heap_elem_t elem;                     /* for exchangig elements */
5
6     pos = (*num_p);                       /* insert at end */
7     *num_p = pos+1;
8     heap[pos].event = ev;
9     heap[pos].t = time;
10    parent = (pos-1)/2;
```

*Big Practical Guide to Computer Simulations*

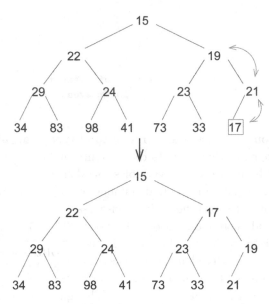

Fig. 6.21 Number 17 is inserted into the heap from Fig. 6.19. Since it is smaller than the value 21 stored at is parent node, numbers 17 and 21 are exchanged. Next, the 17 is exchanged with number 19. Since 17 is not smaller than the number 15 stored at the root, it comes to rest.

```
11    while((pos > 0)&&                        /* move up in heap */
12       (heap[parent].t > heap[pos].t))
13    {
14       elem = heap[parent];                  /* exchange parent/child */
15       heap[parent] = heap[pos];
16       heap[pos] = elem;
17       pos = parent;                         /* move up */
18       parent = (pos-1)/2;
19    }
20    }
```

In lines 6–9 the new element is inserted at the end and the number of elements is increased by one. In lines 10–19 the new element is moved (lines 14–16) towards the root until the current parent is smaller than the new element or until the root is reached (lines 11–12). Note that in case one wants to implement double bookkeeping (see page 194), one would have to update the pointers, which point to an element in the heap from outside, whenever an element is moved. Hence, in this case the data structure which holds the pointers would have to be passed to the function as well.

The next event having been treated, which is always located at the root, it has to be removed from the tree. Furthermore, as explained for the example of the hard spheres above, sometimes arbitrary events might have to be removed from the tree in case they become obsolete. Thus, removing elements, possibly the root, is discussed next: To keep the level-by-level organization, the element to be removed is just replaced by the last element L of the tree, i.e. the element which is located rightmost in the highest level. Now the heap property might be violated again. There are two cases

A) The element L is smaller than the element stored at its current parent node. In this case, element L has to be moved towards the root, until its current parent is smaller. This is similar to inserting an element into the heap, see Fig. 6.22 for an example.
B) The element L is larger than the element stored at the current parent node. Now, L might be also larger than one or both elements stored at its one or two children. Hence L has to be exchanged with the smaller of the elements stored at the children. This is again repeated, until L is smaller than all elements present at its current children, see Fig. 6.23 for an example.

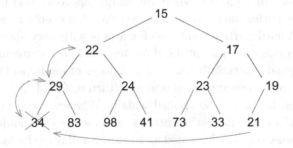

Fig. 6.22 Number 34 is deleted from the heap shown in Fig. 6.21. First, number 34 is replaced by the last element in the heap, number 21. Since 21 is smaller than the number 29, stored at the parent, it starts moving towards the root until the current parent is smaller.

Details of the implementation are left as exercise (9) to the reader. As mentioned above, insertion and removal from the heap can be performed in worst-case running time $\mathcal{O}(\log N)$ for a heap containing $N$ elements. The reason is that the heap is a tree which is always balanced, hence its height grows logarithmically with the number of nodes. This often results in a

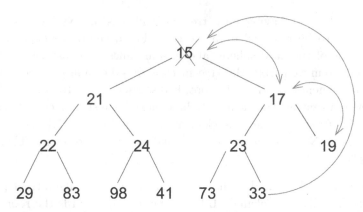

Fig. 6.23   Number 15 is deleted from the heap shown in Fig. 6.22. First, number 15 replaced by the last element in the heap, number 33. Since at least one of the elements located at children is smaller, number 33 starts moving towards the leaves.

much faster simulation compared to the simple approach, hence much larger systems can be treated. A recent example is the study of heat conduction in one-dimensional systems of hard particles [Grassberger et al. (2002)], where the use of heaps led to qualitatively improved results compared to previous work [Dhar (2001)], where the simple approach had been applied.

Note that performing a simulation via collision events is not only possible for real hard particles but also for atoms with very short-ranged interactions, in this case one speaks of *binary-collision approximation*, which has been applied successfully for the simulations of ion impacts on surfaces in molecular-beam epitaxy [Robinson and Torrens (1974)].

In general, heaps can be applied widely. Whenever you have to select "the smallest" or "the largest" from a set of elements, you should always use a priority queue, e.g., implemented as a heap. Similar to the case of lists, in many cases you do not have to implement your own general-purpose priority queue, since it is already contained in the *Standard Template Library*, see Sec. 7.2. Nevertheless, in case you need double bookkeeping for highest performance, you have to have access to the internal heap structure. In this case you have to implement a heap yourself.

Nevertheless, before you change a part of your program and replace some simple selection mechanism by a heap-based selection, you should use **gprof** to analyze where your program spends most of its running time, see Sec. 4.4. Maybe the bulk of running time of your program is spent somewhere else. In this case you should optimize these parts first.

## 6.8 Graphs

For many simulation problems in science, sociology or economy, the underlying model can be represented by graphs. For this reason a short introduction to graph theory is given here. Only the basic definitions, data structures and few algorithms are presented. For more information, the reader should consult a specialized textbook on graph theory, e.g. Refs. [Bolobas (1998); Claiborne (1990); Swamy and Thulasiraman (1991)].

### 6.8.1 *Basic definitions*

Consider a map of a country where several towns, villages or other places are connected by roads or railways. Mathematically, this setting can be represented by a *graph*. A graph consists of *nodes* and *edges*. The nodes represent the towns, villages or other places and the edges describe the roads or railways. Formally, the definition of a graph is given by:

**Definition 6.2** A *graph* $G$ (also often called a *network*) is an ordered pair $G = (V, E)$ where $V$ is a set and $E \subset V \times V$. An element of $V$ is called a *vertex* or *node*. An element $(i, j) \in E$ is called an *edge* or *arc*. In a physical context, where edges represent interactions between particles, edges are often called *bonds*.

If the pairs $(i, j) \in E$ are ordered pairs, i.e. if the edge goes *from i to j*, then $G$ is called a *directed* graph. Otherwise $G$ is called *undirected*, then $(i, j)$ and $(j, i)$ denote the same edge. A graph $G' = (V', E')$ is called *subgraph* of $G$ if it has the properties $V' \subset V$ and $E' \subset E$ ($E' \subset V' \times V'$ by definition). The empty graph $(\emptyset, \emptyset)$ is a subgraph of all graphs. Another special graph is the (undirected) *complete graph* $K_n$ which contains $n$ nodes and all possible $n \times (n-1)/2$ edges, i.e. $K_n = (V, V \times V)$.

First, some further notations are given which apply to both directed and undirected graphs. Some of the definitions are illustrated using an example graph in Fig. 6.24. Usually, we restrict ourselves to *finite* graphs, i.e. the set of nodes and edges are finite. In this case we denote by $n = |V|$ the number of vertices and by $m = |E|$ the number of edges. Let $i \in V$ be a vertex. If $(i, j) \in E$ we call $j$ a *neighbor* of $i$ (and vice versa). Both nodes are *endpoints* of the edge and they are *adjacent* to each other. The set $N(i)$ of neighbors of $i$ is given by $N(i) = \{j \mid (i, j) \in E \lor (j, i) \in E\}$. The degree $d(i)$ of node $i$ is the cardinality of the set of neighbors: $d(i) = |N(i)|$. A vertex with degree 0 is called *isolated*.

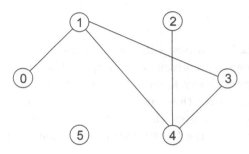

Fig. 6.24 The undirected graph. $G = (\{0, 1, 2, 3, 4, 5\}, \{(0, 1), (1, 3), (3, 4) \ (4, 1), (4, 2)\})$ is shown. The nodes are represented by circles and the edges by lines connecting the circles. The graph has $n = 6$ vertices and $m = 5$ edges; e.g. nodes 3 and 4 are adjacent. The set of neighbors of vertex 1 is $N(1) = \{0, 3, 4\}$. Thus, node 1 has degree 3, while node 2 has only degree 1. Node 5 is isolated. The graph contains the path $0, 1, 4, 3$ from node 0 to node 3 of length 3 and the cycle $1, 3, 4, 1$. Since the nodes 0, 1, 2, 3, and 4 are mutually connected by paths, they form a connected component. There is a second connected component consisting only of node 5.

A *path* from $v_1$ to $v_k$ is a sequence of vertices $v_1, v_2, \ldots, v_k$ which are connected by edges: $(v_r, v_{r+1}) \in E \ \forall r = 1, 2, \ldots, k - 1$. The *length* of the path is $k - 1$. If $v_1 = v_k$ the path is called *closed*. If no node except the first and the last one appears twice in a closed path, it is called a *cycle*. Note that trees, as introduced in Sec. 6.7, are just graphs without cycles. A set of nodes is called a *connected component*, if i) a path exists for each pair of nodes between the nodes within the connected components and ii) no nodes of the graph can be added such that i) still is true. Hence, a connected component is of maximal size. A graph is called *connected* if it has only one connected component.

Now some definitions are given which apply only to directed graphs. For an edge $e = (i, j)$, $i$ is the *head* and $j$ the *tail* of $e$. The edge $e$ is called *outgoing* from $i$ and *incoming* to $j$. Please note that for a directed path it is important that all edges point into the direction of the path, formally the definition is the same as in the case of an undirected graph. A set of nodes is called a *strongly connected component* (SCC), if i) from each of its nodes a directed path to every other node of the set exists within the set and ii) no nodes of the graph can be added such that i) still is true. In a directed graph, the outgoing and incoming edges can be counted separately. The *indegree* is given by $id(i) = |\{j \,|\, (j, i) \in E\}|$ and the *outdegree* is $od(i) = |\{j \,|\, (i, j) \in E\}|$. Obviously, for all vertices $d(i) = id(i) + od(i)$.

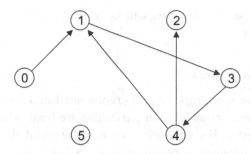

Fig. 6.25 A directed graph. The edges $(i, j)$ are represented by arrows pointing from $i$ to $j$. If one considers the example graph from Fig. 6.24 to be directed, one obtains this figure. Here, the sequence $0, 1, 4, 3$ is not a path since the edges $(4, 1)$ and $(3, 4)$ point in the wrong direction. On the other hand, the graph contains the path $0, 1, 3, 4$. Node 1 has the outdegree $od(1) = 1$ and indegree $id(1) = 2$. The total degree is $d(1) = id(1) + od(1) = 3$ as in the case of the undirected graph.

Graphs appear in many cases where simulation methods have been successfully applied. Here we just list some examples, which have been studied extensively in the literature:

- Citation networks: The nodes represent scientific papers. A directed edge $(i, j)$ means that in article $i$ there is a citation of article $j$.
- Acquaintance networks: Nodes represent persons, an undirected edge $(i, j)$ means that $i, j$ know each other personally (or they have coauthored a scientific article, or they have played in the same team, or ...).
- Protein regulation: Nodes represent proteins. A directed edge $(i, j)$ means that the presence (or absence) of protein $i$ regulates the expression of protein $j$ in a cell.
- World Wide Web: Nodes represent web pages. A directed edge $(i, j)$ means that on page $i$ there is a link (URL) to page $j$.
- Magnetic systems: The nodes represent particles carrying a magnetic moment ("spin") and the edges represent interactions between the spins.

Sometimes functions $f : V \to A$ ($A$ an arbitrary set) on vertices or functions $f : E \to A$ on edges are useful as well. An arbitrary function on vertices or on edges is called *labeling*. A graph together with a labeling is called a *labeled* graph. Typical labelings are distances, costs or capacities. For instance, for a graph where the nodes represent cities connected by roads (edges), labels on the edges can be used to denote distances between neighboring cities.

### 6.8.2 Data structures

Next, we consider how graphs can be represented in a computer program, hence suitable data structures. In particular, we treat adjacency matrices and adjacency lists. For simplicity, we consider unlabeled graphs within the following C example implementations. All of these examples can be extended to labeled graphs in a straightforward way.

The simplest way to represent a graph in a C program is first to use an interval of natural ID numbers $\{0, 1, \ldots, n-1\}$ to represent the nodes and second, for the edges, an *adjacency matrix* $\{a_{ij}\}$ with

$$a_{ij} = \begin{cases} 1 & \exists \text{ edge } (i,j) \\ 0 & \text{else.} \end{cases} \tag{6.8}$$

For the undirected graph shown in Fig. 6.24 and the directed version shown in Fig. 6.25 the matrices look as follows, respectively:

$$\{a_{ij}\} = \begin{pmatrix} 0 & 1 & 0 & 0 & 0 & 0 \\ 1 & 0 & 0 & 1 & 1 & 0 \\ 0 & 0 & 0 & 0 & 1 & 0 \\ 0 & 1 & 0 & 0 & 1 & 0 \\ 0 & 1 & 1 & 1 & 0 & 0 \\ 0 & 0 & 0 & 0 & 0 & 0 \end{pmatrix} \quad \{a_{ij}\} = \begin{pmatrix} 0 & 1 & 0 & 0 & 0 & 0 \\ 0 & 0 & 0 & 1 & 0 & 0 \\ 0 & 0 & 0 & 0 & 0 & 0 \\ 0 & 0 & 0 & 0 & 1 & 0 \\ 0 & 1 & 1 & 0 & 0 & 0 \\ 0 & 0 & 0 & 0 & 0 & 0 \end{pmatrix}$$

Note that in case of an undirected graph, the adjacency matrix is always symmetric. In a C program, an adjacency matrix can be used as simply as every other matrix, see Sec. 1.4. Hence, after allocating an adjacency matrix defined via **short int \*\*adj**, one can add an edge $(i,j)$ to the graph by directly assigning **adj[i][j]=1**. Also the presence of an edge can be tested in $\mathcal{O}(1)$ in the same way by reading **adj[i][j]**. On the other hand, an adjacency matrix requires $\mathcal{O}(n^2)$ memory for a graph with $n$ nodes. Therefore, a lot of memory is wasted for the large sparse graphs, i.e., graphs where the number of edges is less than $\mathcal{O}(n^2)$ (typically only $\mathcal{O}(n)$). In many applications of sparse graphs, a number $n \sim 10^6$ of nodes is present, hence an adjacency matrix would not fit into the main memory. Furthermore, iterating over all neighbors of a given node, a typical operation when performing graph algorithms, requires $\mathcal{O}(n)$ steps even if the average number of neighbors is finite $\mathcal{O}(1)$.

For sparse graphs it is better to store only the edges which are present. This can be done, for example, using a *neighbor list*. Thus, for each node of the graph, again represented by integer

ID numbers $\{0, 1, \ldots, n-1\}$, a list of neighbors is used. Thus, the full graph is represented by an array of lists. For a C implementation, we can use the following data types:

```
1  typedef struct
2  {
3       elem_t *neighbors;    /* pointer to list of neighbors */
4  } gs_node_t;
5
6  typedef struct
7  {
8       int          num_nodes;    /* number of nodes */
9       gs_node_t        *node;    /* array of nodes */
10 } gs_graph_t;
```

Here, a separate type `gs_node_t` (lines 1–4) is introduced for each node, although it currently contains only one element. The reason for this approach is that it can be extended fairly simple, such as for introducing labels on the vertices by just adding another element to the structure. When you use the data structure for lists as presented on page 176, the edges do not carry any label. The `info` is used to store the IDs of the neighboring nodes. In Fig. 6.26 a list representation of the graphs from Figs. 6.24 and 6.25 is shown, respectively. Note that for an undirected graph, an edge $(i, j)$ appears as entry $j$ in the list of node $i$ and as entry $i$ in the list of node $j$.

### 6.8.3 *Generation*

To generate a graph, here the function `gs_graph_generate()` is used, which receives the number of nodes as parameter. Initially no edges are present. The C implementation looks as follows:

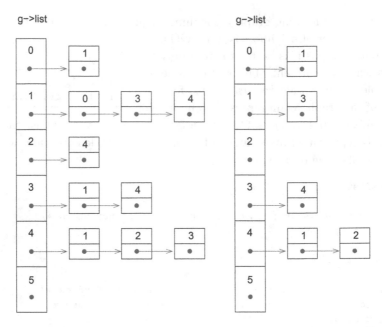

Fig. 6.26   Neighbor-list based representation of the undirected graph shown in Fig. 6.24 (left) and of the directed graph shown in Fig. 6.25 (right).

```
1   gs_graph_t *gs_create_graph(int num_nodes)
2   {
3       gs_graph_t *g;
4       int n;
5
6       g = (gs_graph_t *) malloc(sizeof(gs_graph_t));   /* allocate */
7       g->num_nodes = num_nodes;                        /* initialize */
8       g->node = (gs_node_t *) malloc(num_nodes*sizeof(gs_node_t));
9
10      for(n=0; n<num_nodes; n++)
11        g->node[n].neighbors = NULL;
12
13      return(g);
14  }
```

The function just has to allocate memory of the type `gs_graph_t` (line 6), store the number of nodes in the element `num_nodes` (line 7), and allocate for the element `node` an array of this size (line 8). Finally all neighbor lists are initialized as being empty (lines 10–11).

For an object-oriented design, one needs also a function which destroys a graph. One could split this task into two subtasks: first to remove all edges, i.e., *clear* the graph, and second to free the data structures allocated in `gs_graph_generate()`, i.e., to finally delete a cleared graph. We do not show the details of the implementation, but mention that the source code of these functions is also included in the packages which are available online.

Since it is convenient and failure-safe to hide the graph implementation, next a function for inserting an edge in an undirected graph is presented. The function `gs_insert_edge()` receives the graph (i.e. a pointer) and the IDs of the endpoints of the edge as parameter. The C code looks as follows:

```
1   void gs_insert_edge(gs_graph_t *g, int from, int to)
2   {
3       elem_t *elem1, *elem2, *list;
4
5       list = g->node[from].neighbors;          /* edge exists? */
6       while( list != NULL )
7       {
8           if(list->info == to)                 /* yes */
9               return;
10          list = list->next;
11      }
12
13      elem1 = create_element(to);   /* create neighbor for 'from' */
14      g->node[from].neighbors =
15          insert_element(g->node[from].neighbors, elem1, NULL);
16      elem2 = create_element(from);   /* create neighbor for 'to' */
17      g->node[to].neighbors =
18          insert_element(g->node[to].neighbors, elem2, NULL);
19  }
```

First, it is checked whether the edge is already present in the graph (lines 5–11). If so, nothing has to be done. Note that in lines 6–11, we work directly on the implementation level of the lists. To hide the list implementation, one would have to provide *iterators*, which allow us to access the first and next elements, respectively. Since these examples are of illustrative nature mainly, we do not do this. For powerful general purpose lists, the reader should use available libraries anyway, see Chap. 7. In lines 13–15, the second endpoint is inserted into the neighbor list of the first, while in lines 16–18 the first endpoint is inserted into the neighbor list of the second. Note that the directed version of this function would be even simpler, since one would have to create just one list element containing the second endpoint and insert the list element into the neighbor list of the first one.

The functions presented so far can be used to generate graphs. Here, we consider a particular example, which has been studied quite a lot during the past decade, a graph generated by *preferential attachment* [Albert and Barabási (2002); Newman (2003)].

These graphs are *grown* step by step, starting from some small initial graph. The basic idea is that the graph growth should reflect processes in nature, society or technology, where existing members are contacted more often if they have already many contacts. For example, for an acquaintance network, people becoming members of some group are more likely to get in contact to group members who already know many people. For a network describing the growth of the Internet, one might assume that a well known Internet page is more likely to be referred to by new web pages. This is also called the *Matthäus principle* "Give to those more who have already".

One simple algorithm to create such a graph works as follows: One starts with a complete graph $K_{m+1}$ having $m + 1$ nodes. This means each node has initially $d(i) = m$ neighbors. Then iteratively new nodes are added and connected to exactly $m$ different nodes, which are randomly chosen among the currently existing nodes. The probability to connect to a certain node $i$ shall be proportional to its current degree $d(i)$. In Fig. 6.27, an example graph generated by preferential attachment is shown.

Note that initially all $m + 1$ nodes have the same degree. But through random fluctuations, some nodes will acquire more neighbors during the iteration, hence they have a higher probability to acquire even more neighbors. In this way few *hubs* are created which have a lot of neighbors.

The following C function creates random graphs through preferential attachment. The function receives as parameters a graph (intended to have no edges) and the number $m$. The main idea is to hold an array `pick`, where each node $i$ is contained exactly $d(i)$ times at any stage of the iteration. When for a new node the $m$ neighbors are chosen randomly, they are just picked randomly from `pick`. This ensures that the probability to pick a node is proportional to $d(i)$. The C code looks as follows:

```
1   void gs_preferential_attachment(gs_graph_t *g, int m)
2   {
3     int t;
4     int n1, n2;
5     int *pick;              /* array which holds for each edge (n1,n2) */
6                             /* the IDs n1 and n2. Used for picking */
7                             /* nodes proportional to its current degree */
8     int num_pick;           /* number of entries in 'pick' so far */
```

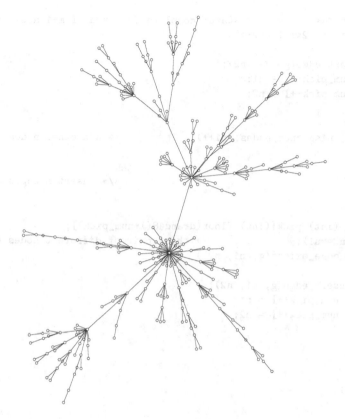

Fig. 6.27  Random graph generated by preferential attachment $(m = 1)$ with $n = 300$ nodes. The graph was generated using `gs_preferential_attachment()` and drawn using the `fdb` program from the `dot` package, see Sec. 9.2.3.

```
9     int max_pick;                              /* maximum number of entries */
10
11    if(g->num_nodes < m+1)
12    {
13        printf("graph too small (at least %d edges per node)!\n", m);
14        exit(1);
15    }
16    max_pick = 2*m*g->num_nodes- m*(m+1);
17    pick = (int *) malloc(max_pick*sizeof(int));
18    num_pick=0;
19
```

```
20    for(n1=0; n1<m+1; n1++) /* start: complete subgraph of m+1 nodes */
21      for(n2=n1+1; n2<m+1; n2++)
22      {
23        gs_insert_edge(g, n1, n2);
24        pick[num_pick++] = n1;
25        pick[num_pick++] = n2;
26      }
27
28    for(n1=m+1; n1<g->num_nodes; n1++)                /* add other nodes */
29    {
30      t=0;
31      while(t<m)                                      /* insert m edges */
32      {
33        do
34          n2 = (int) pick[(int) floor(drand48()*num_pick)];
35        while(n2==n1);                /* chose pair of different nodes */
36        if(!gs_edge_exists(g, n1, n2))
37        {
38          gs_insert_edge(g, n1, n2);
39          pick[num_pick++] = n1;
40          pick[num_pick++] = n2;
41          t++;
42        }
43      }
44    }
45    free(pick);
46 }
```

First (lines 11–15), it is verified that the graph is large enough to have at least $m$ neighbors per node. Next (lines 16–18), the `pick` array is allocated. The graph is initialized as the complete graph $K_{m+1}$ in lines 20–26. In the final part, for each of the remaining nodes (line 28), $m$ new edges are inserted (lines 30–43), such that the neighbor is picked from `pick` (lines 33–35). The function `drand48()` generates a (pseudo) random number between 0 and 1 and `floor()` generates the largest integral number not greater than the argument. Note that `drand48()` might not be defined on all systems. In this case you can use `(double) rand()/RAND_MAX` instead, which is standard C. More details on generation of (pseudo) random numbers are presented in Chap. 8.

### 6.8.4  *Connected Components*

So far, we have been concerned with the storage and the generation of graphs. Next, we present a particular algorithm for a simple, edge-traversal based analysis of graphs. As pointed out in Sec. 6.8.1, graphs may exhibit several *connected components*, such that nodes belonging to the same component are connected by paths, while for nodes belonging to different components, there exist no path between them. For example, for a graph describing a communication network, nodes belonging to the same component can communicate among each other, while nodes belonging to different components cannot.

Next, a simple algorithm is discussed which determines the connected components of a graph. This discussion be performed in three steps: First the basic idea is explained. Next the C code is given. Finally, the operation of the algorithm for a small sample graph is discussed. The basic idea is to start with any ("seed") node. It is considered as first node of the first component. Subsequently, all neighbors of the "seed" node, i.e. all nodes connected by an ede to it, are attributed to the same component. In the same way, the neighbors of these neighbors belong to the same component, and so on. However, note that for a correct determination of the components, each node does not need to be processed more than once. For example, the "seed" node is processed first and there is no need to process it again when the neighbours of any of its neighbors are considered. To accomplish this, for each node an entry in the array `comp` is kept, which indicates whether the node has not been encountered so far during the determination of the components (then `comp[node]` $= -1$). Conveniently, this variable will also store an ID, identifying the component to which the node belongs. Initially this variable will be set to $-1$ for all nodes, indicating that each node has not been considered so far. Since for each node one has to consider all neighbors once, the neighbors will be stored in a "bag" once they are encountered, actually implemented as an array `candidate`, During the component construction, the algorithm takes one node after the other from the bag to check their neighbors. In this way, each node connected to the seed node by a path will be considered and added to the component at some point. Note that the order in which the nodes in the bag are considered does not matter here. If the order is *last-in first-out* (LIFO), as for a stack, then, after the first neighbor of the seed has been considered, in turn its first neighbor will be considered, etc., before the second neighbor of the seed node will be considered. Thus, the algorithm will tend to go as far as possible directly. This is called *depth-first search*. If on the other hand the bag is treated in *first-in first-out* (FIFO) fashion, then all neighbors of the seed are taken from the bag before the neighbors of neighbors are considered, etc. In this case the strategy is called *breadth-first search*.

When no nodes are left in the bag, the construction of the respective component has been completed. Thereafter, the algorithm continues by iterating over the nodes until it encounters one which is not part of any component yet (i.e. comp is -1). Thus, the construction of the next component starts. After the iteration over all nodes, with the purpose of finding seeds of individual components, has finished, the component construction is completed.

Next, the C code implementing the above procedure is discussed. It utilizes the graph data structures introduced in Sec. 6.8.2. The function gs_components() receives as arguments the graph g and the array comp where the IDs of the components to which a node belongs will be stored. Initially all entries are set to $-1$ (lines 10–11). The function returns the number of components found. Among the local variables, there is in particular the array candidate, which stores the (IDs of) nodes for which the neighbors have still to be checked, i.e. the bag. The size of the array is equal to the number $n$ of nodes (line 9), because in principle a node can be connected to all other nodes (actually $n - 1$ would be sufficient). The variable num_candidates counts the number of nodes which are currently stored in candidate. The variable neighb is a pointer to a node (see data structure on page 205), which is used to iterate over all neighbors of a node taken from candidate.

```
1   int gs_components(gs_graph_t *g, int *comp)
2   {
3     int n1, n2, nn;                    /* nodes/loop counters */
4     int *candidate;                    /* nodes still to treat */
5     int num_candidates;
6     int num_components = 0;      /* number of components so far found */
7     elem_t *neighb;                  /* for iterating over neighbors */
8
9     candidate = (int *) malloc(g->num_nodes * sizeof(int));
10    for(n1=0; n1<g->num_nodes; n1++)
11       comp[n1] = -1;                 /* initializes as non-assigned */
```

The main part of the function consists of a loop over all nodes. If a node has not yet been assigned to a component (line 14), a new component is found and the component construction performed (lines 16–35). Thus, the node is put into the candidate array. The component construction continues while there are still candidates left (line 19). One after the other the candidates are taken (line 21) and a loop over all neighbors is performed (lines 22–32). For each neighbor it is checked (line 26) whether it has not been assigned to a component yet. If not, it is added to the component and to the bag of candidates (line 27–30).

```
12    for(n1=0; n1<g->num_nodes; n1++) /* all possible component seeds */
13    {
14      if(comp[n1] == -1)                /* not yet part of a component ? */
15      {
16        comp[n1] = num_components;
17        candidate[0] = n1;
18        num_candidates = 1;
19        while(num_candidates > 0)  /* still nodes in current comp. ? */
20        {
21          n2 = candidate[--num_candidates];        /* next candidate */
22          neighb = g->node[n2].neighbors;
23          while(neighb != NULL)          /* go through all neighbors */
24          {
25            nn = neighb->info;
26            if(comp[nn] == -1)           /* not yet part of component ? */
27            {
28              comp[nn] = num_components;        /* add to candidates */
29              candidate[num_candidates++] = nn;
30            }
31            neighb = neighb->next;
32          }
33        }
34        num_components++;
35      }
36    }
37
38    free(candidate);
39    return(num_components);
40  }
```

To understand how the algorithm operates, we consider the small example graph from Fig. 6.24. Initially all comp values are set to −1. The variable num_components is initialized as 0. The change of the variables during the execution of the algorithm is shown in Fig. 6.28. Within the loop over all nodes, first node 0 is encountered. Is has not been assigned to a component, thus, a new component is started with ID 0. Node 0 is added to candidate. Hence, during the iteration of the loop over all candidates

```
n1=0: comp[0]=0
                    candidate={0}   n2=0:
                                        comp[1]=0
                    candidate={1}   n2=1:
                                        comp[3]=0
                                        comp[4]=0
                    candidate={3,4} n2=4:
                                        comp[2]=0
                    candidate={3,2} n2=2:
                    candidate={3}   n2=3:
n1=1:
n1=2:
n1=3:
n1=4:
n1=5: comp[5]=1
                    candidate={5}   n2=5:
```

Fig. 6.28 Operation of gs_components() for the sample graph of Fig. 6.24. The graphs consists of two components with IDs 0 (nodes 0,1,2,3,4) and 1 (node 5). The first column shows the current node n1 of the main loop. Each entry in the second column represents the start of a new component and the corresponding assignment of the component ID to the seed node n1 (line 16 of the code). In the third column, the state of the candidate array at the beginning of each iteration of the loop in line 19 is displayed. The fourth column shows the current node n2 taken from the candidates. In the fifth column, all neighbors of n2 which are added to the component are indicated via the corresponding component ID assignment (line 28).

(lines 19–33), first node 0 is considered. It exhibits one neighbor, node 1, which is added to the component. Therefore, next node 1 is considered in the candidates loop. Node 1 has three neighbors, nodes 0, 3 and 4. Node 0 is already part of the component (comp[2]=0, i.e., $\neq -1$), thus nothing happens here. Nodes 3 and 4 are added to the component and to the candidates. Next, node 4 is considered. Its neighbors 1 and 3 are already part of the component. Hence, only its neighbor node 2 is added to the component and to the bag of candidates. Due to the FIFO operation on the candidates, next node 2 is considered. Its only neighbor node 4 is already part of the component, nothing happens. The same holds for node 3, which has neighbors nodes 1 and 4. Now, the candidate array is empty and the construction of the first component is finished. During the main loop, nodes 1, 2, 3, and 4 are encountered next. Since they have been already assigned to a component, no further component identification procedure is started. Finally node 5 is encountered. It is the seed of a new component with ID 1. Since it has no neighbors, i.e., the condition in line 23 of the code is immediately FALSE, no other nodes will be added to the component. At this point, the main loop terminates and the full component identification procedure is completed.

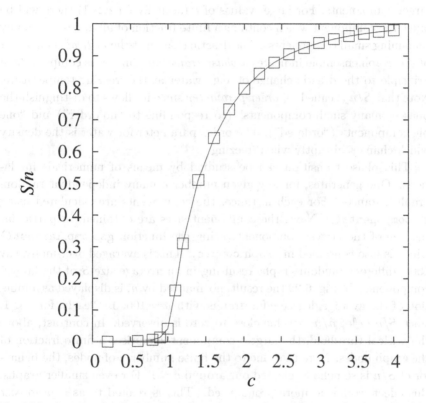

Fig. 6.29 Average size $S$ of the largest component of a random graphs, normalized by the number $n$ of nodes, as a function of the average connectivity $c$. Here, $n = 10000$. The average was taken over 1000 independent random graph instances for each value of $c$. The lines are guides to the eyes only.

As an example application of how the determination of components is used in conjunction with graph models, we study *Erdős-Rényi (ER) random graphs*. These are graphs, which here have $n$ nodes and $m$ edges, which connect randomly and independently chosen pairs of nodes. Thus, each node has an average degree $c = 2m/n$.[6] If $c$ is very small, like 0.2 neighbors per node on average, the nodes are only sparsely connected, hence the number of components will be large. Consequently, the size of the components, even of the largest component, will be small. Actually, one can show that the size of the largest components grows logarithmically in the number $n$ of nodes. If $c$ is increased, more and more edges join the graph. Thus, small components get to be connected, i.e., merge to form larger components. For larger values of $c$ (actually for $c > 1$) there will be one large component which contains a finite fraction of all nodes plus many remaining small components. This drastic change at the *critical value* $c = 1$ of the graph ensemble indicates a *phase transition*, since it is comparable in principle to the drastic change of, e.g. water at the freezing temperature. Note that $S/n$ is called an *order parameter* since it allows to distinguish the phases "many small components" (corresponding to "no order") and "one big component" ("ordered"). The order parameter for water is the density which changes abruptly when freezing.

This phase transition can be studied by means of numerical simulations. One generates, for any given number $c$ many independent random graph instances. For each instance, the components are calculated using `gs_components()`. Next, the component sizes are obtained, in particular the size of the largest component (using the function `gs_comp_largest()` which is also contained in `graph_comp.c`). This is averaged over many, say 1000, different random graphs resulting in an average size $S$ of the largest components. In Fig. 6.29 the result, normalized by $n$, is displayed as a function of the average degree $c$ for graphs with $n = 10000$. Clearly, for $c < 1$, since $S/n \sim \log n/n$, a value close to zero is observed. In contrast, above the critical threshold, the largest component comprises a finite fraction of the graph nodes. Note that due to the finite number $n$ of nodes, the behavior of $S/n$ is somehow smeared our around $c = 1$. For even smaller graphs, this effect would be more pronounced. This is refered to as a *finite-size effect*. For $n \to \infty$, $S/n$ is exactly zero below $c = 1$ and rises monotonously to one for $c > 1$.

With this, we close our small introduction to graphs. For an exercise concerning graph-related functions, see exercise (10).

---

[6]Alternatively, and more standard in mathematics, one can define ER graphs such that for each pair of nodes an edge appears with some probability $p$, here $p = c/(N-1)$. In the limit $n \to \infty$ the two definitions of ER random graphs show the same typical behavior.

# Exercises

(solutions: can be downloaded from http://www.worldscientific.com/r/9019-supp)

## (1) Permutations

Design, implement and test a recursive function **permute()** for obtaining and printing all permutations of an integer array **a[]** *in place*, i.e. no further array is needed.

| SOLUTION SOURCE CODE |
|---|
| DIR: algorithms |
| FILE(S): |
| permutation.c |

The function prototype reads as follows:

```
/****************** permutation() ************************/
/** Obtains all permutations of positions 0..n-1 of a    **/
/** given array 'a' of numbers and prints them,          **/
/** including the higher index entries (from 0..n_max-1). **/
/**                                                       **/
/** Parameters: (*) = return parameter                   **/
/**             n: current range                         **/
/**         n_max: size of array                         **/
/**             a: array                                 **/
/** Returns:                                             **/
/**         (nothing)                                     **/
/*********************************************************/
void permutation(int n, int n_max, int *a)
```

The basic idea is as follows: To solve the problem, for range **n** (initially **n_max**), one puts into the last current element **a[n-1]**, one after the other, all elements from the indices **0...n-1**, and calls each time recursively the function for range **n-1**.

Hint: Do not forget to put back the numbers to their initial positions, either right after the recursive call or at the end.

## (2) Power of a number

Design, implement and test the function **fast_power()** for calculating a power $a^n$ ($n$ being an integer). Use a divide-and-conquer approach by reducing the calculation of $a^n$ to the calculation of $a^{n/2}$ if $n > 2$. This leads to a running time $\mathcal{O}(\log n)$ instead of $\mathcal{O}(n)$ for the trivial approach.

| SOLUTION SOURCE CODE |
|---|
| DIR: algorithms |
| FILE(S): power.c |

The function prototype reads as follows:

```
/****************** fast_power() ************************/
/** Calculates the a^n (n being a natural number) using  **/
/** a divide-and-conquer approach.                       **/
/**                                                      **/
/** Parameters: (*) = return parameter                   **/
/**             a: base                                  **/
/**             n: power                                 **/
/** Returns:                                             **/
/**         a^n                                          **/
/********************************************************/
double fast_power(double a, int n)
```

Hint: Design your function first for $n \geq 0$ and then reduce the case $n < 0$ to the case $n > 0$.

(3) **Number Partitioning**

The *number-partitioning problem* (NPP) is defined as follows: Given a set of $n$ non-negative integers $A = \{a[0]...a[n-1]\}$, we want to partition it into two subsets $A_1 \subset A$

| SOLUTION SOURCE CODE |
| --- |
| DIR: algorithms |
| FILE(S): partition.c |

and $A_2 \subset A$ such that the difference $d = |\sum_{a \in A_1} a - \sum_{a' \in A_2} a'|$ is minimized. For example, $A = \{2, 3, 4, 10\}$ can be partitioned into two subsets $A_1 = \{2, 3, 4\}$ and $A = \{10\}$ with $d = 1$. Solve this problem, i.e. find the minimum difference, using a dynamic programming approach. Design, implement and test the resulting program.

Hint: Introduce the two-dimensional array `part[][]`, where `part[i][s]=1` if *some* subset of $\{a[0]...a[i]\}$ has sum `s` (with `i=0,...,n − 1` and `s=0,...,`$\sum_{a \in A} a$). The array `part` can be calculated (with some suitable initialization for `part[0][s]`) using `part[i][s]=1` if `part[i-1][s]==1` or `part[i-1][s-a[i]]==1`. Then, using $|\sum_{a \in A_1} a - \sum_{a' \in A_2} a'| = |\sum_{a \in A} a - 2\sum_{a' \in A_2} a'|$, the minimal difference can be obtained from `part[n-1][s]` (`s= 0..`$\sum_{a \in A} a$). Note that this approach does not directly yield the two subsets, just the minimum difference.

Technical hint: Pass the integers as parameters to your program.

(4) **Tour of a knight**

On a chessboard, a knight is allowed to move from its current position into one of up to eight possible positions in its neighborhood. The possible moves of a knight are shown in Fig. 6.30 (left).

| SOLUTION SOURCE CODE |
| --- |
| DIR: algorithms |
| FILE(S): knight.c |

The *Knight's tour problem* is to find a tour for a $n \times n$ chessboard, starting in the upper left corner of the board, such that each square is visited exactly once. A possible solution for a $8 \times 8$ board is shown in Fig. 6.30 (right).

Write a C program solving this problem using a backtracking approach.

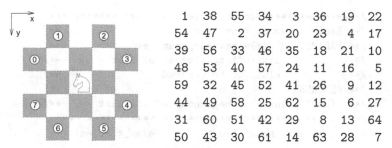

| 1 | 38 | 55 | 34 | 3 | 36 | 19 | 22 |
| 54 | 47 | 2 | 37 | 20 | 23 | 4 | 17 |
| 39 | 56 | 33 | 46 | 35 | 18 | 21 | 10 |
| 48 | 53 | 40 | 57 | 24 | 11 | 16 | 5 |
| 59 | 32 | 45 | 52 | 41 | 26 | 9 | 12 |
| 44 | 49 | 58 | 25 | 62 | 15 | 6 | 27 |
| 31 | 60 | 51 | 42 | 29 | 8 | 13 | 64 |
| 50 | 43 | 30 | 61 | 14 | 63 | 28 | 7 |

Fig. 6.30 (left) The 8 possible moves of a knight. (right) A possible solution of the knight's tour problem for a 8 × 8 chessboard.

Design, implement and test the resulting program. Hints: Use an array visit[][], where initially visit[x][y]=0 everywhere, and where visit[x][y]> 0 indicates at which stage of the tour the square is visited. Furthermore, use arrays dx[] and dy[] which store the possible move, e.g. (see left of Fig. 6.30) dx[0]=-2 and dy[0]=-1. Finally, you could use a function, which places exactly one knight and calls itself for placing the next knight. The function prototype could look as follows:

```
/****************** move_knight() *************************/
/** Calculates recursively via backtracking a tour for    **/
/** the knight, such that all squares with visit[x][y]=0  **/
/** are visited once.                                      **/
/**                                                        **/
/** Parameters: (*) = return parameter                     **/
/**           n: size of board                             **/
/**        step: how many sites already visited?           **/
/**         x,y: current position of knight                **/
/**       visit: states whether and when square was visited **/
/**       dx,dy: possible move directions                  **/
/** Returns:                                               **/
/**       (nothing)                                        **/
/***********************************************************/
void move_knight(int n, int step, int x, int y,
                 short int **visit, int *dx, int *dy)
```

(5) **Remove element from list**

Design, implement and test the function remove_node() for removing elements from a list, as described on page 181. The function prototype reads as follows:

SOLUTION SOURCE CODE

DIR: algorithms
FILE(S): list_rm_e.c

```
/************** remove_element() ****************/
/** Remove element from list                  **/
/** PARAMETERS: (*)= return-parameter         **/
/**             list: first element of list   **/
/**             elem: to be removed           **/
/** RETURNS:                                  **/
/**   (new) pointer to beginning of the list  **/
/************************************************/
elem_t *remove_node(elem_t *list, elem_t *elem)
```

(6) **Mergesort for lists**

Design, implement and test a function for sorting a given list using the mergesort algorithm (see Sec. 6.3). The main approach is as follows: First you have to split the given list into two lists of roughly half the size. Then

| SOLUTION SOURCE CODE |
| --- |
| DIR: algorithms |
| FILE(S): |
| list_mergesort.c |

the function is called recursively for each of the two lists. Finally, one has to build the resulting list via merging the two sorted lists. (Good exercise for pointer arithmetic !!) The function prototype reads as follows:

```
/**************** mergesort_list() **************/
/** Sorts list using mergesort algorithm      **/
/** PARAMETERS: (*)= return-parameter         **/
/**             list: first element of list   **/
/** RETURNS:                                  **/
/**   (new) pointer to beginning of the list  **/
/************************************************/
elem_t *mergesort_list(elem_t *list)
```

(7) **Find value in tree**

Design, implement and test a function for testing whether a given value for the key is contained in the tree. The function prototype reads as follows:

| SOLUTION SOURCE CODE |
| --- |
| DIR: algorithms |
| FILE(S): tree_find.c |

```
/********************* find_node() ****************/
/** Locates a node with a given 'key' in a tree. **/
/** If the 'key' is not contained in the tree,   **/
/** NULL is returned                             **/
/** PARAMETERS: (*)= return-parameter            **/
/**          tree: pointer to root of tree       **/
/**          key: to be located                  **/
/** RETURNS:                                      **/
/** pointer to node (NULL if key is not exisiting) **/
/**************************************************/
node_t *find_node(node_t *tree, int key)
```

## (8) Remove node from tree

Design, implement and test a function for removing a node from a tree which has a given value for the key. For the design, follow the outline as given in Sec. 6.7.

```
SOLUTION SOURCE CODE
DIR: algorithms
FILE(S): tree_rm.c
```

The function prototype reads as follows:

```
/****************** remove_node() *****************/
/** Removes node containung the 'key_rem' from  the  **/
/** tree.                                             **/
/** If the 'key' is not contained in the tree,        **/
/** nothing happens.                                  **/
/** PARAMETERS: (*)= return-parameter                 **/
/**          tree: pointer to root of tree            **/
/**       key_rem: to be removed                      **/
/**    (*) node_p: address of ptr to removed node     **/
/*+             (NULL if not found)                   **/
/** RETURNS:                                          **/
/** (new) pointer to the root                         **/
/****************************************************/
node_t *remove_node(node_t *tree, int key_rem, node_t **node_p)
```

## (9) Remove node from heap

Design, implement and test a function for removing a node from a heap. For the design, follow the outline as given in Sec. 6.7.1.
The function prototype reads as follows:

```
SOLUTION SOURCE CODE
DIR: algorithms
FILE(S):
heap_remove.c
```

```
/****************** heap_remove() *****************/
/** Deletes event at position 'pos' from heap        **/
/** PARAMETERS: (*)= return-parameter                 **/
/**      (*) heap: array containing heap              **/
/**      (*) num_p: ptr to: number of elements        **/
/**             pos: of element to be removed          **/
/** RETURNS:                                          **/
/**    (nothing)                                       **/
/****************************************************/
void heap_remove(heap_elem_t *heap, int *num_p, int pos)
```

(10) **Random graph**

Design, implement and test a function for generating an Erdős-Rényi random graph. This is a graph containing $n$ nodes and $m$ edges. The edges are randomly drawn

| SOLUTION SOURCE CODE |
|---|
| DIR: algorithms |
| FILE(S): graph_r.c |

with equal probability from all possible pairs $i, j \in V$. The function is similar to `gs_preferential_attachment()`, but simpler. The function prototype reads as follows:

```
/****************** gs_random_graph() ******************/
/** Function adds exactly m randomly chosen edges to   **/
/** the graph.                                         **/
/** No self loops are allowed. No edge is allowed to   **/
/*+ appear twice!                                      **/
/** PARAMETERS: (*)= return-parameter                  **/
/**          (*) g: graph                              **/
/**              m: number of edges to be added per node **/
/** RETURNS:                                           **/
/**     (nothing)                                      **/
/*******************************************************/
void gs_random_graph(gs_graph_t *g, int m)
```

# Chapter 7

# Libraries

Libraries are collections of data types and functions which can be used in other programs. There are libraries for numerical methods such as integration or solving differential equations, for storing, sorting and accessing data, for fancy data types like lists or trees, for generating colorful graphics and for thousands of other applications. Some libraries can be obtained for free, while other, usually specialized libraries have to be purchased. The use of libraries speeds up the software development process enormously, because you do not have to implement every standard method by yourself. Hence, you should always check whether someone has done the jobs for you already, before starting to write a program.

You have already learned about the basics of libraries in Chap. 1. For instance, the library for mathematical functions was discussed, which is included with all C/C++ compilers. Here, only few example libraries are presented, which are freely available: The *standard C library*, the *standard template (C++) library* (STL), and the *GNU scientific library*. Examples for other useful libraries, not discussed in detail here, are the *boost* library [Boost; Karlsson (2005)] and the *LEDA* library [Mehlhorn and Näher (1999)]. The boost library is also freely available and is in fact a collection of many libraries, which are based on the STL. There are, for example, libraries for graphs, hash tables, interval arithmetics, parallel programming (MPI), statistical distributions and random numbers, respectively. The LEDA library, on the other hand, does not directly build on the STL. It offers also data types for graphs, in particular many very efficient graph algorithms, several container types, arbitrary-precision arithmetics, data compression, encryption, and support for two- and three-dimensional geometry. Unfortunately, the full version of the LEDA library is quite expensive. Nevertheless, you can get a free version, which contains a comprehensive subset of the main functionality, including many container data types like priority queues or trees.

Although many libraries are available, sometimes it is inevitable to implement some specialized but basic functions by yourself. In this case, after the code has been proven to be reliable and useful for some time, you can put it in a self-created library. How to create libraries is explained in the last part of this chapter.

Many libraries are *dynamically linked*. This means, when the linker assembles the executable, no code will be included from the libraries. Only when the program is executed, the executable functions from the corresponding library will be loaded to the memory. This helps to keep the sizes of the executables of your simulation programs small. Nevertheless, sometimes you have to force the linker to include the library functions in the executable already when the program is linked. This is called *static linking*. This may happen when you want to run your simulations on another machine where the library is not installed. You can tell the linker to link the executable code for the library function by using the linker/compiler option `-static`.

## 7.1   Standard C library

You have already encountered many standard C library functions in Chap. 1, for example for allocating memory via `malloc()`, for calling shell functions via `system()`, or converting strings to integers via `atoi()`. To use functions from this library properly, you have to use the directive

```
#include <stdlib.h>
```

Note that you do *not* have to use a `-l` option for linking executables of functions for the standard C library, because they are linked always.

Here, we are discussing only one more function, which is used quite frequently in the context of simulations. It is `qsort()`, which allows to sort in ascending order arrays of arbitrary data types. The function implements the efficient *quicksort* algorithm [Sedgewick (1990); Cormen et al. (2001)], which sorts $n$ elements in typically $\mathcal{O}(n \log n)$ comparison steps. The order is determined by some sorting criterion, which has to be provided by the program which calls `qsort()`.

The function prototype of `qsort()` looks as follows:

```
void qsort(void *base, size_t nel, size_t width,
            int (*compare)(const void *, const void *));
```

Thus, it expects as argument a pointer **base** to the beginning of the array to be sorted and the number **nel** of elements in the array. The third argument specifies the number of bytes of one element. The last argument provides the sorting criterion. Technically, it is a pointer to a function, which itself has the prototype

```
int compare(const void *p1, const void *p2);
```

This function has to be provided by the program which calls **qsort**. It should take two pointers p1, p2 as arguments, which point to two array elements which have to be compared. The function **compare()** should return the value -1, if the element indicated by **p1** is "smaller" than the element indicated by **p2**. In the opposite case, +1 should be returned, and 0 if the two elements are equal. What "smaller" means is determined by the function; hence, if you want to sort in descending order, you can return -1 if actually the second element is smaller. Note that the pointers are pointers to **void**; thus, the **compare()** function usually has to perform an explicit type conversion using a cast (see Sec. 1.1.2) to access the content of the array elements. Also, the function **compare()** is not allowed to modify the content of the array, indicated by the **const** qualifier (see Sec. 1.2).

As example, sorting an array of numbers of type **double** in ascending order is considered. In this case, the function to compare two numbers may look like

| GET SOURCE CODE |
|---|
| DIR: **libraries** <br> FILE(S): **sorting.c** |

```
1  int compare_double(const void *p1, const void *p2)
2  {
3    if( *(double *)p1 < *(double *)p2)
4      return (-1);
5    else if( *(double *)p1 == *(double *)p2 )
6      return (0);
7    else
8      return(1);
9  }
```

Here, the casts to access the two elements indicated by p1 and p2 are directly included in the comparison expression. For more complex **compare()** functions, it is usually more convenient (and readable) to create two variables of the corresponding type and assign the content of the two pointers to the variables.

Provided that you have an array `val[]` of n numbers of type `double`, the call to `qsort()` should look as follows:

```
qsort(val, n, sizeof(double), compare_double);
```

Note that the sorting is performed *in place*. Thus, the original order within the array is lost. You are asked to write a slightly more complicated `compare()` function in exercise (1). In exercise (6) of Chap. 8, `qsort()` is used to calculate a confidence interval from measured sample points.

## 7.2 *Standard Template Library*

The *Standard (Template) Library* (STL) [STL; Josuttis (1999)] is part of the "standard" C++ library. The STL provides three main ingredients: *containers* to store objects, *iterators*, which allow the programmer to access the containers in different ways, and *algorithms* which perform more complicated tasks such as sorting. Example containers are vectors, lists and sets. You can store objects of nearly arbitrary type in these containers, one type of objects per individual container. Thus, you can have vectors of "something", lists of "something", or sets of "something". This means that the types of member data or of arguments in member functions for the containers cannot be determined in advance but have to be defined in each case by the actual data type you want to use. This is achieved via *templates*. In this section, it will be explained how to use templates by discussing below two data structures from the STL, *priority queues* and *maps*. If you want to learn how to implement your own templates, please refer to specialized literature [Stroustrup (2000)].

In Sec. 6.7.1 priority queues were introduced. They allow objects to be stored such that always the "smallest" object can be retrieved. This is possible, because the objects are kept partially sorted in the priority queue. In Sec. 6.7.1, a C implementation of events being stored in a priority queue is presented, which is based on heaps. Unfortunately, that implementation is for a specific data structure `heap_elem_t`. If you want to store other objects of type (say) `your_elem_t` on a heap, you have to copy the code and exchange all occurrences of `heap_elem_t` by `your_elem_t`. Furthermore, you have to provide some basic operations, such as a function to compare two objects of type `your_elem_t` to find out which is the smaller one. This work is facilitated for you by using templates.

Let us start with a simple priority queue of strings using the STL. Note that the STL contains also a class **string** for strings. To use the string class and the template class for priority queues, you have to write in your program

GET SOURCE CODE

DIR: libraries
FILE(S): pqueue0.cpp

```
#include <queue>
#include <string>
```

The following simple part of a program shows how a priority queue of strings is defined and used:

```
1    priority_queue<string> q2;
2    q2.push("hello");                   // fill priority queue
3    q2.push("you");
4    q2.push("are");
5    q2.push("welcome");
6
7    while(!q2.empty())                   // get events from queue
8    {
9      string s;
10     s = q2.top();                      // get "smallest" string
11     q2.pop();                          // remove it
12     cout << s << endl;
13   }
```

In line 1, the priority queue **q2** is defined, and initialized automatically as empty queue. Behind the class name **priority_queue**, the template arguments are given in angular brackets **< >**, here the type **string** of the queue elements. In this way, for all templates the types and further arguments (see below) are specified. In line 2, an auxiliary string variable is declared. Some strings are put into the queue using the member function **push()**, which receives as argument the object to be put into the queue (technically it receives a reference to the object). In lines 8–13, the priority queue is read out. Here, the following three other member functions are used: **empty()** tests, whether the queue is empty. The function **top()** returns a reference to the currently first element in the queue. Using **pop()**, the currently first element is removed. For completeness, we note that the member function **size()** returns the number of elements being currently stored in the queue.

When running this program, the output will read:

```
you
welcome
hello
are
```

Hence, the strings are returned by the queue in decreasing lexicographic order. For simple predefined data types, one can specify other sorting criteria as template parameters for the queue. We only consider here the most general case: The sorting criterion is completely self-defined, which will usually be the case for priority queues of more complex objects.

Let us assume that you want to create a priority queue for events. Each event describes a collision of two atoms in a simulation of a gas. Within an event-driven simulation (see

| GET SOURCE CODE |
| --- |
| DIR: `libraries` |
| FILE(S): `pqueue.cpp` |

Sec. 6.7.1), one event will be treated after the other, in the order of increasing event times. Thus, each event can be described by the IDs of the two participating atoms and a time, where the collision takes place. The following data structure is used here:

```
struct event_t
{
  int part1, part2;    // IDs of particles colliding
  double time;         // time of collision
};
```

A structure is used, which is a class where by default all members are `public`. Everything shown in the following would work also if we defined a class for events with data capsuling, member functions for access etc.

For the priority queue, we have to provide a function to compare two events. For this purpose, no standard comparison operator can be used; therefore, we have to provide one. To achieve maximum flexibility, this comparison function must be a *function object* (also called *functor*). This is an object `obj` of a class which provides the operator () as member function. Consequently, applying this operator to the object, i.e. writing `obj()`, implements exactly a call to the function. The advantage over defining an ordinary function is that object functions are C++ objects as well. Thus, they can have all class properties like member data etc. In this way, one can have different function objects of the same class, which behave in a different way, because their internal data looks different. The following example, which provides a comparison for events, should make it clear how classes providing function objects are defined:

```
class EventSortCriterion
{
 public:
 bool operator() (const event_t &e1, const event_t &e2)
   {
     return(e1.time > e2.time);
   }
};
```

This looks like a standard class definition, without member data here, and one member function is the () operator. The operator receives two arguments and returns a variable of type `bool`, which is a special C++ type. The type `bool` has the possible values `true` and `false`, being equivalent to 0 and non-zero, compatible with standard C. In front of the arguments `const` is written, which prohibits that the values are modified by the operator.

Next, it is shown how a priority queue of events is defined and used:

```
1    priority_queue<event_t, vector<event_t>, EventSortCriterion>  q;
2    event_t e;                           // one event
3    int i;                               // loop counter
4    int num_events = 20;
5
6    srand48(1000);
7    for(i=0; i<num_events; i++)          // put events into queue
8    {
9      e.part1 = i;
10     e.part2 = i+10;
11     e.time = 1000*drand48();
12     q.push(e);
13   }
14
15   while(!q.empty())                    // get events from queue
16   {
17     e = q.top();                       // get smallest event
18     q.pop();                           // remove it
19     cout << "(" << e.part1 << "," << e.part2 << ") at t="
20          << e.time << endl;
21   }
```

The most important line is the first line. Here, the priority queue is defined. In comparison to the queue of strings example above, here two additional template arguments are given. The second is the data type, which is used to implement the queue. Here, a vector is used, which stores elements of type `event_t`. Note that this is also a template, as visible from the fact that the type is written inside angular brackets. One can use other classes to implement priority queues. Possible are container classes with random access which provide member functions front(), push_back() and pop_back(), such as a *deque* [Josuttis (1999)], which is defined in the STL as well. The third and final template argument is the name of the class which provides the sorting criterion via the operator (). The remaining part of the program uses the standard priority queue member functions push(), empty(), top() and pop() as in the above example: In lines 6–13, events are initialized randomly and pushed into the queue, while in lines 15–20, the events are retrieved from the queue in increasing order.

Note that the STL priority queue does not allow objects to be easily removed from within the queue, only the top object can be removed. This is by purpose, because it must be guaranteed that the partial sorting of the data remains. Nevertheless, this makes rescheduling difficult, i.e. the change of event times during the simulation. If you need this, here two options are stated. First, you can use your own implementation, like that presented in Sec. 6.7.1. The STL implementation is also given completely in 39 lines of code in [Josuttis (1999)]. It can be easily modified such that you can access the underlying data structure directly. Nevertheless, this is a bit laborious and not elegant, because you effectively write your own class.

Second, the container used for an object of the class priority_queue is accessible as **protected** member data. Hence, you can derive a child class from priority_queue, which enables you to use all standard random-access modes of the container, which is actually based on a heap for the STL. You can use for example the erase() member function to remove objects. After you have completed all changes, you have to make sure by calling the function make_heap() that the partial sorting is restored. For details, please refer to the STL documentation.

As next example, we consider a *map*, which allows $N$ *value objects* of arbitrary but fixed type to be stored via a *key*. Each *element* in a map is a (key, value) pair. Using the key, it is possible to access the stored value objects in almost random-access fashion. This means, one does not need to iterate over $\mathcal{O}(N)$ value objects stored in the container to find a value object with a specific key. Internally, the value objects are stored in the current implementation in a balanced binary search tree. This means, all important operations can be performed in $\mathcal{O}(\log N)$ steps. Nevertheless, iteration over all stored value objects is also possible, as we will see below.

Here, a basic example is considered. The simulation consists of some "agents" which are distributed in the x-y plane. Each agent has a position, serving as key. Furthermore, the data

GET SOURCE CODE

DIR: `libraries`
FILE(S): `maps.cpp`

associated with each agent consists of some "capacity" and of some "interaction range", the exact meaning is not important here. This is implemented via the following two type definitions for `agent_t` and `agentkey_t`:

```
struct agent_t
{
  int capacity;    // how much an agent can process
  int range;       // interaction range
};

struct agentkey_t
{
  int x,y;         // position of agent
};
```

Since the implementation is based on search trees, one has to supply a comparison criterion. Here, we assume that the x coordinates determine the order at first. If the x coordinates of two agents are the same, the y coordinates are compared next. The comparison is implemented as corresponding function object class, similar to the priority queue shown above:

```
class AgentkeySortCriterion
{
 public:
 bool operator() (const agentkey_t &k1, const agentkey_t &k2) const
   {
    return( (k1.x < k2.x)||( (k1.x == k2.x)&&(k1.y < k2.y) ));
   }
};
```

In the `main()` function, a map `m1` is defined via

```
1    typedef map<agentkey_t, agent_t, AgentkeySortCriterion> agentmap_t;
2    agentmap_t m1;                               // stores all agents
```

The first template argument signifies the type of the keys, the second refers to the type of the value objects to be stored, while the third is the sorting criterion. The third argument is optional, if the operator `<` is defined for the keys, as for example for integer keys.

For inserting and removing elements, you can use the member functions insert() and erase(). The insertion is a bit inconvenient. It is more convenient to use an array notation, as in the following example:

```
3    int num_agents = 20;                           // number of agents
4    int sizex=10, sizey= 10;                       // size of system
5    agentkey_t k1;                                 // one key
6    agent_t a1;                                    // one agent
7
8    for(int i=0; i<num_agents; i++)                // distrib. agents
9    {                                              // random positions
10     k1.x = static_cast<int>(floor(sizex*drand48()));
11     k1.y = static_cast<int>(floor(sizey*drand48()));
12     a1.capacity = 100;
13     a1.range = 1 + static_cast<int>(floor(3*drand48()));
14     m1[k1] = a1;
15   }
16
```

Two variables k1 and a1 are used (lines 5–6) to store the keys and agents, respectively. The agent positions and range values are initialized randomly (lines 10–11). Note that a *static cast* is used. This is a C++ operator, which performs some checks to be performed by the compiler and prevents meaningless type conversions. The agents are actually stored in line 14, using the array notation. You can access the map elements in the same way for reading, *but*, unlike for an array, if you ask for an element with a key value, which does not exist, the element will be created with default values.

For this reason, we use the find() member function in the following example to test whether an element with a specific key exists. The code loops over all elements in the map. For this purpose, one needs a special object, an *iterator*, which is something like an internal pointer in the map. There are some predefined iterators accessible via public member functions: begin() and end(). The first one returns always an iterator to the currently first element of the map. The second one returns an iterator which points behind the last element.

```
17    agentmap_t::iterator pos, pos2;                  // to iterate in map
18    for(pos=m1.begin(); pos != m1.end(); pos++)      // print all agents
19    {
20      int r = pos->second.range;
21      cout << "Agent (" << pos->first.x << "," << pos->first.y <<
22        "): cap=" << pos->second.capacity << ", range=" << r << endl;
23      printf("neighbors:\n");
24      for(k1.x=pos->first.x-r; k1.x<=pos->first.x+r; k1.x++)
25        for(k1.y=pos->first.y-r; k1.y<=pos->first.y+r; k1.y++)
26          if( ((k1.x != pos->first.x)||(k1.y != pos->first.y))&&
27              (k1.x>=0)&&(k1.x<sizex)&&(k1.y>=0)&&(k1.y<sizey))
28          {                               // look for neighbors near agent
29            pos2 = m1.find(k1);
30            if(pos2 != m1.end())
31              cout << "(" << pos2->first.x << "," << pos2->first.y
32                << ")\n";
33          }
34    }
```

The iterator `pos` is declared in line 17. The actual iteration takes place in line 18 and looks very similar to a normal `for` loop. An iterator `pos` points always to an element consisting of a key and a value. The key of the element can be accessed via `pos->first` (see for example line 24), while the value is accessible via `pos->second` (line 20). Here, we iterate via `k1` over a square of positions around the position of the current agent (lines 24–25). If `k1` contains a valid position inside the system $[0,$ `sizex`$] \times [0,$ `sizey`$]$ (lines 26–27), then it is tested whether at this position there exists another agent via the `find()` member function (lines 29–30). If yes, the neighbor agent is printed (line 31). With these examples you should have enough background to use maps efficiently. For more details, please consult the online documentation [STL] also present on the website: http://www.worldscientific.com/r/9019-supp.

The STL is under continuous development and extension [Karlsson (2005); Becker (2007); Wilson (2007)]. In particular, some elements of the *boost* library [Boost; Karlsson (2005)] have already found its way to the technical report *TR1*, where future extensions of the C++ standard library are scheduled. In TR1, you can find, for example, *hash tables* and *smart pointers.*. Hash tables implement nearly constant-time access to objects with arbitrary keys. Smart pointers are aware of the objects they point to, which facilitates memory management and helps to avoid memory leaks. For instance, if a pointer of type `auto_ptr` is deleted, the memory where it points at is automatically deallocated.

## 7.3 *GNU scientific library*

The *GNU Scientific library* (GSL) [Galassi et al. (2006)] contains a huge number of subroutines that can be used to solve standard numerical problems. Amongst other things, the library features:

- complex numbers and functions
- polynomials
- evaluation, differentiation and integration of functions
- performing interpolations
- Fourier, wavelet and Hankel transforms
- minimizing functions
- diagonalization of matrices
- solving linear equations (through matrix inversion)
- random numbers
- statistics/histograms
- fitting data
- solving nonlinear equations
- solving ordinary and partial differential equations
- Monte Carlo simulation and simulated annealing

Some of the functionality comes through a C interface to the BLAS (basic linear algebra subroutines) package, which is originally implemented using Fortran. The library is under continuous development. Note that it comes under the *GNU General Public License* (GPL). This means the library is free, including the source code. You are allowed to extend and to modify the library. You can include it into other software, provided it is also free (GPL) if it is distributed as well.

We start to explain the usage via a simple example: We want to print a table of the probability mass function of the Binomial distribution, defined in Sec. 8.1.1. For this and related purposes, the GSL offers functions for different distributions. When using any of these, one has to include the header file `gsl/gsl_randist.h`. Similar header files exist for the other sections of the GSL. The function to calculate the probability mass function of the Binomial distribution has the following prototype:

```
double gsl_ran_binomial_pdf(const unsigned int x, const double p,
                            const unsigned int n);
```

The first argument contains the value, for which the probability is to be calculated. The second and third arguments contain the values of the parameters $p$ and $x$ (with $0 \leq p \leq 1$ and

```
GET SOURCE CODE
DIR: libraries
FILE(S): binomial.c
```

$x = 0, 1, \ldots, n$). This and all other functions of the GSL are explained in the user guide, which can be downloaded from the web page or bought as printed copy [Galassi et al. (2006)]. Next, the small sample C program named `binomial.c` is shown, which reads in $p$ and $n$ from the command line and prints a table with all feasible values of $p(x)$:

```
1   #include <stdio.h>
2   #include <gsl/gsl_randist.h>
3
4   int main(int argc, char *argv[])
5   {
6       int n;                          /* parameters of distribution */
7       double p;
8       int x;                                      /* argument of pdf */
9       double prob;         /* resulting probability and sum of all probs */
10      int argz = 1;             /* for treating command line arguments */
11
12      sscanf(argv[argz++], "%lf", &p);
13      n = atoi(argv[argz++]);                        /* get arguments */
14
15      for(x=0; x<=n; x++)        /* iterate to print full distribution */
16      {
17          prob = gsl_ran_binomial_pdf(x, p, n);
18          printf("%d %f\n", x, prob);
19      }
20      return(0);
21  }
```

Line 2 contains the inclusion of the suitable GSL header file. Lines 6–10 contain the variable declarations. In lines 12 and 13, the command-line arguments are read in, see Sec. 1.3. Lines 15–19 contain the main loop, where the evaluation of the distribution is called (line 17) and the result is printed to the standard output (line 18). The program can be compiled using:

```
cc -o binomial  binomial.c -lgsl -lgslcblas -lm
```

For linking, in addition to the GSL (`-lgsl`) also an auxiliary library (`-lgslcblas`) and the standard math library (`-lm`) have to be stated. Note that we assume that the library is installed in standard places, such that compiler and linker can find the needed header files and libraries. Otherwise, corresponding paths have to be stated using the −I and −L options.

The program has to be called with parameter $\langle p \rangle$ as first and $\langle n \rangle$ as second argument, e.g.

```
binomial 0.4 10
```

This call was used to generate the plot on page 254 via the *gnuplot* program (see Sec. 8.4.1). The GSL offers also a function to generate random numbers, which are distributed according to the Binomial distribution, see also Sec. 8.2.2. Two ingredients are needed:

First, the basis are general pseudo random number generators, see Sec. 8.2.1. There are different high-quality generators provided by the GSL. It is now explained how to use them. For

| GET SOURCE CODE |
| --- |
| DIR: libraries |
| FILE(S): randnum.c |

all dynamically generated data structures, one first has to allocate a generator, then one can use it. Finally, one has to deallocate the generator. This is illustrated in the following example:

```
1   #include <stdio.h>
2   #include <gsl/gsl_rng.h>
3
4   int main()
5   {
6     gsl_rng *rng;              /* pointer to a random number generator */
7     int n = 100;                      /* number of random numbers */
8     int i;                                  /* loop counter */
9     double r;                           /* a random number */
10
11    rng = gsl_rng_alloc(gsl_rng_mt19937);    /* allocate generator */
12    gsl_rng_set(rng, 1000);                  /* set seed to 1000 */
13
14    for(i=0; i<n; i++)                       /* generate numbers */
15    {
16      r = gsl_rng_uniform(rng);
17      printf("%4.3f\n", r);
18    }
19
20    gsl_rng_free(rng);                       /* delete generator */
21
22    return(0);
23  }
```

All generators are of type `gsl_rng`, a corresponding pointer variable is declared in line 6. The generator is actually allocated in line 11 using the function `gsl_rng_alloc()`, which expects an argument of type `gsl_rng_type *`. Many different generators are predefined in the GSL, see below. Here, the generator "`gsl_rng_mt19937`" is used, which is also the default generator. All random number generators can be initialized with a seed using the function `gsl_rng_set()`, which expects a generator and an `int` number as arguments. Here, the seed value 1000 is used (line 12). The actual generation of the numbers is performed in the loop in lines 14–18. The generation of numbers is performed always by means of auxiliary functions. They receive a pointer to the generator as argument, sometimes also parameters of the corresponding distribution, and return a random number. Here, we generate numbers which are uniformly distributed in the interval $[0, 1)$ via the function `gsl_rng_uniform()`, see line 16. Therefore, the value 0 is included, but the value 1 is excluded. Finally, the generator is destroyed (line 20).

The GSL offers several high-quality generators. Also other generators are included for back compatibility. In total more than 27 different generator algorithms are implemented. The best known are

- `gsl_rng_mt19937`
  The MT19937 generator [Matsumoto and Nishimura (1998)]. It has a period of about $10^{6000}$ and passed the *Diehard* statistical tests [Marsaglia].
- `gsl_rng_gfsr4`
  is a lagged-Fibonacci generator [Ziff (1998)]. In the present implementation, the generator has a period of about $10^{2017}$.
- `gsl_rng_ranlux`
  The *ranlux* 24 bit generator [Lüscher (1994)]. It has a period of $10^{171}$.
- `gsl_rng_rand48`
  is the Unix linear congruential random number generator `rand48()`. It is also included for compatibility and has lower quality compared to modern generators, e.g. a period of about $10^{14}$. But this is sufficient for small applications without the need for high-quality statistics.
- `gsl_rng_rand`
  is the standard Unix (BSD) linear congruential random number generator `rand()`. It is included for compatibility (i.e. you can use it on non-Unix system) and has a very bad quality compared to modern generators, e.g. a period of about $10^9$.

Note that it is not necessary to hard-code the type of the random number generator in the program. One can define an environment variable GSL_RNG_TYPE and call in the program the function `gsl_rng_env_setup()` (no arguments). Then the global variable `gsl_rng_default` of type `gsl_rng_type *` will contain the current generator and can be used as argument when calling `gsl_rng_alloc()`. If the environment variable is not assigned, `gsl_rng_default` will contain `gsl_rng_mt19937`.

In addition to the functions discussed so far, the GSL enables us to investigate the properties of random number generators, copy generators, store the state of generators in files, and read them back in.

Now, to actually generate data which is distributed according to a Binomial distribution $\text{Bin}(n, p)$, the following function is provided:

```
unsigned int gsl_ran_binomial(const gsl_rng * r, double p,
                              unsigned int n);
```

The first argument is a function pointer to a GSL random number generator as described above. One has to pass the parameters $p$ and $n$ as second and third arguments. Thus, to generate 10000 numbers which are distributed according to a Binomial (10,0.4) distribution, one has to replace line 16 in the above program by

```
r = (double) gsl_ran_binomial (rng, 0.4, 10);
```

Apart from the Binomial distribution, the GSL implements many different distributions. Usually the probability density function (or the probability mass function for discrete variables) and a function for generating (pseudo) random numbers according to the distribution are available. Examples are the Gaussian distribution, the exponential distribution, the Levy distribution, the Chi-squared distribution, the Pareto distribution, the Weibull distribution, and the hyper-geometric distribution. See the GSL documentation for more details.

To record histograms of the generated random numbers and of any other data, the GSL offers histograms via the `gsl_histogram` data type. Included are functions for creating, updating, accessing, printing, analyzing and deleting histograms. Interestingly, the histograms can have bins of non-uniform size, which is useful for logarithmic binning. Again, see the GSL documentation for more details and exercise (2).

We close this section by showing how one can solve a system of linear equations using the GSL. For this example already a couple of different data types and functions are needed: To represent vectors and matrices, GSL offers the special data types `gsl_vector` and `gsl_matrix`. They are not simply arrays, because they store the dimensions of the represented objects as well. In the GSL manual, the internal structures of the data type is explained, but we restrict ourselves here to use the objects through defined access functions. Furthermore, for solving systems of linear equations, one needs *permutations* as auxiliary data, which are just arrays of $n$ integers in the range 0 to $n-1$ (data type `gsl_permutation`). Permutations are needed to describe how rows and columns of matrices are permuted to solve the linear system. The basic steps needed in the program to solve the system $\mathbf{A}\underline{x} = \underline{b}$ are

- allocation of the matrix $\mathbf{A}$ and vectors $\underline{x}$ and $\underline{b}$,
- initializing the values of $\mathbf{A}$ and $\underline{b}$,
- performing an LU decomposition of $\mathbf{A}$ such that $\mathbf{PA} = \mathbf{LU}$ where $\mathbf{P}$ is a permutation matrix, $\mathbf{L}$ a lower triangular matrix and $\mathbf{U}$ a upper diagonal matrix,
- solving the system via forward and back substitution,
- and finally freeing all allocated memory.

As illustration, we solve the following system of linear equations:

$$
\begin{array}{rrrrr}
x_1 & +2.5x_2 & +3x_3 & & = 15 \\
2x_1 & +5x_2 & +x_3 & +2x_4 & = 23 \\
3.5x_1 & +3x_2 & & +3x_4 & = 21.5 \\
& 2x_2 & +x_3 & +4x_4 & = 15
\end{array}
$$

This is done by the following small sample program

```
1   #include <stdio.h>
2   #include <gsl/gsl_matrix.h>
3   #include <gsl/gsl_vector.h>
4   #include <gsl/gsl_permutation.h>
5   #include <gsl/gsl_linalg.h>
6
7   int main()
8   {
9     gsl_matrix *A;                /* matrix */
10    int i, j;                     /* row/column index */
11    gsl_vector *b, *x;            /* vectors */
12    gsl_permutation *perm;        /* a permutation */
13    int signum;                   /* sign of permutation */
14
```

```
15      double values[4*4] = {1.0, 2.5, 3.0, 0.0,              /* data */
16                            2.0, 5.0, 1.0, 2.0,
17                            3.5, 3.0, 0.0, 3.0,
18                            0.0, 2.0, 1.0, 4.0};
19      double values2[4] = {15.0, 23.0, 21.5, 15.0};
20
21      A = gsl_matrix_alloc(4,4);            /* allocate and initialize */
22      for(i=0; i<4; i++)
23        for(j=0; j<4; j++)
24          gsl_matrix_set(A, i, j, values[i*4+j]);
25      x = gsl_vector_alloc(4);
26      b = gsl_vector_alloc(4);
27      for(i=0; i<4; i++)
28        gsl_vector_set(b, i, values2[i]);
29      perm = gsl_permutation_alloc(4);
30
31      gsl_linalg_LU_decomp(A, perm, &signum);    /* solve equation */
32      gsl_linalg_LU_solve(A, perm, b, x);
33
34      gsl_vector_fprintf(stdout, x, "%f");
35
36      gsl_matrix_free(A);                        /* free memory */
37      gsl_vector_free(b);
38      gsl_vector_free(x);
39      gsl_permutation_free(perm);
40      return(0);
41    }
```

The variables are declared in lines 8–12. The parameters describing the equations through **A** and $\underline{b}$ are first put in two one-dimensional arrays **values** and **values2** (lines 14–18), be-

| GET SOURCE CODE |
|---|
| DIR: **libraries** |
| FILE(S): **lin_eq.c** |

cause in this way the matrices can be initialized more easily. The allocation and initialization of the matrix, of the vectors and of the permutation is performed in lines 20–28. Note that the write access to matrix and vector elements is through the functions `gsl_matrix_set()` and `gsl_vector_set()`, respectively. These functions perform also checks for bounds, i.e. whether an access outside the matrix dimensions is attempted. To read data, the GSL offers also corresponding functions `gsl_matrix_get()` and `gsl_vector_get()`, respectively, which are not used in this example.[1]

---

[1] Note that one can access the $i$'th element of the vector b directly via b->data[i
* b->stride] and the $i, j$'th element of a matrix **mat** directly via mat->data[i *

The actual calculation is performed in lines 30 and 31. Note that the system is passed to `gsl_linalg_LU_decomp()` via the first argument, which will also contain the resulting LU decomposition upon return in the left lower and right upper triangles of the matrix; hence, the original content is overwritten. The second argument points to the permutation where **P** is stored. The function `gsl_linalg_LU_solve()` is used to perform the forward and back substitution to actually obtain the solution which is stored in the vector passed as last (pointer) argument. The result vector is printed by `gsl_vector_fprintf()` line by line. Finally (lines 35-38) the used memory is freed.

For more complex usage of the GSL, such as to solve differential equations, to optimize functions or to perform Monte Carlo simulations, please refer to the GSL documentation.

## 7.4 Creating your own libraries

Although many useful libraries are available, sometimes you have to write some code by yourself. Over the years you will collect many functions and data structures, which – if properly designed – can be included in other programs, in which case it is convenient to put these subroutines into a library. Then you do not have to include the object file every time you compile one of your programs. If your self-created library is put into a standard search path, you can access it like a system library, you even do not have to remember where the object file is stored.

To create a library you must have an object file, for example `tasks.o`, and a header file such as `tasks.h` where all data types and function prototypes are defined. Furthermore, to facilitate the use of the library, you should write a *man* page, which is not necessary for technical reasons but results in a more convenient usage of your library, particularly if other people want to benefit from it. To learn how to write a *man* page you should consult `man man` and have a look at the source code of some *man* pages, they are stored, for example, in `/usr/man`.

A library is created with the UNIX command `ar`. To include `tasks.o` in your library `libmy.a`, you have to enter

```
ar r libmy.a tasks.o
```

---

`mat->tda + j]`, please refer to the GSL documentation for details.

In a library several object files may be collected. The option "r" replaces the given object files, if they already belong to the library, otherwise they are added. If the library does not exist yet, it is created. For more options, please refer to the *man* page of `ar`.

After including an object file, you have to update an internal object table of the library. This is done by

```
ar s libmy.a
```

Now you can compile a program `prog.c` using your library via

```
cc -o prog prog.c libmy.a
```

In case `libmy.a` contains several object files, it saves some typing by just writing `libmy.a`. Furthermore you do not have to remember the names of all your object files.

To make the handling of the library more comfortable, you can create a directory, e.g. `~/lib` and put your libraries there. Additionally, you should create the directory `~/include` where all personal header files can be collected. Then your compile command may look like this:

```
cc -o prog prog.c -I$HOME/include -L$HOME/lib -lmy
```

The option `-I` states the search path for additional header files, the `-L` option tells the linker where your libraries are stored and via `-lmy` the library `libmy.a` is actually included. Please note that the prefix `lib` and the postfix `.a` are omitted with the `-l` option. Finally, it should be pointed out, that the compiler command given above works for all working directories, once you have set up the library structure as explained. Consequently, you do not have to remember directories or names of object files.

# Exercises

(solutions: can be downloaded from http://www.worldscientific.com/r/9019-supp)

(1) **Sorting vectors**

Design, implement and test a program, which generates an array of two-dimensional vectors $(x, y)$ sorts the vectors by ascending lengths $\sqrt{(x^2 + y^2)}$ and finally prints the vectors.

| SOLUTION SOURCE CODE |
|---|
| DIR: **libraries** |
| FILE(S): **vectorsort.c** |

Use the following type for the vectors:

```
typedef struct
{
    double x,y;                   /* elements of vector */
} vector_t;
```

Use the qsort() function from the standard C library. For this purpose, write a function compare_vector(), which should have the following prototype:

```
/****************** compare_vector() ****************/
/** Auxiliary function which compares two vectors   **/
/** by its length.                                  **/
/** Used to call qsort.                             **/
/** PARAMETERS: (*)= return-paramter                **/
/**    p1, p2: pointers to the two vectors          **/
/** RETURNS:                                        **/
/**    -1 if *p1<*p2, 0 if *p1=*p2, +1 else         **/
/****************************************************/
int compare_vector(const void *p1, const void *p2)
```

Hint: It is sufficient to calculate $x^2 + y^2$; thus, you do not need the mathematical library.

(2) **Histogram of random numbers**

Design, implement and test a program, which draws using the GSL $n$ numbers from an exponential distribution with probability density function $p(x) = \frac{1}{\lambda} \exp(-x/\lambda)$, puts the sample numbers into a histogram and prints the histogram.

| SOLUTION SOURCE CODE |
|---|
| DIR: **libraries** |
| FILE(S): |
| **exponential.c** |

Hints:

- View the program **randnum.c** (page 236) as blueprint.
- Use the function **gsl_ran_exponential()** to generate exponentially distributed random numbers.

- Consult the histogram section of the GSL manual. Use the gsl_histogram data structure and corresponding GSL functions.
- Implement logarithmic binning with bins $[0, x_0)$, $[x_0, x_0 f)$, $[x_0 f, x_0 f^2)$, ...for $f > 1$. Use the function gsl_histogram_set_ranges() to setup the logarithmic binning.

Run the program for $n = 10^6$ and $\lambda = 2.0$. Use 30 bins defined by $x_0 = 0.2$ and $f = 1.2$. View the histogram with a suitable program, e.g. *gnuplot*, see Sec. 8.4.1.

Additional exercise:

The GSL function gsl_histogram_fprintf(), does not allow the result to be printed as (normalized) probability density function. For this purpose design, implement and test a function print_norm_histogram() for printing the probability density function corresponding to a histogram. The function prototype reads as follows:

```
/************* print_norm_histogram() ****************/
/** Prints histogram as probability density function. **/
/** For each bin [l,u) its mid point                **/
/** (l+u)/2 and its count normalized by the         **/
/** total count and by the bin width (u-l) is shown. **/
/**                                                  **/
/** Parameters: (*) = return parameter              **/
/**        h: histogram                             **/
/*                                                  **/
/** Returns:                                        **/
/**        (nothing)                                **/
/****************************************************/
void print_norm_histogram(gsl_histogram *h)
```

Hints:

- Histograms allow for negative bin counts; hence, the function should ignore them.
- For each bin $i$, you can print the midpoint $(l_i + u_i)/2$ between its lower end $l_i$ and upper end $u_i$. The normalized value for bin $i$ is $c_i/(N(u_i - l_i))$, where $c_i$ is the count of bin $i$ and $N$ is the sum of all nonzero counts.
- It is not necessary to access the internal data of a histogram. You can use the functions gsl_histogram_bins() to obtain the number of bins, gsl_histogram_get() to get the count of a bin, and gsl_histogram_get_range() to get the ranges of a bin; see the GSL manual for descriptions.

Run the program as above and compare the result using *gnuplot* to the analytical formula for the probability density function.

# Chapter 8

# Randomness and Statistics

In this chapter, we are concerned with statistics in a very broad sense. This involves generation of (pseudo) random data, display/plotting of data and the statistical analysis of simulation results.

Frequently, a simulation involves the explicit generation of random numbers, for instance, as auxiliary quantity for stochastic simulations. In this case it is obvious that the simulation results are random as well. Although there are many simulations which are explicitly not random, the resulting behavior of the simulated systems may appear also random, for example the motion of interacting gas atoms in a container. Hence, methods from statistical data analysis are necessary for almost all analysis of simulation results.

This chapter starts (Sec. 8.1) by an introduction to randomness and statistics. In Sec. 8.2 the generation of pseudo random numbers according to some given probability distribution is explained. Basic analysis of data, i.e., the calculation of mean, variance, histograms and corresponding error bars, is covered in Sec. 8.3. Next, in Sec. 8.4, it is shown how data can be represented graphically using suitable plotting tools, *gnuplot* and *xmgrace*. Hypothesis testing and how to measure or ensure independence of data is treated in Sec. 8.5. How to fit data to functions is explained in Sec. 8.6. In the concluding section, a special technique is outlined which allows to cope with the limitations of simulations due to finite system sizes.

Note that some examples are again presented using the C programming language. Nevertheless, there exist very powerful freely available programs like R [R], where many analysis (and plotting) tools are available as additional packages.

## 8.1 Introduction to probability

Here, a short introduction to concepts of probability and randomness is given. The presentation here should be concise concerning the subjects presented in this book. Nevertheless, more details, in particular proofs, examples and exercises, can be found in standard textbooks [Dekking et al (2005); Lefebvre (2006)]. Here often a sloppy mathematical notation is used for brevity, e.g. instead of writing "a function $g : X \to Y$, $y = g(x)$", we often write simply "a function $g(x)$".

A *random experiment* is an experiment which is truly random (like radioactive decay or quantum mechanical processes) or at least unpredictable (like tossing a coin or predicting the position of a certain gas atom inside a container which holds a hot dense gas).

**Definition 8.1** The *sample space* $\Omega$ is a set of all possible outcomes of a random experiment.

For the coin example, the sample space is $\Omega = \{\text{head, tail}\}$. Note that a sample space can be in principle infinite, like the possible $x$ positions of an atom in a container. With infinite precision of measurement we have $\Omega^{(x)} = [0, L_x]$, where the container shall be a box with linear extents $L_x$ ($L_y, L_z$ in the other directions, see below).

For a random experiment, one wants to know the probability that certain events occur. Note that for the position of an atom in a box, the probability to find the atom *precisely at* some $x$-coordinate $x \in \Omega^{(x)}$ is zero if one assumes that measurements result in real numbers with infinite precision. For this reason, one considers probabilities $P(A)$ of subsets $A \subset \Omega$ (in other words $A \in 2^{\Omega}$, $2^{\Omega}$ being the *power set* which is the set of all subsets of $\Omega$). Such a subset is called an *event*. Therefore $P(A)$ is the probability that the outcome of a random experiment is inside $A$, i.e. one of the elements of $A$. More formally:

**Definition 8.2** A *probability function* $P$ is a function $P : 2^{\Omega} \longrightarrow [0, 1]$ with

$$P(\Omega) = 1 \tag{8.1}$$

and for each finite or infinite sequence $A_1, A_2, A_3, \ldots$ of mutual disjoint events ($A_i \cap A_j = \emptyset$ for $i \neq j$) we have

$$P(A_1 \cup A_2 \cup A_3 \cup \ldots) = P(A_1) + P(A_2) + P(A_3) + \ldots \tag{8.2}$$

For a fair coin, both sides would appear with the same probability, hence one has $P(\emptyset) = 0$, $P(\{\text{head}\}) = 0.5$, $P(\{\text{tail}\}) = 0.5$, $P(\{\text{head, tail}\}) = 1$. For the hot gas inside the container, we assume that no external forces act on the atoms. Then the atoms are distributed uniformly. Thus, when measuring the $x$ position of an atom, the probability to find it inside the region $A = [x, x + \Delta x] \subset \Omega^{(x)}$ is $P(A) = \Delta x / L_x$.

The usual set operations applies to events. The *intersection* $A \cap B$ of two events is the event which contains elements that are both in $A$ and $B$. Hence $P(A \cap B)$ is the probability that the outcome of an experiment is contained in both events $A$ and $B$. The *complement* $A^c$ of a set is the set of all elements of $\Omega$ which are not in $A$. Since $A^c$, $A$ are disjoint and $A \cup A^c = \Omega$, we get from Eq. (8.2):

$$P(A^c) = 1 - P(A). \tag{8.3}$$

Furthermore, one can show for two events $A, B \subset \Omega$:

$$P(A \cup B) = P(A) + P(B) - P(A \cap B) \tag{8.4}$$

**Proof.** $P(A) = P(A \cap \Omega) = P(A \cap (B \cup B^c)) = P((A \cap B) \cup (A \cap B^c)) \overset{(8.2)}{=} P(A \cap B) + P(A \cap B^c)$. If we apply this for $A \cup B$ instead of $A$, we get $P(A \cup B) = P((A \cup B) \cap B) + P((A \cup B) \cap B^c)) = P(B) + P(A \cap B^c)$. Eliminating $P(A \cap B^c)$ from these two equations gives the desired result. $\qquad\square$

Note that Eqs. (8.2) and (8.3) are special cases of this equation.

If a random experiment is repeated several times, the possible outcomes of the repeated experiment are tuples of outcomes of single experiments. Thus, if you throw the coin twice, the possible outcomes are (head,head), (head,tail), (tail,head), and (tail,tail). This means the sample space is a power of the single-experiment sample spaces. In general, it is also possible to combine different random experiments into one. Hence, for the general case, if $k$ experiments with sample spaces $\Omega^{(1)}, \Omega^{(2)}, \ldots, \Omega^{(k)}$ are considered, the sample space of the combined experiment is $\Omega = \Omega^{(1)} \times \Omega^{(2)} \times \ldots \times \Omega^{(k)}$. For example, one can describe the measurement of the position of the atom in the hot gas as a combination of the three independent random experiments of measuring the $x$, $y$, and $z$ coordinates, respectively.

If we assume that the different experiments are performed *independently*, then the total probability of an event for a combined random experiment is the product of the single-experiment probabilities: $P(A^{(1)}, A^{(2)}, \ldots, A^{(k)}) = P(A^{(1)})P(A^{(2)}) \ldots P(A^{(k)})$.

For tossing the fair coin twice, the probability of the outcome (head,tail) is $P(\{(\text{head,head})\}) = P(\{\text{head}\})P(\{\text{head}\}) = 0.5 \cdot 0.5 = 0.25$. Similarly, for the experiment where all three coordinates of an atom inside the container are measured, one can write $P([x, x+\Delta x] \times [y, y+\Delta y] \times [z, z+\Delta z]) = P([x, x + \Delta x])P([y, y + \Delta y])P([z, z + \Delta z]) = (\Delta x/L_x)(\Delta y/L_y)(\Delta z/L_z) = \Delta x \Delta y \Delta z/(L_x L_y L_z)$.

Often one wants to calculate probabilities which are restricted to special events $C$ among all events, hence relative or *conditioned* to $C$. For any other event $A$ we have $P(C) = P((A \cup A^c) \cap C) = P(A \cap C) + P(A^c \cap C)$, which means $\frac{P(A \cap C)}{P(C)} + \frac{P(A^c \cap C)}{P(C)} = 1$. Since $P(A \cap C)$ is the *joint* probability of an outcome in $A$ and $C$ and because $P(C)$ is the probability of an outcome in $C$, the fraction $\frac{P(A \cap C)}{P(C)}$ gives the probability of an outcome $A$ and $C$ relative to $C$, i.e. the probability of event $A$ given $C$, leading to the following

**Definition 8.3**   The *probability of A under the condition C* is

$$P(A|C) = \frac{P(A \cap C)}{P(C)}. \tag{8.5}$$

As we have seen, we have the natural normalization $P(A|C) + P(A^c|C) = 1$. Rewriting Eq. (8.5) one obtains $P(A|C)P(C) = P(A \cap C)$. Therefore, the calculation of $P(A \cap C)$ can be decomposed into two parts, which are sometimes easier to obtain. By symmetry, we can also write $P(C|A)P(A) = P(A \cap C)$. Combining this with Eq. (8.5), one obtains the famous *Bayes' rule*

$$P(C|A) = \frac{P(A|C)P(C)}{P(A)}. \tag{8.6}$$

This means one of the conditional probabilities $P(A|C)$ and $P(C|A)$ can be expressed via the other, which is sometimes useful if $P(A)$ and $P(C)$ are known. Note that the denominator in the Bayes' rule is sometimes written as $P(A) = P(A \cap (C \cup C^c)) = P(A \cap C) + P(A \cap C^c) = P(A|C)P(C) + P(A|C^c)P(C^c)$.

If an event $A$ is *independent* of the condition $C$, its conditional probability should be the same as the unconditional probability, i.e., $P(A|C) = P(A)$. Using $P(A \cap C) = P(A|C)P(C)$ we get $P(A \cap C) = P(A)P(C)$, i.e., the probabilities of independent events have to be multiplied. This was used already above for random experiments, which are conducted as independent subexperiments.

So far, the outcomes of the random experiments can be anything like the sides of coins, sides of a dice, colors of the eyes of randomly chosen people or states of random systems. In mathematics, it is often easier to handle numbers instead of arbitrary objects. For this reason one can represent the outcomes of random experiments by numbers which are assigned via special functions:

**Definition 8.4** For a sample space $\Omega$, a *random variable* is a function $X : \Omega \longrightarrow \mathbb{R}$.

For example, one could use $X(\text{head})=1$ and $X(\text{tail}) = 0$. Hence, if one repeats the experiments $k$ times independently, one would obtain the number of heads by $\sum_{i=1}^{k} X(\omega^{(i)})$, where $\omega^{(i)}$ is the outcome of the $i$'th experiment.

If one is interested only in the values of the random variable, the connection to the original sample space $\Omega$ is not important anymore. Consequently, one can consider random variables $X$ as devices, which output a random number $x$ each time a random experiment is performed. Note that random variables are usually denoted by upper-case letters, while the actual outcomes of random experiments are denoted by lower-case letters.

Using the concept of random variables, one deals only with numbers as outcomes of random experiments. This enables many tools from mathematics to be applied. In particular, one can combine random variables and functions to obtain new random variables. This means, in the simplest case, the following: First, one performs a random experiment, yielding a random outcome $x$. Next, for a given function $g$, $y = g(x)$ is calculated. Then, $y$ is the final outcome of the random experiment. This is called a *transformation* $Y = g(X)$ of the random variable $X$. More generally, one can also define a random variable $Y$ by combining *several* random variables $X^{(1)}, X^{(2)}, \ldots, X^{(k)}$ via a function $\tilde{g}$ such that

$$Y = \tilde{g}\left(X^{(1)}, X^{(2)}, \ldots, X^{(k)}\right). \tag{8.7}$$

In practice, one would perform random experiments for the random variables $X^{(1)}, X^{(2)}, \ldots, X^{(k)}$, resulting in outcomes $x^{(1)}, x^{(2)}, \ldots, x^{(k)}$. The final number is obtained by calculating $y = \tilde{g}(x^{(1)}, x^{(2)}, \ldots, x^{(k)})$. A simple but the most important case is the linear combination of random variables $Y = \alpha_1 X^{(1)} + \alpha_2 X^{(2)} + \ldots + \alpha_k X^{(k)}$, which will be used below. For all examples considered here, the random variables $X^{(1)}, X^{(2)}, \ldots, X^{(k)}$ have the same properties, which means that the same random experiment is repeated $k$ times. Nevertheless, the most general description which allows for different random variables will be used here.

The behavior of a random variable is fully described by the probabilities of obtaining outcomes smaller or equal to a given parameter $x$:

**Definition 8.5**    The *distribution function* of a random variable $X$ is a function $F_X : \mathbb{R} \longrightarrow [0, 1]$ defined via

$$F_X(x) = P(X \le x) \tag{8.8}$$

The index $X$ is omitted if no confusion arises. Sometimes the distribution function is also named *cumulative* distribution function. One also says, the distribution function defines a *probability distribution*. Stating a random variable or stating the distribution function are fully equivalent methods to describe a random experiment.

For the fair coin, we have, see left of Fig. 8.1

$$F(x) = \begin{cases} 0 & x < 0 \\ 0.5 & 0 \le x < 1 \\ 1 & x \ge 1 \end{cases} . \tag{8.9}$$

For measuring the $x$ position of an atom in the uniformly distributed gas we obtain, see right of Fig. 8.1

$$F(x) = \begin{cases} 0 & x < 0 \\ x/L_x & 0 \le x < L_x \\ 1 & x \ge L_x \end{cases} . \tag{8.10}$$

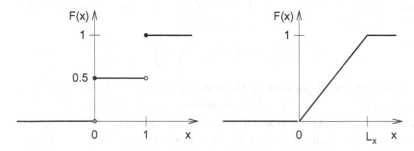

Fig. 8.1   Distribution function of the random variable for a fair coin (left) and for the random $x$ position of a gas atom inside a container of length $L_x$.

Since the outcomes of any random variable are finite, there are *no* possible outcomes $X \le x$ in the limit $x \to -\infty$. Also, *all* possible outcomes fulfill $X \le x$ for $x \to \infty$. Consequently, one obtains for all random variables $\lim_{x \to -\infty} F(x) = 0$ and $\lim_{x \to +\infty} F(x) = 1$. Furthermore, from Def. 8.5, one obtains immediately:

$$P(x_0 < X \le x_1) = F_X(x_1) - F_X(x_0) \tag{8.11}$$

Therefore, one can calculate the probability to obtain a random number for any arbitrary interval, hence also for unions of intervals.

The distribution function, although it contains all information, is sometimes less convenient to handle, because it gives information about cumulative probabilities. It is more obvious to describe the outcomes of the random experiments directly. For this purpose, we have to distinguish between *discrete* random variables, where the number of possible outcomes is denumerable or even finite, and *continuous* random variables, where the possible outcomes are non-denumerable. The random variable describing the coin is discrete, while the position of an atom inside a container is continuous.

### 8.1.1  *Discrete random variables*

We first concentrate on discrete random variables. Here, an alternative but equivalent description to the distribution function is to state the probability for each possible outcome directly:

**Definition 8.6**  For a discrete random variable $X$, the *probability mass function* (pmf) $p_X : \mathbb{R} \to [0,1]$ is given by

$$p_X(x) = P(X = x). \tag{8.12}$$

Again, the index $X$ is omitted if no confusion arises. Since a discrete random variable describes only a denumerable number of outcomes, the probability mass function is zero almost everywhere. In the following, the outcomes $x$ where $p_X(x) > 0$ are denoted by $\tilde{x}_i$. Since probabilities must sum up to one, see Eq. 8.1, one obtains $\sum_i p_X(\tilde{x}_i) = 1$. Sometimes we also write $p_i = p_X(\tilde{x}_i)$. The distribution function $F_X(x)$ is obtained from the pmf via summing up all probabilities of outcomes smaller or equal to $x$:

$$F_X(x) = \sum_{\tilde{x}_i \le x} p_X(\tilde{x}_i) \tag{8.13}$$

For example, the pmf of the random variable arising from the fair coin Eq. (8.9) is given by $p(0) = 0.5$ and $p(1) = 0.5$ ($p(x) = 0$ elsewhere). The generalization to a possibly unfair coin, where the outcome "1" arises with probability $p$, leads to:

**Definition 8.7** The *Bernoulli distribution with parameter* $p$ $(0 < p \le 1)$ describes a discrete random variable $X$ with the following probability mass function

$$p_X(1) = p, \quad p_X(0) = 1 - p. \tag{8.14}$$

Performing a Bernoulli experiment means that one throws a generalized coin and records either "0" or "1" depending on whether one gets head or tail.

There are a couple of important characteristic quantities describing the pmf of a random variable. Next, we describe the most important ones for the discrete case:

**Definition 8.8**

- The *expectation value* is

$$\mu \equiv \mathrm{E}[X] = \sum_i \tilde{x}_i P(X = \tilde{x}_i) = \sum_i \tilde{x}_i p_X(\tilde{x}_i) \tag{8.15}$$

- The *variance* is

$$\sigma^2 \equiv \mathrm{Var}[X] = \mathrm{E}[(X - \mathrm{E}[X])^2] = \sum_i (\tilde{x}_i - \mathrm{E}[X])^2 p_X(\tilde{x}_i) \tag{8.16}$$

- The *standard deviation*

$$\sigma \equiv \sqrt{\mathrm{Var}[X]} \tag{8.17}$$

The expectation value describes the "average" one would typically obtain if the random experiment is repeated very often. The variance is a measure for the spread of the different outcomes of random variable. As example, the Bernoulli distribution exhibits

$$\mathrm{E}[X] = 0p(0) + 1p(1) = p \tag{8.18}$$

$$\mathrm{Var}[X] = (0 - p)^2 p(0) + (1 - p)^2 p(1)$$
$$= p^2(1 - p) + (1 - p)^2 p = p(1 - p) \tag{8.19}$$

One can calculate expectation values of functions $g(x)$ of random variables $X$ via $\mathrm{E}[g(X)] = \sum_i g(\tilde{x}_i) p_X(\tilde{x}_i)$. For the calculation here, we only need that the calculation of the expectation value is a linear operation. Hence, for numbers $\alpha_1, \alpha_2$ and, in general, two random variables $X_1, X_2$ one has

$$E[\alpha_1 X_1 + \alpha_2 X_2] = \alpha_1 \mathrm{E}[X_1] + \alpha_2 \mathrm{E}[X_2]. \tag{8.20}$$

In this way, realizing that $E[X]$ is a number, one obtains:

$$\sigma^2 = \text{Var}(X) = E[(X - E[X])^2] = E[X^2] - 2E[X\,E[X]] + E[E[X]^2]$$
$$= E[X^2] - E[X]^2 = E[X^2] - \mu^2 \tag{8.21}$$
$$\Leftrightarrow \quad E[X^2] = \sigma^2 + \mu^2 \tag{8.22}$$

The variance is *not linear*, which can be seen when looking at a linear combination of two *independent* random variables $X_1, X_2$ (implying $E[X_1 X_2] = E[X_1]\,E[X_2]$ $(\star)$)

$$
\begin{aligned}
\sigma^2_{\alpha_1 X_1 + \alpha_2 X_2} &= \text{Var}[\alpha_1 X_1 + \alpha_2 X_2] \\
&\overset{(8.21)}{=} E[(\alpha_1 X_2 + \alpha_2 X_2)^2] - E[\alpha_1 X_1 + \alpha_2 X_2]^2 \\
&\overset{(8.20)}{=} E[\alpha_1^2 X_1^2 + 2\alpha_1\alpha_2 X_1 X_2 + \alpha_2^2 X_2^2] \\
&\quad - (\alpha_1 E[X_1] + \alpha_2 E[X_2])^2 \\
&\overset{(8.20),(\star)}{=} \alpha_1^2 E[X_1^2] + \alpha_2^2 E[X_2^2] - \alpha_1^2 E[X_1]^2 + \alpha_2^2 E[X_2]^2 \\
&\overset{(8.21)}{=} \alpha_1^2 \text{Var}[X_1] + \alpha_2^2 \text{Var}[X_2] \tag{8.23}
\end{aligned}
$$

The expectation values $E[X^n]$ are called the *$n$'th moments* of the distribution. This means that the expectation value is the first moment and the variance can be calculated from the first and second moments.

Next, we describe two more important distributions of discrete random variables. First, if one repeats a Bernoulli experiment $n$ times, one can measure how often the result "1" was obtained. Formally, this can be written as a sum of $n$ random variables $X^{(i)}$ which are Bernoulli distributed: $X = \sum_{i=1}^{n} X^{(i)}$ with parameter $p$. This is a very simple example of a transformation of a random variable, see page 249. In particular, the transformation is linear. The probability to obtain $x$ times the result "1" is calculated as follows: The probability to obtain exactly $x$ times a "1" is $p^x$, the other $n - x$ experiments yield "0" which happens with probability $(1-p)^{n-x}$. Furthermore, there are $\binom{n}{x} = n!/(x!(n-x)!)$ different sequences with $x$ times "1" and $n - x$ times "0". Hence, one obtains:

**Definition 8.9** The *binomial distribution* with parameters $n \in \mathbb{N}$ and $p$ $(0 < p \leq 1)$ describes a random variable $X$ which has the pmf

$$p_X(x) = \binom{n}{x} p^x (1 - p)^{n-x} \quad (0 \leq x \leq n) \tag{8.24}$$

A common notation is $X \sim B(n, p)$.

Note that the probability mass function is assumed to be zero for argument values that are not stated. A sample plot of the distribution for parameters $n = 10$ and $p = 0.4$ is shown in the left of Fig. 8.2. The Binomial distribution has expectation value and variance

$$E[X] = np \tag{8.25}$$

$$\text{Var}[X] = np(1 - p) \tag{8.26}$$

(without proof here). The distribution function cannot be calculated analytically in closed form.

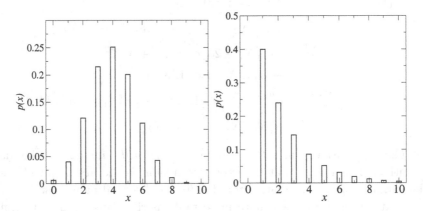

Fig. 8.2 (Left) Probability mass function of the binomial distribution for parameters $n = 10$ and $p = 0.4$. (Right) Probability mass function of the geometric distribution for parameter $p = 0.4$.

In the limit of a large number of experiments $(n \to \infty)$, constrained such that the expectation value $\mu = np$ is kept fixed, the pmf of a Binomial distribution is well approximated by the pmf of the *Poisson distribution*, which is defined as follows:

**Definition 8.10** The *Poisson distribution* with parameter $\mu > 0$ describes a random variable $X$ with pmf

$$p_X(x) = \frac{\mu^x}{x!} e^{-\mu} \tag{8.27}$$

Indeed, as required, the probabilities sum up to 1, since $\sum_x \frac{\mu^x}{x!}$ is the Taylor series of $e^\mu$. The Poisson distribution exhibits $E[X] = \mu$ and $\text{Var}[X] = \mu$. Again, a closed form for the distribution function is not known.

Furthermore, one could repeat a Bernoulli experiment just until the first time a "1" is observed, without limit for the number of trials. If a "1" is observed for the first time after exactly $x$ times, then the first $x-1$ times the outcome "0" was observed. This happens with probability $(1-p)^{x-1}$. At the $x$'th experiment, the outcome "1" is observed which has the probability $p$. Therefore one obtains

**Definition 8.11** The *geometric distribution* with parameter $p$ ($0 < p \leq 1$) describes a random variable $X$ which has the pmf

$$p_X(x) = (1-p)^{x-1}p \quad (x \in \mathbb{N}) \tag{8.28}$$

A sample plot of the pmf (up to $x = 10$) is shown in the right of Fig. 8.2. The geometric distribution has (without proof here) the expectation value $E[X] = 1/p$, the variance $\text{Var}[X] = (1-p)/p^2$ and the following distribution function:

$$F_X(x) = \begin{cases} 0 & x < 1 \\ 1 - (1-p)^m & m \leq x < m+1 \quad (m \in \mathbb{N}) \end{cases}$$

### 8.1.2 *Continuous random variables*

As stated above, random variables are called continuous if they describe random experiments where outcomes from a subset of the real numbers can be obtained. One may describe such random variables also using the distribution function, see Def. 8.5. For continuous random variables, an alternative description is possible, equivalent to the pmf for discrete random variables: The probability density function states the probability to obtain a certain number per unit:

**Definition 8.12** For a continuous random variable $X$ with a continuous distribution function $F_X$, the *probability density function* (pdf) $p_X : \mathbb{R} \to [0, 1]$ is given by

$$p_X(x) = \frac{dF_X(x)}{dx} \tag{8.29}$$

Consequently, one obtains, using the definition of a derivative and using Eq. (8.11)

$$F_X(x) = \int_{-\infty}^{x} d\tilde{x}\, p_X(\tilde{x}) \tag{8.30}$$

$$P(x_0 < X \le x_1) = \int_{x_0}^{x_1} d\tilde{x}\, p_X(\tilde{x}) \tag{8.31}$$

Below some examples for important continuous random variables are presented. First, we extend the definitions Def. 8.13 of expectation value and variance to the continuous case:

**Definition 8.13**

- The *expectation value* is

$$E[X] = \int_{-\infty}^{\infty} dx\, x\, p_X(x) \tag{8.32}$$

- The *variance* is

$$\mathrm{Var}[X] = E[(X - E[X])^2] = \int_{\infty}^{-\infty} dx\, (x - E[X])^2 p_X(x) \tag{8.33}$$

Expectation value and variance have the same properties as for the discrete case, i.e., Eqs. (8.20), (8.21), and (8.23) hold as well. Also the definition of the n'th moment of a continuous distribution is the same.

Another quantity of interest is the *median*, which describes the central point of the distribution. It is given by the point such that the cumulative probabilities left and right of this point are both equal to 0.5:

**Definition 8.14**   The *median* $x_{\mathrm{med}} = \mathrm{Med}[X]$ is defined via

$$F_X(x_{\mathrm{med}}) = 0.5 \tag{8.34}$$

The simplest distribution is the uniform distribution, where the probability density function is nonzero and constant in some interval $[a, b)$:

**Definition 8.15**   The *uniform distribution*, with real-valued parameters $a < b$, describes a random variable $X$ which has the pdf

$$p_X(x) = \begin{cases} 0 & x < a \\ \frac{1}{b-a} & x \le x < b \\ 0 & x \ge b \end{cases} \tag{8.35}$$

One writes $X \sim U(a, b)$.

The distribution function simply rises linearly from zero, starting at $x = a$, till it reaches 1 at $x = b$, see for example Eq. 8.10 for the case $a = 0$ and $b = L_x$. The uniform distribution exhibits the expectation value $E[X] = (a + b)/2$ and variance $Var[X] = (b - a)^2/12$. Note that via the linear transformation $g(X) = (b - a) * X + a$ one obtains $g(X) \sim U(a, b)$ if $X \sim U(0, 1)$. The uniform distribution serves as a basis for the generation of (pseudo) random numbers in a computer, see Sec. 8.2.1. All distributions can be in some way obtained via transformations from one or several uniform distributions, see Secs. 8.2.2–8.2.5.

Probably the most important continuous distribution in the context of simulations is the Gaussian distribution:

**Definition 8.16** The *Gaussian distribution*, also called *normal distribution*, with real-valued parameters $\mu$ and $\sigma > 0$, describes a random variable $X$ which has the pdf

$$p_X(x) = \frac{1}{\sqrt{2\pi\sigma^2}} \exp\left(-\frac{(x - \mu)^2}{2\sigma^2}\right) \tag{8.36}$$

One writes $X \sim N(\mu, \sigma^2)$.

The Gaussian distribution has expectation value $E[X] = \mu$ and variance $Var[X] = \sigma^2$. A sample plot of the distribution for parameters $\mu = 5$ and $\sigma = 3$ is shown in the left of Fig. 8.3. The Gaussian distribution for $\mu = 0$ and $\sigma = 1$ is called *standard normal distribution* $N(0, 1)$. One can obtain any Gaussian distribution from $X_0 \sim N(0, 1)$ by applying the transformation $g(X_0) = \sigma X_0 + \mu$. Note that the distribution function for the Gaussian distribution cannot be calculated analytically. Thus, one uses usually numerical integration or tabulated values of $N(0, 1)$.

The *central limit theorem* describes how the Gaussian distribution arises from a sum of random variables:

**Theorem 8.1** *Let* $X^{(1)}, X^{(2)}, \ldots, X^{(n)}$ *be independent random variables, which follow all the same distribution exhibiting expectation value $\mu$ and variance $\sigma^2$. Then*

$$X = \sum_{i=1}^{n} X^{(i)} \tag{8.37}$$

*is in the limit of large $n$ approximately Gaussian distributed with mean $n\mu$ and variance $n\sigma^2$, i.e. $X \sim N(n\mu, n\sigma^2)$.*

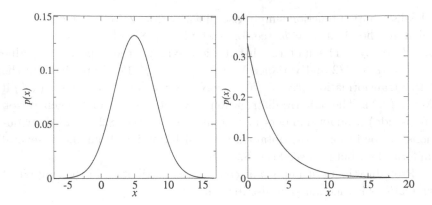

Fig. 8.3 (Left) Probability density function of the Gaussian distribution for parameters $\mu = 5$ and $\sigma = 3$. (Right) Probability density function of the exponential distribution for parameter $\mu = 3$.

*Equivalently, the suitably normalized sum*

$$Z = \frac{\frac{1}{n}\sum_{i=1}^{n} X^{(i)} - \mu}{\sigma/\sqrt{n}} \tag{8.38}$$

*is approximately standard normal distributed $Z \sim N(0,1)$.*

For a proof, please refer to standard text books on probability. Since sums of random processes arise very often in nature, the Gaussian distribution is ubiquitous. For instance, the movement of a "large" particle swimming in a liquid called *Brownian motion* is described by a Gaussian distribution.

Another common probability distribution is the exponential distribution.

**Definition 8.17** The *exponential distribution*, with real-valued parameter $\mu > 0$, describes a random variable $X$ which has the pdf

$$p_X(x) = \begin{cases} 0 & x < 0 \\ \frac{1}{\mu}\exp\left(-x/\mu\right) & x \geq 0 \end{cases} \tag{8.39}$$

A sample plot of the distribution for parameter $\mu = 3$ is shown in the right of Fig. 8.3. The exponential distribution has expectation value $E[X] = \mu$ and variance $Var[X] = \mu^2$. The distribution function can be obtained analytically and is given by

$$F_X(x) = 1 - \exp\left(-x/\mu\right) \tag{8.40}$$

The exponential distribution arises under circumstances where processes happen with certain *rates*, i.e., with a constant probability per time unit. Very often, waiting queues or the decay of radioactive atoms are modeled by such random variables. Then the time duration till the first event (or between two events if the experiment is repeated several times) follows Eq. (8.39).

Next, we discuss a distribution, which has attracted recently [Newman (2003); Newman et al. (2006)] much attention in various disciplines like sociology, physics and computer science. Its probability distribution is a power law:

**Definition 8.18** The *power-law distribution*, also called *Pareto distribution*, with real-valued parameters $\gamma > 0$ and $\kappa > 0$, describes a random variable $X$ which has the pdf

$$p_X(x) = \begin{cases} 0 & x < 1 \\ \frac{\gamma}{\kappa}(x/\kappa)^{-\gamma+1} & x \geq 1 \end{cases} \tag{8.41}$$

A sample power-law distribution is shown in Fig. 8.4. When plotting a power-law distribution with double-logarithmic scale, one sees just a straight line.

A discretized version of the power-law distribution appears for example in empirical social networks. The probability that a person has $x$ "close friends" follows a power-law distribution. The same is observed for computer networks for the probability that a computer is connected to $x$ other computers. The power-law distribution has a finite expectation value only if $\gamma > 1$, i.e. if it falls off quickly enough. In that case one obtains $E[X] = \gamma\kappa/(\gamma-1)$. Similarly, it exhibits a finite variance only for $\gamma > 2$: $Var[X] = \frac{\kappa^2\gamma}{(\gamma-1)^2(\gamma-2)}$. The distribution function can be calculated analytically:

$$F_X(x) = 1 - (x/\kappa)^{-\gamma} \quad (x \geq 1) \tag{8.42}$$

In the context of extreme-value statistics, the Fisher-Tippett distribution (also called log-Weibull distribution) plays an important role.

**Definition 8.19** The *Fisher-Tippett distribution*, with real-valued parameter $\lambda > 0$, describes a random variable $X$ which has the pdf

$$p_X(x) = \lambda e^{-\lambda x} e^{-e^{-\lambda x}} \tag{8.43}$$

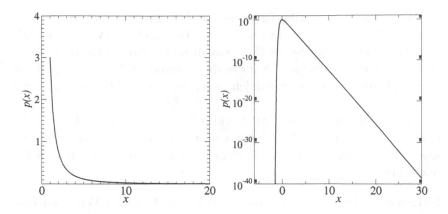

Fig. 8.4    (Left) Probability density function of the power-law distribution for parameters $\gamma = 3$ and $\kappa = 1$. (Right) Probability density function of the Fisher-Tippett distribution for parameter $\lambda = 3$ with logarithmically scaled $y$-axis.

In the special case of $\lambda = 1$, the Fisher-Tippett distribution is also called *Gumbel* distribution.

A sample Fisher-Tippett distribution is shown in the right part of Fig. 8.4. The function exhibits a maximum at $x = 0$. This can be shifted to any value $\mu$ by replacing $x$ by $x - \mu$. The expectation value is $\mathrm{E}[X] = \nu/\lambda$, where $\nu \equiv 0.57721\ldots$ is the *Euler-Mascheroni constant*. The distribution exhibits a variance of $\mathrm{Var}[X] = \frac{\pi}{\sqrt{6}\lambda}$. Also, the distribution function is known analytically:

$$F_X(x) = e^{-e^{-\lambda x}} \tag{8.44}$$

Mathematically, one can obtain a Gumbel ($\lambda = 1$) distributed random variable from $n$ standard normal $N(0,1)$ distributed variables $X^{(i)}$ by taking the maximum of them and performing the limit $n \to \infty$, i.e. $X = \lim_{n\to\infty} \max\left\{X^{(1)}, X^{(2)}, \ldots, X^{(n)}\right\}$. The Gumbel distribution arises by normalizing $X$ to variance 1 and having the maximum probability at $x = 0$. This is also true for some other "well-behaved" random variables like exponential distributed ones, if they are normalized such that the maximum is at $x = 0$ and the variance is one. The Fisher-Tippett distribution can be obtained from the Gumbel distribution via a linear transformation.

For the estimation of confidence intervals (see Secs. 8.3.2 and 8.3.3) one needs the chi-squared distribution and the $F$ distribution, which are presented next for completeness.

**Definition 8.20** The *chi-squared distribution*, with $\nu > 0$ *degrees of freedom* describes a random variable $X$ which has the probability density function (using the *Gamma function* $\Gamma(x) = \int_0^\infty t^{x-1}e^{-t}\,dt$)

$$p_X(x) = \frac{1}{2^{\nu/2}\Gamma(\nu/2)}x^{\frac{\nu-2}{2}}e^{-\frac{x}{2}} \quad (x > 0) \tag{8.45}$$

and $p_X(x) = 0$ for $x \le 0$.

Distribution function, mean and variance are not stated here. A chi-squared distributed random variable can be obtained from a sum of $\nu$ squared standard normal distributed random variables $X_i$: $X = \sum_{i=1}^\nu X_i^2$. The chi-squared distribution is implemented in the *GNU scientific library* (see Sec. 7.3).

**Definition 8.21** The *F distribution*, with $d_1, d_2 > 0$ *degrees of freedom* describes a random variable $X$ which has the pdf

$$p_X(x) = d_1^{d_1/2}d_2^{d_2/2}\frac{\Gamma(d_1/2+d_2/2)}{\Gamma(d_1/2)\Gamma(d_2/2)}\frac{x^{d_1/2-1}}{(d_1x+d_2)^{d_1/2+d_2/2}} \quad (x > 0) \tag{8.46}$$

and $p_X(x) = 0$ for $x \le 0$.

Distribution function, mean and variance are not stated here. An F distributed random variable can be obtained from a chi-squared distributed random variable $Y_1$ with $d_1$ degrees of freedom and a chi-squared distributed random variable $Y_2$ with $d_2$ degrees of freedom via $X = \frac{Y_1/d_1}{Y_2/d_2}$. The F distribution is implemented in the *GNU scientific library* (see Sec. 7.3).

Finally, note that also discrete random variables can be described using probability density functions if one applies the so-called *delta function* $\delta(x - x_0)$. For the purpose of computer simulations this is not necessary. Consequently, no further details are presented here.

## 8.2 Generating (pseudo) random numbers

For many simulations in science, economy or social sciences, random numbers are necessary. Quite often the model itself exhibits random parameters which remain fixed throughout the simulation; one speaks of *quenched disorder*. A famous example in the field of condensed matter physics are *spin glasses*, which are random alloys of magetic and non-magnetic materials. In this case, when one performs simulations of small systems, one has to perform an average over different disorder realizations to obtain physical quantities. Each realization of the disorder consists of randomly chosen positions of the magnetic and non-magnetic particles. To generate a disorder realization within the simulations, random numbers are required.

But even when the simulated system is not inherently random, very often random numbers are required by the algorithms, e.g. to realize a finite-temperature ensemble or when using randomized algorithms. In summary, the application of random numbers in computer simulations is ubiquitous.

In this section an introduction to the generation of random numbers is given. First it is explained how they can be generated at all on a computer. Then, different methods are presented for obtaining numbers which obey a target distribution: the *inversion method*, the *rejection method* and *Box-Müller method*. More comprehensive information about these and similar techniques can be found in Refs. [Morgan (1984); Devroye (1986); Press et al. (1995)]. In this section it is assumed that you are familiar with the basic concepts of probability theory and statistics, as presented in Sec. 8.1.

## 8.2.1 *Uniform (pseudo) random numbers*

First, it should be pointed out that standard computers are deterministic machines. Thus, it is completely impossible to generate true random numbers directly. One could, for example, include interaction with the user. It is, for example, possible to measure the time interval between successive keystrokes, which is randomly distributed by nature. But the resulting time intervals depend heavily on the current user which means the statistical properties cannot be controlled. On the other hand, there are external devices, which have a true random physical process built in and which can be attached to a computer [Qantis; Westphal] or used via the internet [Hotbits]. Nevertheless, since these numbers are really random, they do not allow to perform stochastic simulations in a controlled and reproducible way. This is important in a scientific context, because spectacular or unexpected results are often tried to be reproduced by other research groups. Also, some program bugs turn up only for certain random numbers. Hence, for debugging purposes it is important to be able to run exactly the same simulation again. Furthermore, for the true random numbers, either the speed of random number generation is limited if the true random numbers are cheap, or otherwise the generators are expensive.

This is the reason why *pseudo random numbers* are usually taken. They are generated by deterministic rules. As basis serves a number generator function **rand()** for a uniform distribution. Each time **rand()** is called, a new (pseudo) random number is returned. (Now the "pseudo" is omitted for convenience) These random numbers should "look like" true random numbers and should have many of the properties of them. One says they should be "good". What "look like" and "good" means, has to be specified: One would like to have a random number generator such that each possible number has indeed the same probability of occurrence. Additionally, if two generated numbers $r_i, r_k$ differ only slightly, the random numbers $r_{i+1}, r_{k+1}$ returned by the respective subsequent calls should differ substantially, hence consecutive numbers should have a low correlation. There are many ways to specify a correlation, hence there is no unique criterion. Below, the simplest one will be discussed.

The simplest methods to generate pseudo random numbers are *linear congruential generators*. They generate a sequence $x_1, x_2, \ldots$ of integer numbers between 0 and $m - 1$ by a recursive rule:

$$x_{n+1} = (ax_n + c)\mathrm{mod}\, m. \qquad (8.47)$$

The initial value $x_0$ is called *seed*. Here we show a simple C implementation **lin_con()**. It stores the current number in the local variable x which is declared as **static**, such that it is remembered, even when the function is terminated (see page 42 of Sec. 1.2). There are two arguments. The first argument **set_seed** indicates whether one wants to set a seed. If yes, the new seed should be passed as second argument, otherwise the value of the second argument is ignored. The

| GET SOURCE CODE |
| --- |
| DIR: **randomness** <br> FILE(S): **rng.c** |

function returns the seed if it is changed, or the new random number. Note that the constants $a$ and $c$ are defined inside the function, while the modulus $M$ is implemented via a macro **RNG_MODULUS** to make it visible outside **lin_con()**:

```
1  #define RNG_MODULUS  32768                    /* modulus */
2
3  int lin_con(int set_seed, int seed)
4  {
5    static int x = 1000;        /* current random number */
6    const int a = 12351;                      /* multiplier */
7    const int c = 1;                              /* shift */
8
```

```
9    if(set_seed)                          /* new seed ? */
10     x = seed;
11   else                          /* new random number ? */
12     x = (a*x+c) % RNG_MODULUS;
13
14   return(x);
15 }
```

If you just want to obtain the next random number, you do not care about the seed. Hence, we use for convenience `rn_lin_con()` to call `lin_con()` with the first argument being 0:

```
1 int rand_lin_con()
2 {
3   return(lin_con(0,0));
4 }
```

If we want to set the seed, we also use for convenience a special trivial function `seed_lin_con()`:

```
1 void srand_lin_con(int seed)
2 {
3   lin_con(1, seed);
4 }
```

To generate random numbers $r$ distributed in the interval $[0, 1)$ one has to divide the current random number by the modulus $m$. It is desirable to obtain equally distributed outcomes in the interval, i.e. a uniform distribution. Random numbers generated from this distribution can be used as input to generate random numbers distributed according to other, basically arbitrary, distributions. Below, you will see how random numbers obeying other distributions can be generated. The following simple C function generates random numbers in $[0, 1)$ using the macro `RNG_MODULUS` defined above:

```
1 double drand_lin_con()
2 {
3   return( (double) lin_con(0,0) / RNG_MODULUS);
4 }
```

One has to choose the parameters $a, c, m$ in a way that "good" random numbers are obtained, where "good" means "with less correlations". Note that in the past several results from simulations have been proven to be wrong because of the application of bad random number generators [Ferrenberg et al. (1992); Vattulainen et al. (1994)].

**Example 8.1** To see what "bad generator" means, consider as an example the parameters $a = 12351, c = 1, m = 2^{15}$ and the seed value $I_0 = 1000$. 10000 random numbers are generated by dividing each of them by $m$. They are distributed in the interval $[0, 1)$. In Fig. 8.5 the distribution of the random numbers is shown.

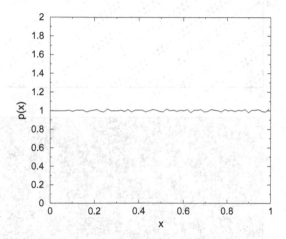

Fig. 8.5 Distribution of random numbers in the interval $[0, 1)$ obtained from converting a histogram into a pdf, see Sec. 8.3.3. The random numbers are generated using a linear congruential generator with the parameters $a = 12351, c = 1, m = 2^{15}$.

The distribution looks rather flat, but by taking a closer look some regularities can be observed. These regularities can be studied by recording $k$-tuples of $k$ successive random numbers $(x_i, x_{i+1}, \ldots, x_{i+k-1})$. A good random number generator, exhibiting no correlations, would fill up the $k$-dimensional space uniformly. Unfortunately, for linear congruential generators, instead the points lie on $(k - 1)$-dimensional planes. It can be shown that there are *at most* of the order $m^{1/k}$ such planes. A bad generator has much fewer planes. This is the case for the example studied above, see top part of Fig. 8.6

The result for $a = 123450$ is even worse: only 15 different "random" numbers are generated (with seed 1000), then the iteration reaches a fixed point (not shown in a figure).

If instead $a = 12349$ is chosen, the two-point correlations look like that shown in the bottom half of Fig. 8.6. Obviously, the behavior is much more

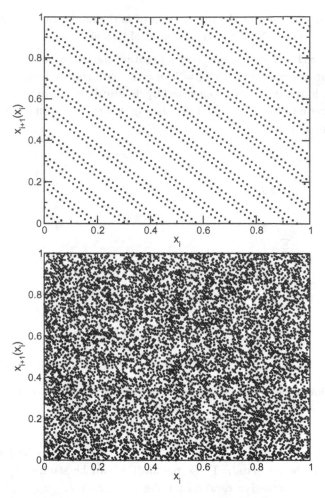

Fig. 8.6   Two point correlations $x_{i+1}(x_i)$ between successive random numbers $x_i, x_{i+1}$. The top case is generated using a linear congruential generator with the parameters $a = 12351, c = 1, m = 2^{15}$, the bottom case has instead $a = 12349$.

irregular, but poor correlations may become visible for higher $k$-tuples.

A generator which has passed several empirical tests is $a = 7^5 = 16807$, $m = 2^{31} - 1$, $c = 0$. When implementing this generator you have to be careful, because during the calculation numbers are generated which do not fit into 32 bit. A clever implementation is presented in Ref. [Press et al. (1995)]. Finally, it should be stressed that this generator, like all linear congruential generators, has the low-order bits much less random than the high-order bits. For that reason, when you want to generate integer numbers in an interval [1,N], you should use

```
r = 1+(int) (N*x_n/m);
```

instead of using the modulo operation as with `r=1+(x_n % N)`.

In standard C, there is a simple built-in random number generator called `rand()` (see corresponding documentation), which has a modulus $m = 2^{15}$, which is very poor. On most operating systems, also `drand48()` is available, which is based on $m = 2^{48}$ ($a = 25214903917$, $c = 11$) and needs also special arithmetics. It is already sufficient for simulations which no not need many random numbers and do not require highest statistical quality. In recent years, several high-standard random number generators have been developed. Several very good ones are included in the freely available *GNU scientific library* (see Sec. 7.3). Hence, you do not have to implement them yourself.

So far, it has been shown how random numbers can be generated which are distributed uniformly in the interval $[0, 1)$. In general, one is interested in obtaining random numbers which are distributed according to a given probability distribution with some density $p(x)$. In the next sections, several techniques suitable for this task are presented.

### 8.2.2 *Discrete random variables*

In case of discrete distributions with finite number of possible outcomes, one can create a table of the possible outcomes together with their probabilities $p_i = p_X(x_i)$ ($i = 1, \ldots, i_{max}$), assuming that the $x_i$ are sorted in ascending order. To draw a number, one has to draw a random number $u$ which is uniformly distributed in $[0, 1)$ and take the entry $j$ of the table such that for the sum $s_j \equiv \sum_{i=1}^{j} p_X(x_i)$ of the probabilities the condition $s_{j-1} < u < s_j$ holds. For example, consider a discrete random variable with $p_1 = 1/8$, $p_2 = 1/4$, $p_3 = 1/2$ and $p_4 = 1/8$. Using this approach, e.g, if the random number is contained in the interval $]1/8, 3/8]$, the second outcome will be selected, see Fig. 8.7. Note that one can search the array quickly by *bisection search*: The array is iteratively divided into two halves and each time continued in that half where the corresponding entry $j$ is contained. In this way, generating a random number has a time complexity which grows only logarithmically with the number $i_{max}$ of possible outcomes. This pays off if the number of possible outcomes is very large.

In exercise (1) you are asked to write a function to sample from the

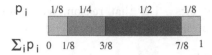

Fig. 8.7 A discrete distribution with four outcomes with probabilities $p_1 = 1/8$, $p_2 = 1/4$, $p_3 = 1/2$ and $p_4 = 1/8$. The probabilities are represented in the interval $[0, 1]$ by sub intervals which have lengths equal to the probabilities, respectively. This allows to draw random numbers according the distribution.

probability distribution of a discrete variable, in particular for a Poisson distribution.

Although just a logarithmic growth with the number $N$ of possible outcomes is already pretty efficient, there even exist an approach [Walker (1977)] where the time to draw a random number is constant, independent of $N$. This is explained next, following the implementation [Fukui and Todo (2009)]:

We will consider as an example $N = 5$ possible outcomes with $p_1 = 0.1$, $p_2 = 0.25$, $p_3 = 0.45$, $p_4 = 0.1$, $p_5 = 0.1$. The basic idea is to distribute the different outcomes to $N$ "packages". Each package represents a weight $p_{\text{avg}} = 1/N$. The main point is that each package contains only up to two outcomes. Outcomes with small probabilities $p_i \leq p_{\text{avg}}$ will be represented in just one package, while outcomes with higher probabilities will be contained in more than one package. For each package $i$ one stores which outcomes $a_i$ (and maybe also $b_i$) are represented and what fraction $q_i$ belongs to outcome $a_i$ (outcome $b_i$ correspond to the fraction $1 - q_i$). This information is stored in a table. Once the table is set up, drawing a random outcome is now easy: First one selects randomly the package $i$, each with the same probability $1/N$. For this purpose one $U(0, 1)$ uniformly distributed random number is required. Next a second $U(0, 1)$ random number $r$ is drawn. If $r < q_i$ the outcome $a_i$ selected else $b_i$.

One only has to set up the table before the actual simulation starts. This works as follows: One starts by assigning event $i$ in package $i$ by setting $a_i = i$ and $q_i = p_i$. During the computation of the table $q_i$ states how much of the probability for outcome $a_i$ is still represented in package $i$. For the example, this initial situation is shown in Fig. 8.8. Next, the outcomes are partially sorted such that all outcomes with $p_i > p_{\text{avg}}$ come first (to the left), the other outcomes ($p_i \leq p_{\text{avg}}$) come next, to the right. The partial order can be generated conveniently by using two index counters $t_0$ and $t_1$ which are put initially to the left at package 1 (entry 0 for a C array) and to the right at package $N$ (entry $N - 1$ for a C array). These counters

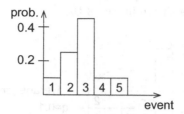

Fig. 8.8 The initial representation of the table for the example, just containing the probabilities of the outcomes. The heights of the bars correspond to the probabilities $q_i$, the numbers written in the bars to the outcomes $a_i$.

are moved stepwise right and left, respectively, until each them points to an elements such that the partial order is not fulfilled, i.e. $q_{t_0} > p_{avg}$ and $q_{t_1} < p_{avg}$. Those two elements are exchanged, hence the partial order is restored and the pointers can move on. This is repeated until the two index counters have passed each other, i.e. until $t_0 > t_1$ holds. After that $t_1$ will point to the rightmost outcome where the probability is larger than $p_{avg}$. For our example, the situation is shown in Fig. 8.9.

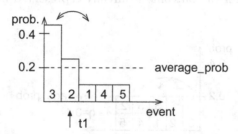

Fig. 8.9 After the table has been rearranged, all outcomes with $q_i > 1/N$ are located to the left, all other outcomes to the right.

Now the final phase of the table setup starts. Starting from the right, the packages are filled up such that each one represents a probability $p_{avg}$. For this purpose a corresponding share will be taken from package $t_1$, i.e. $q_{t_1}$ will be reduced by an amount $(p_{avg} - q_t)$ and $b_t = a_{t_1}$ is assigned.

Since $q_{t1} > p_{avg}$, the amount remaining in package $t_1$ will be positive, but it might now be smaller than $p_{avg}$. In this case, the package from now on belongs to those which have to be filled up. Thus, pointer $t_1$ will be moved one position to the left, i.e., $t_1 = t_1 - 1$ and then again point to a package representing an amount larger than $p_{avg}$. For the example, the situation will look like shown in Fig. 8.10.

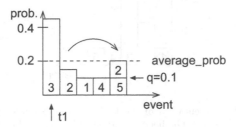

Fig. 8.10  Situation after the rightmost package $t$ has been filled up to an amount $p_{avg} = 1/N$ by taking the missing amount from the package that $t_1$ pointed to. The corresponding outcome is stored in $b_t$, represented as number "2" on the top of the rightmost package. Since the amount in package $t_1$ has fallen below $p_{avg} = 1/N$, $t_1$ is moved one position to the left.

This is repeated until all packages have been filled up to level $p_{avg} = 1/N$. This results in the situation shown in Fig. 8.11 (with $q_1 = 0.2$, $q_2 = 0.15$ and $q_3 = q_4 = q_5 = 0.1$). Note that the leftmost package will always just contain the outcome it already represented at the beginning, hence $q_1 = 1/N$.

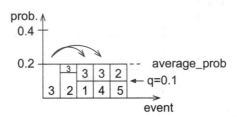

Fig. 8.11  Result for the table after all packages have been arranged to represent an amount $p_{avg} = 1/N$.

So far, the values of $q_i$ represent global probabilities for the outcomes $b_i$. To allow for drawing the random numbers as explained above, they have to be turned into conditional probabilities, representing only the relative fraction of outcome $a_i$ in package $i$. Since each package represents an amount $p_{avg} = 1/N$, due to Eq. (8.5) all values of $q_i$ have to be divided by $p_{avg}$, i.e., multiplied by $N$.

You are asked to implement this approach in exercise (2). Although drawing random numbers using this approach takes only a constant amount of computation steps, it requires generating two (pseudo) random numbers each time. Thus, the implementation is more efficient in comparison to the straightforward approach, see the beginning of this section, only for large numbers $N$ of possible outcomes.

Since all possible discrete distributions can be generated using these approaches, we concentrate on techniques for generating continuous random variables in the following.

### 8.2.3 Inversion Method

Given is a random number generator `drand()` which is assumed to generate random numbers $U$ which are distributed uniformly in $[0, 1)$. The aim is to generate random numbers $Z$ with probability density $p_Z(z)$. The corresponding distribution function is

$$F_Z(z) \equiv P(Z \leq z) \equiv \int_{-\infty}^{z} dz' p_Z(z') \tag{8.48}$$

The target is to find a function $g(u)$, such that after the transformation $Z = g(U)$ the outcomes of $Z$ are distributed according to (8.48). It is assumed that $g$ can be inverted and is strongly monotonically increasing. Then one obtains

$$F_Z(z) = P(Z \leq z) = P(g(U) \leq z) = P(U \leq g^{-1}(z)) \tag{8.49}$$

Since the distribution function $F_U(u) = P(U \leq u)$ for a uniformly distributed variable is just $F_U(u) = u$ ($u \in [0, 1]$), one obtains $F_Z(z) = g^{-1}(z)$. Thus, one just has to choose $g(z) = F_Z^{-1}(z)$ for the transformation function in order to obtain random numbers, which are distributed according to the probability distribution $F_Z(z)$. Of course, this only works if $F_Z$ can be inverted. If this is not possible, you may use the methods presented in the subsequent sections, or you could generate a table of the distribution function, which is in fact a discretized approximation of the distribution function, and use the methods for generating discrete random numbers as shown in Sec. 8.2.2. This can be even refined by using a linearized approximation of the distribution function. Here, we do not go into further details, but present an example where the distribution function can be indeed inverted.

**Example 8.2**   Let us consider the exponential distribution with parameter $\mu$, with distribution function $F_Z(z) = 1 - \exp(-z/\mu)$, see page

258.   Therefore, one can obtain exponentially distributed random numbers $Z$ by generating uniform distributed random numbers $u$ and choosing $z = -\mu \ln(1 - u)$.

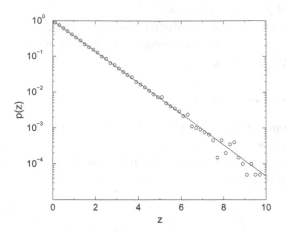

Fig. 8.12   Histogram pdf (see page 293) of random numbers generated according to an exponential distribution ($\mu = 1$) compared with the probability density function (straight line) in a logarithmic plot.

The following simple C function generates a random number which is exponentially distributed. The parameter $\mu$ of the distribution is passed as argument.

| GET SOURCE CODE |
| --- |
| DIR: random<br>FILE(S): expo.c |

```
1  double rand_expo(double mu)
2  {
3    double randnum;              /* random number U(0,1) */
4    randnum = drand48();
5
6    return(-mu*log(1-randnum));
7  }
```

Note that we use in line 4 the simple `drand48()` random number generator, which is included in the C standard library and works well for applications with moderate statistical requirements. For more sophisticated generates, e.g. see the *GNU scientific library* (see Sec. 7.3).

In Fig. 8.12 a histogram pdf (see page 293) for $10^5$ random numbers generated in this way and the exponential probability function for $\mu = 1$ are shown with a logarithmically scaled $y$-axis. Only for larger values are deviations visible. They are due to statistical fluctuations since $p_Z(z)$ is very small there.

### 8.2.4 Rejection Method

As mentioned above, the inversion method works only when the distribution function $P$ can be inverted analytically. For distributions not fulfilling this condition, sometimes this problem can be overcome by drawing several random numbers and combining them in a clever way.

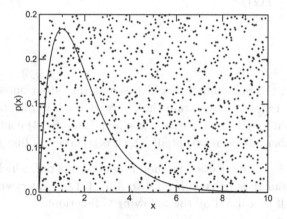

Fig. 8.13 The rejection method: Points $(x, y)$ are scattered uniformly over a bounded rectangle. The probability that $y \leq p(x)$ is proportional to $p(x)$.

First the *simple rejection method* is presented which works for random variables where the pdf $p(x)$ fits into a box $[x_0, x_1] \times [0, y_{\max})$, i.e., $p(x) = 0$ for $x \notin [x_0, x_1]$ and $p(x) \leq y_{\max}$. A generalization to pdfs which cannot be boxed is given below. For the simple approach, the basic idea of generating a random number distributed according to $p(x)$ is to generate random pairs $(x, y)$, which are distributed uniformly in $[x_0, x_1] \times [0, y_{\max}]$ and accept only those numbers $x$ where $y \leq p(x)$ holds, i.e., the pairs which are located below $p(x)$, see Fig. 8.13. Therefore, the probability that $x$ is drawn is proportional to $p(x)$, as desired.

The following C function realizes the rejection method for an arbitrary pdf. It takes as arguments the boundaries of the box `y_max`, `x0` and `x1` as well as a pointer `pdf` to the function

| GET SOURCE CODE |
| --- |
| DIR: **randomness** <br> FILE(S): `reject.c` |

realizing the pdf. For an explanation of function pointers, see page 59.

```
1  double reject(double y_max, double x0, double x1,
2                 double (* pdf)(double))
3  {
4    int found;                    /* flag if valid number has been found */
5    double x,y;                   /* random points in [x0,x1]x[0,p_max] */
6    found = 0;
7
8    while(!found)                 /* loop until number is generated */
9    {
10     x = x0 + (x1-x0)*drand48();          /* uniformly on [x0,x1] */
11     y = y_max *drand48();                /* uniformly in [0,p_max] */
12     if(y <= pdf(x))                           /* accept ? */
13       found = 1;
14   }
15   return(x);
16 }
```

In lines 10–11 the random point, which is uniformly distributed in the box, is generated. Lines 12–13 contain the check whether a point below the pdf curve has been found. The search in the loop (lines 8–14) continues until a random number has been accepted, which is returned in line 15.

**Example 8.3** The rejection method is applied to a pdf, which has density 1 in $[0, 0.5)$ and rises linearly from 0 to 4 in $[1, 1.5)$. Everywhere else it is zero. This pdf is realized by the following C function:

```
1  double pdf(double x)
2  {
3    if( (x<0)||
4        ((x>=0.5)&&(x<1))||
5        (x>1.5) )
6      return(0.0);
7    else if((x>=0)&&(x<0.5))
8      return(1.0);
9    else
10     return(4.0*(x-1));
11 }
```

The resulting empirical histogram pdf is shown in Fig. 8.14.

The simple rejection method can always be applied if the probability density is boxed, but it has the drawback that more random numbers have to be generated than can be used: If $A = (x_1 - x_0)y_{max}$ is the area of the box, one has on average to generate $2A$ auxiliary random numbers to obtain one random number of the desired distribution. If this leads to a

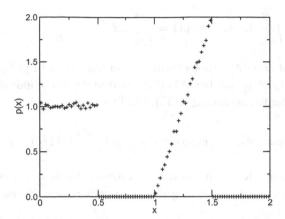

Fig. 8.14   Histogram pdf (see page 293) of $10^5$ random numbers generated using the rejection method for an artificial pdf.

very poor efficiency, as a simple solution you can consider to use several boxes for different parts of the pdf.

It is even better not to border the pdf $p(x)$ by a box but by a shape which resembles $p(x)$ as close as possible. Again, the only requirement is that one can distribute points uniformly within the shape. This is always possible if the shape is given by a function $cq(x)$ and $q(x)$ is another pdf for which the generation of random numbers can be done easily, e.g. using the inversion method. The bordering condition $p(x) \leq cq(x)$ must be fulfilled for the entire support of the pdf $p(x)$. Note that this approach also allows in many cases to generate random numbers for pdfs where the support stretches to infinity. In detail, the approach works as follows: First a random number $x$ according the pdf $q(x)$ is generated. Next, a random number $y$ is generated, which is uniformly distributed in $[0, cq(x)]$, i.e. via drawing a number uniformly in $[0, 1]$ and multiplying it by $cq(x)$. Finally, the number $x$ is accepted if $y < p(x)$. Thus, the probability for the acceptance is given by $p(x)/(cq(x))$. Since the probability density for the generation of $x$ in the first step is $q(x)$, the joint probability that $x$ is generated and accepted is

$$p(x, \text{accept}) = \frac{p(x)}{cq(x)} q(x) = \frac{p(x)}{c} .$$

Thus, the probability is proportional to $p(x)$. Again, the numbers $x$ which are not accepted are just ignored, i.e. the process is repeated until a number is accepted. The probability that any number $x$ is accepted is simple the integral over $p(x, \text{accept})$, i.e.

$$p_{\text{accept}} = \int_{-\infty}^{\infty} dx\, p(x, \text{accept}) = \int_{-\infty}^{\infty} dx\, \frac{p(x)}{c} = \frac{1}{c} \int_{-\infty}^{\infty} dx\, p(x) = \frac{1}{c},$$

where the last equality follows from the normalization of pdf $p(x)$. Finally, the probability that number $x$ is generated under the condition that it is accepted evaluates according to Eq. (8.5) as

$$p(x|\text{accept}) = p(x, \text{accept})/p_{\text{accept}} = \frac{p(x)}{c}/(1/c) = p(x),$$

as desired. Since for each iteration 2 random numbers $x$ and $y$ are generated, the average number of generated numbers per accepted number is $2/p_{\text{accept}} = 2c$.

As an example, we will consider the Gaussian distribution as shown in Eq. (8.36), here for the choices of mean $\mu = 0$ and variance $\sigma = 1$. The generalization to arbitrary values of $\mu$ and $\sigma > 0$ is explained in Sec. 8.2.5, where also a rejection-free approach specifically for Gaussian numbers is presented. Here, we consider the case that a positive number $x$ according the Gaussian distribution is to be generated. For the general case, due to the symmetry of the Gaussian distribution, one finally draws a uniformly in $[0, 1]$ distributed random number $s$ and negates $x$ if $s < 1/2$.

To border the Gaussian, we use for $q(x)$ an exponential distribution with parameter $\mu$, the density given by Eq. (8.39). The value of the parameter $\mu$ will be determined below. The exponential was chosen because the generation of exponentially distributed random numbers is particularly simple using the rejection method as shown in Sec. 8.2.3. The multiplier $c$ must be chosen such that the pdf of the Gaussian distribution lies completely below $cq(x)$, see Fig. 8.15. To maximize the efficiency, $c$ should be made as small as possible. From inspection of Fig. 8.15 we read off that the best choice of $c$ is such that $p(x)$ and $cq(x)$ touch exactly in one point $x$ (and neither cross in two points nor do not touch at all). This leads to the following condition:

$$p(x) = cq(x)$$

$$\Rightarrow \quad \frac{1}{\sqrt{2\pi}} \exp\left(-x^2/2\right) = \frac{c}{\mu} \exp\left(-x/\mu\right)$$

$$\Rightarrow \quad -x^2/2 = \log(\sqrt{2\pi}c/\mu) - x/\mu$$

$$\Rightarrow \quad (x - 1/\mu)^2 = 1/\mu^2 - 2\log(\sqrt{2\pi}c/\mu)$$

$$\Rightarrow \quad x = 1/\mu \pm \sqrt{1/\mu^2 - 2\log(\sqrt{2\pi}c/\mu)}$$

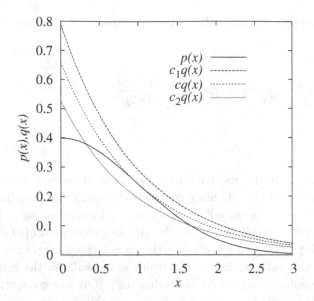

Fig. 8.15 The pdf $p(x)$ of the Gaussian distribution Eq. (8.36) together with a bordering exponential distribution $cq(x)$, where $q(x)$ is given by Eq. (8.39). The constant $c$ is chosen such that $p(x)$ and $cq(x)$ touch in one point, thus $p(x)$ is bordered by $cq(x)$ and the area under $cq(x)$ is minimized, see text. For a larger constant $c_1 > c$ $p(x)$ is also bordered by $c_1 q(x)$ but the area is larger, thus the method is less efficient. For $c_2 < c$, $p(x)$ is not bordered by $c_2 q(x)$, hence the approach does not work for this case.

Such that there is exactly one point where $p(x)$ and $cq(x)$ agree, the expression under the square root must be identically zero,[1] i.e.

$$1/\mu^2 = 2\log(\sqrt{2\pi}c/\mu)$$
$$\Rightarrow \quad \exp(1/(2\mu^2)) = \sqrt{2\pi}c/\mu$$
$$\Rightarrow \quad c = \frac{\mu}{\sqrt{2\pi}}\exp(1/(2\mu^2)). \qquad (8.50)$$

Still, $c$ depends on $\mu$, which is the parameter of the exponential. Since the efficiency of the random number generation is maximal if the area under

---

[1] If the expression is larger than zero, there are two distinct intersections. If the expression is smaller than zero, there is no intersection, i.e., the scaled exponential lies above the Gaussian.

the function $cq(x)$ is minimal, we have to minimize $c$ with respect to $\mu$, i.e.

$$\frac{dc}{d\mu} \stackrel{!}{=} 0$$

$$\Rightarrow \quad \frac{1}{\sqrt{2\pi}} \exp(1/(2\mu^2)) + \frac{\mu}{\sqrt{2\pi}} \frac{-2}{2\mu^3} \exp(1/(2\mu^2)) = 0$$

$$\Rightarrow \quad 1 - \frac{1}{\mu^2} = 0$$

$$\Rightarrow \quad \mu = \pm 1 \,.$$

Since $\mu$ is restricted to positive values (otherwise the exponential is not normalized), we obtain $\mu = 1$. Since $c(\mu)$ tends to infinity for both limits $\mu \to 0$ and $\mu \to \infty$, $\mu = 1$ is indeed the optimum choice minimizing $c$. Hence, for the bordering function we obtain the simple exponential $q(x) = \exp(-x)$ and from Eq. (8.50) we get via inserting $\mu = 1$ the constant $c = \sqrt{e/(2\pi)}$. Note that this calculation and the result are specific to the generation of Gaussian random numbers via bordering the pdf by an exponential. Nevertheless, the general approach of finding a suitable function $q(x)$ is similar: One has to choose a function $q(x)$ such that random numbers can be easily generated. Also one has to determine the constant $c$ such that $cq(x)$ borders the pdf as close as possible. Usually this involves minimizing $c$ with respect to one or several parameters of $q(x)$.

To illustrate the results, in Fig. 8.16, the Gaussian is shown together with the optimum bordering function $\exp(1/2 - x)/\sqrt{2\pi}$ and some points uniformly distributed under the the bordering function. In Exercise (4) you are asked to implement a C function which realizes the rejection method for the Gaussian.

The rejection method can also be generalized to higher dimensions. As example, here the generation of random numbers is considered which are distributed uniformly in a hypersphere or on it's surface. Therefore, we consider a circle in two dimensions or a sphere in three dimensions. This is useful, e.g, when studying random movements in real space, like in *random walk* models. One can, in principle, also use the inversion method via suitable coordinate transformations. This is rather simple in two dimensions, slightly more difficult in three dimensions, but can become quite cumbersome in even higher dimensions. In contrast, the present approach is very simple and suitable for all dimensions $d$. Here we consider the unit hypersphere, i.e., with radius one. The basic idea is to generate uniform vectors $\vec{x} = (x_1, x_2, \ldots, x_d)$ in the hypercube $[-1, 1]^d$, calculate the squared length

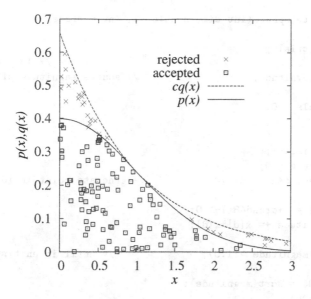

Fig. 8.16 The pdf $p(x)$ of the standard Gaussian for the positive $x$-axis together with it's bordering function $cq(x) = \exp(1/2 - x)/\sqrt{2\pi}$ and 100 points $(x, y)$ which are scattered with uniform probability under the bordering function. Points below the curve $p(x)$ are accepted, i.e., the $x$-coordinate will be taken as sample result.

$|\vec{x}|^2 = \sum_i x_i^2$ and reject all vectors where $|\vec{x}|^2 > 1$. Thus, the hypersphere is "cut out" of the hypercube. Since the distribution in the hypercube is uniform, so is the distribution in the hypersphere. If you aim at a uniform distribution on the surface of the hypersphere, you can project all values inside the sphere onto the surface, by taking $\vec{x}' = \vec{x}/|\vec{x}|$, among those which are not rejected.

Below, a C function generating random numbers in or on the surface of the hypersphere is
shown. It takes the dimension of the system,

| GET SOURCE CODE |
| --- |
| DIR: **randomness** |
| FILE(S): **sphere.c** |

the vector $\vec{x}$ to be generated (i.e., a pointer) and a flag **surface** as arguments. When **surface** is 0, the points are generated inside the hypersphere,
else on the surface of the hypersphere. The function returns the number of
trials needed to generate the returned vector.[2]

---

[2]Note that this can be used to numerically estimate the value of $\pi$: In two dimensions, the area of the unit circle is $\pi$, while the area of the square is 4. Thus, averaged over many calls, the fraction of successful trials should converge to $\pi/4 \approx= 0.7854$.

```
1   int generate_sphere(int dim, double *x, int surface)
2   {
3     int num_trials;
4     int d;                                /* loop counter */
5     double magnitude;                     /* square magnitude of vector */
6
7     num_trials = 0;
8     do
9     {
10      num_trials++;
11      magnitude = 0;
12      for(d=0; d<dim; d++)                /* generate point in [0,1]^dim */
13      {
14        x[d] = 2*drand48()-1.0;
15        magnitude += x[d]*x[d];
16      }
17    } while(magnitude > 1.0);             /* until in unit sphere ? */
18
19    magnitude = sqrt(magnitude);
20    if(surface)
21      for(d=0; d<dim; d++)                /* normalize point to length 1 */
22        x[d] /= magnitude;
23    return(num_trials);
24  }
```

The main loop lines 8–17 is performed until a point $\vec{x}$ inside the hypersphere is obtained. Therein, the loop (lines 12–16) generates a uniformly distributed point inside the hypercube $[-1, 1]^d$ and calculates the length of the vector on the fly. In the final part (lines 19–22) the projection onto the surface of the hypersphere is performed in case the flag surface is set.

As numerical experiment, 1000 points were generated inside the unit circle and 200 points on it's circumference. The resulting scatter plot of the points is shown in Fig. 8.17.

### 8.2.5   *The Gaussian Distribution*

In case neither the distribution function can be inverted nor the probability fits into a box, special methods have to be applied. As an example, a method for generating random numbers distributed according to a Gaussian distribution is considered. Other methods and examples of how different techniques can be combined are collected in [Morgan (1984)].

The probability density function for the Gaussian distribution with mean $\mu$ and variance $\sigma^2$ is shown in Eq. (8.36), see also Fig. 8.18. It is,

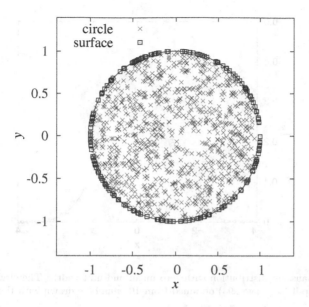

Fig. 8.17   Distribution of 1000 randomly generated points inside a circle and 200 points on its circumference.

apart from uniform distributions, the most common distribution occurring in simulations.

Here, the case of a standard Gaussian distribution ($\mu = 0$, $\sigma = 1$) is considered. If you want to realize the general case, you have to draw a standard Gaussian distributed number $z$ and then use $\sigma z + \mu$ which is distributed as desired.

Since the Gaussian distribution extends over an infinite interval and because the distribution function cannot be inverted, the methods from above are not applicable. The simplest technique to generate random numbers distributed according to a Gaussian distribution makes use of the central limit theorem 8.1. It tells us that any sum of $K$ independently distributed random variables $U_i$ (with mean $\mu$ and variance $v$) will converge to a Gaussian distribution with mean $K\mu$ and variance $Kv$. If again $U_i$ is taken to be uniformly distributed in $[0, 1)$ (which has mean $\mu = 0.5$ and variance $v = 1/12$), one can choose $K = 12$ and the random variable $Z = \sum_{i=1}^{K} U_i - 6$ will be distributed approximately according to a standard Gaussian distribution. The drawbacks of this method are that 12 random numbers are needed to generate one final random number and that numbers larger than 6 or smaller than -6 will never appear.

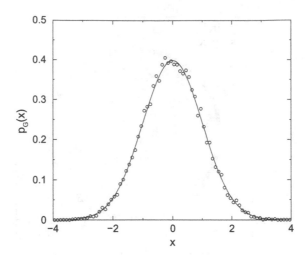

Fig. 8.18 Gaussian distribution with zero mean and unit width. The circles represent a histogram pdf (see page 293) obtained from $10^4$ numbers drawn with the Box-Müller method.

In contrast to this technique the *Box-Müller method* is exact. You need two random variables $U_1, U_2$ uniformly distributed in $[0, 1)$ to generate two independent Gaussian variables $N_1, N_2$. This can be achieved by generating $u_1, u_2$ from $U_1, U_2$ and assigning

$$n_1 = \sqrt{-2\log(1 - u_1)} \cos(2\pi u_2)$$
$$n_2 = \sqrt{-2\log(1 - u_1)} \sin(2\pi u_2)$$

A proof that $n_1$ and $n_2$ are indeed distributed according to (8.36) works as follows: Let us write $n_1, n_2$ in polar coordinates $(r, \theta)$, i.e., $(r, \theta) = f(n_1, n_2)$. Thus, the inverse is:

$$n_1 = r\cos(\theta)$$
$$n_2 = r\sin(\theta). \tag{8.51}$$

Now we have to find the probability density for $(r, \theta)$.

In general, for any two random variables $R, \Theta$ connected to two other random variables $N_1, N_2$ via the transformation $(R, \Theta) = f(N_1, N_2)$, $p_{N_1, N_2}$ being the (joint) pdf for $(N_1, N_1)$, we have

$$p_{R,\Theta}(r,\theta) = p_{N_1,N_2}(f^{-1}(r,\theta))|\mathbf{J}^{-1}|, \qquad (8.52)$$

where $|\mathbf{J}^{-1}|$ ist the Jacobi determinant of the inverse transformation. Therefore, we have to calculate the Jacobi determinant of the inverse transformation given in Eq. (8.51), for which we obtain:

$$|\mathbf{J}^{-1}| = \begin{vmatrix} \frac{\partial n_1}{\partial r} & \frac{\partial n_1}{\partial \theta} \\ \frac{\partial n_2}{\partial r} & \frac{\partial n_2}{\partial \theta} \end{vmatrix} = \begin{vmatrix} \cos(\theta) & -r\sin(\theta) \\ \sin(\theta) & r\cos(\theta) \end{vmatrix} = r\cos^2(\theta) + r\sin^2(\theta) = r.$$

Since $n_1, n_2$ should be Gaussian distributed we obtain from Eq. (8.52)

$$p_{R,\Theta}(r,\theta) = \frac{r}{2\pi}e^{-n_1^2/2 - n_2^2/2} = \frac{r}{2\pi}e^{-r^2/2}. \qquad (8.53)$$

This factorizes. We assume that $\theta$ is uniformly distributed in $[0, 2\pi)$, i.e., with density $1/2\pi$ (generated by setting $\theta = 2\pi u_2$). It remains $p_R(r) = re^{-r^2/2}$. On can easily see that the distribution function of this pdf is $F_R(r) = 1 - \exp(-r^2/2)$ $(r \geq 0)$. You can check this by calculating the derivative of $F_R(r)$. Thus, one can use the inversion method to generate number distributed according the pdf $p_R(r)$. For this we obtain $r = \sqrt{-2\log(1 - u_1)}$. The rules for generating $\theta$ and $r$, inserted into the inverse transformation Eq. (8.51), result exactly in the formulas given by the Box-Muller approach.

There exist other methods for generating Gaussian random numbers, some even more efficient, see Refs. [Press et al. (1995); Morgan (1984)]. A method which is based on the simulation of particles in a box is explained in [Fernandez and Criado (1999)]. In Fig. 8.18 a histogram pdf of $10^4$ random numbers drawn with the Box-Müller method is shown. Note that you can find an implementation of the Box-Müller method in the solution of Exercise (5).

## 8.3 Basic data analysis

The starting point is a *sample* of $n$ measured points $\{x_0, x_1, \ldots, x_{n-1}\}$ of some quantity, as obtained from a simulation. Examples are the density of a gas, the transition time between two conformations of a molecule, or the price of a stock. We assume that formally all measurements can be described by random variables $X_i$ representing the same random variable $X$ and that all measurements are statistically independent of each other (treating statistical dependencies is treated in Sec. 8.5). Usually, one does not know the underlying probability distribution $F(x)$, having density $p(x)$, which describes $X$.

### 8.3.1   *Estimators*

Thus, one wants to obtain information about $X$ by looking at the sample $\{x_0, x_1, \ldots, x_{n-1}\}$. In principle, one does this by considering *estimators* $h = h(x_0, x_1, \ldots, x_{n-1})$. Since the measured points are obtained from random variables, $H = h(X_0, X_1, \ldots, X_{n-1})$ is a random variable itself. Estimators are often used to estimate parameters $\theta$ of random variables, e.g. moments of distributions. The most fundamental estimators are:

- The *mean*

$$\overline{x} \equiv \frac{1}{n} \sum_{i=0}^{n-1} x_i \qquad (8.54)$$

- The *sample variance*

$$s^2 \equiv \frac{1}{n} \sum_{i=0}^{n-1} (x_i - \overline{x})^2 \qquad (8.55)$$

The sample standard deviation is $s \equiv \sqrt{s^2}$.

As example, next a simple C function is shown, which calculates the mean of $n$ data points. The function obtains the number $n$ of data points and an array containing the data as arguments. It returns the average:

| GET SOURCE CODE |
| --- |
| DIR: **randomness** |
| FILE(S): **mean.c** |

```
1   double mean(int n, double *x)
2   {
3       double sum = 0.0;                        /* sum of values */
4       int i;                                   /* counter */
5
6       for(i=0; i<n; i++)            /* loop over all data points */
7           sum += x[i];
8       return(sum/n);
9   }
10
```

You are asked to write a similar function for calculating the variance in exercise (5).

The sample mean can be used to estimate the expectation value $\mu \equiv E[X]$ of the distribution. This estimate is *unbiased*, which means that the expectation value of the mean, for any sample sizes $n$, is indeed the expectation value of the random variable. This can be shown quite easily. Note that formally the random variable from which the sample mean $\bar{x}$ is drawn is $\overline{X} = \frac{1}{n}\sum_{i=0}^{n-1} X_i$:

$$\mu_{\overline{X}} \equiv E[\overline{X}] = E\left[\frac{1}{n}\sum_{i=0}^{n-1} X_i\right] = \frac{1}{n}\sum_{i=0}^{n-1} E[X_i] = \frac{1}{n}n\,E[X] = E[X] = \mu \quad (8.56)$$

Here again the linearity of the expectation value was used. The fact that the estimator is unbiased means that if you repeat the estimation of the expectation value via the mean several times, on average the correct value is obtained. This is independent of the sample size. In general, the estimator $h$ for a parameter $\theta$ is called unbiased if $E[h] = \theta$.

Contrary to what you might expect due to the symmetry between Eqs. (8.16) and (8.55), the sample variance is *not* an unbiased estimator for the variance $\sigma^2 \equiv Var[X]$ of the distribution, but is *biased*. The fundamental reason is, as mentioned above, that $\overline{X}$ is itself a random variable which is described by a distribution $P_{\overline{X}}$. As shown in Eq. (8.56), this distribution has mean $\mu$, independent of the sample size. On the other hand, the distribution has the variance

$$\sigma_{\overline{X}}^2 \equiv Var[\overline{X}] = Var\left[\frac{1}{n}\sum_{i=0}^{n-1} X_i\right] \overset{(8.23)}{=} \frac{1}{n^2}\sum_{i=0}^{n-1} Var[X_i]$$

$$= \frac{1}{n^2}n\,Var[X] = \frac{\sigma^2}{n} \quad (8.57)$$

Thus, the distribution of $\overline{X}$ gets narrower with increasing sample size $n$. This has the following consequence for the expectation value of the sample variance which is described by the random variable $S^2 = \frac{1}{n}\sum_{i=0}^{n-1}(X_i - \overline{X})^2$:

$$
\begin{aligned}
E[S^2] &= E\left[\frac{1}{n}\sum_{i=0}^{n-1}(X_i - \overline{X})^2\right] = E\left[\frac{1}{n}\sum_{i=0}^{n-1}(X_i^2 - 2X_i\overline{X} + \overline{X}^2)\right] \\
&= \frac{1}{n}\left(\sum_{i=0}^{n-1}E[X_i^2] - nE[\overline{X}^2]\right) \overset{(8.22)}{=} \frac{1}{n}\left(n(\sigma^2 + \mu^2) - n(\sigma_{\overline{X}}^2 + \mu_{\overline{X}}^2)\right) \\
&\overset{(8.57)}{=} \frac{1}{n}\left(n\sigma^2 + n\mu^2 - n\frac{\sigma^2}{n} - n\mu^2\right) = \frac{n-1}{n}\sigma^2
\end{aligned}
\tag{8.58}
$$

This means that, although $s^2$ is biased, $\frac{n}{n-1}s^2$ is an unbiased estimator for the variance of the underlying distribution of $X$. Nevertheless, $s^2$ also becomes unbiased for $n \to \infty$.[3]

For some distributions, for instance a power-law distribution Eq. (8.41) with exponent $\gamma \le 2$, the variance does not exist. Numerically, when calculating $s^2$ according Eq. (8.55), one observes that it will not converge to a finite value when increasing the sample size $n$. Instead one will observe occasionally jumps to higher and higher values. One says the estimator is *not robust*. To get still an impression of the spread of the data points, one can instead calculate the *average deviation*

$$
D \equiv \frac{1}{n}\sum_{i=0}^{n-1}|x_i - \overline{x}|
\tag{8.59}
$$

In general, an estimator is the less robust, the higher the involved moments are. Even the sample mean may not be robust, for instance for a power-law distribution with $\gamma \le 1$. In this case one can use the *sample median*, which is the value $x_m$ such that $x_i \le x_m$ for half the sample points, i.e. $x_m$ is the $(n+1)/2$'th sample point if they are sorted in ascending order.[4] The sample median is clearly an estimator of the median (see Def. 8.14). It is more robust, because it is less influenced by the sample points in the tail. The simplest way to calculate the median is to sort all sample points in ascending order and take the sample point at the $(n/2+1)$'th position.

---

[3]Sometimes the sample variance is defined as $S^* = \frac{1}{n-1}\sum_{i=0}^{n-1}(x_i - \overline{x})^2$ to make it an unbiased estimator of the variance.

[4]If $n$ is even, one can take the average between the $n/2$'th and the $(n+1)/2$'th sample point in ascending order.

This process takes a running time $\mathcal{O}(n \log n)$. Nevertheless, there is an algorithm [Press et al. (1995); Cormen et al. (2001)] which calculates the median even in linear running time $\mathcal{O}(n)$.

### 8.3.2 Confidence intervals

In the previous section, we have studied estimators for parameters of a random variable $X$ using a sample obtained from a series of independent random experiments. This is a so-called *point estimator*, because just one number is estimated.

Since each estimator is itself a random variable, each estimated value will be usually off the true value $\theta$. Consequently, one wants to obtain an impression of how far off the estimate might be from the real value $\theta$. This can be obtained for instance from:

**Definition 8.22** The *mean squared error* of a point estimator $H = h(X_0, X_1, \ldots, X_{n-1})$ for a parameter $\theta$ is

$$
\begin{aligned}
\mathrm{MSE}(H) &\equiv \mathrm{E}[(H - \theta)^2] = \mathrm{E}[(H - \mathrm{E}[H] + \mathrm{E}[H] - \theta)^2] \\
&= \mathrm{E}[(H - \mathrm{E}[H])^2] + \mathrm{E}[2(H - \mathrm{E}[H])(\mathrm{E}[H] - \theta)] + \mathrm{E}[(\mathrm{E}[H] - \theta)^2] \\
&= \mathrm{E}[(H - \mathrm{E}[H])^2] + 2 \underbrace{(\mathrm{E}[H] - \mathrm{E}[H])}_{=0}(\mathrm{E}[H] - \theta) + (\mathrm{E}[H] - \theta)^2 \\
&= \mathrm{Var}[H] + (\mathrm{E}[H] - \theta)^2
\end{aligned}
\tag{8.60}
$$

If an estimator is unbiased, i.e., if $\mathrm{E}[H] = \theta$, the mean squared error is given by the variance of the estimator. Hence, if for independent samples (each consisting of $n$ sample points) the estimated values are close to each other, the estimate is quite accurate. Unfortunately, usually only *one* sample is available (how to circumvent this problem rather ingeniously, see Sec. 8.3.4). Also the mean squared error does not immediately provide a probabilistic interpretation of how far the estimate is away from the true value $\theta$.

Nevertheless, one can obtain an estimate of the error in a probabilistic sense. Here we want to calculate a so-called *confidence interval* also sometimes named *error bar*.

**Definition 8.23** For a parameter $\theta$ describing a random variable, two estimators $l_\alpha = l_\alpha(x_0, x_1, \ldots, x_{n-1})$ and $u_\alpha = u_\alpha(x_0, x_1, \ldots, x_{n-1})$ which are obtained from a sample $\{x_0, x_1, \ldots, x_{n-1}\}$ provide a *confidence interval* if, for given *confidence level* $1 - \alpha \in (0, 1)$ we have

$$P(l_\alpha < \theta < u_\alpha) = 1 - \alpha \qquad (8.61)$$

The value $\alpha \in (0, 1)$ is called conversely *significance level*.

This means, the true but unknown value $\theta$ is contained in the interval $(l, u)$, which is itself a random variable as well, with probability $1 - \alpha$. Typical values of the confidence level are 0.68, 0.95 and 0.99 ($\alpha = 0.32$, 0.05, 0.01, respectively), providing increasing confidence. The more one wants to be sure that the interval really contains the true parameter, i.e. the smaller the value of $\alpha$, the larger the confidence interval will be.

Next, it is quickly outlined how one arrives at the confidence interval for the mean, for details please consult the specialized literature. First we recall that according to its definition the mean is a sum of independent random variables. For computer simulations, one can assume that usually (see below for a counterexample) a sufficiently large number of experiments is performed.[5] Therefore, according to the central limit theorem 8.1 $\overline{X}$ should exhibit (approximately) a pdf $f_{\overline{X}}$ which is Gaussian with an expectation value $\mu$ and some variance $\sigma_{\overline{X}}^2 = \sigma^2/n$. This means, the probability $\alpha$ that the sample means fall *outside* an interval $I = [\mu - z\sigma_{\overline{X}}, \mu + z\sigma_{\overline{X}}]$ can be easily obtained from the standard normal distribution. This situation is shown in the Fig. 8.19. Note that the interval is symmetric about the mean $\mu$ and that its width is stated in multiples $z = z(\alpha)$ of the standard deviation $\sigma_{\overline{X}}$. The relation between significance level $\alpha$ and half interval width $z$ is just $\int_{-z}^{z} dx \, f_{\overline{X}}(x) = 1 - \alpha$. Hence, the weight of the standard normal distribution *outside* the interval $[-z, z]$ is $\alpha$. This relation can be obtained from any table of the standard Gaussian distribution or from the function `gsl_cdf_gaussian_P()` of the *GNU scientific library* (see Sec. 7.3). Usually, one considers integer values $z = 1, 2, 3$ which correspond to significance levels $\alpha = 0.32$, 0.05, and 0.003, respectively. So far, the confidence interval $I$ contains the unknown expectation value $\mu$ and the

---

[5]This is different for many empirical experiments, for example, when testing new treatments in medical sciences, where often only a very restricted number of experiments can be performed. In this case, one has to consider special distributions, like the *Student distribution*.

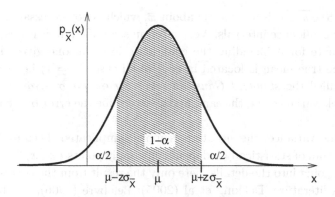

Fig. 8.19 Probability density function of the sample mean $\overline{X}$ for large enough sample sizes $n$ where the distribution becomes Gaussian. The true expectation value is denoted by $\mu$ and $\sigma_{\overline{X}}$ is the variance of the sample mean. The probability that a random number drawn from this distribution falls outside the symmetric interval $[\mu - z\sigma_{\overline{X}}, \mu + z\sigma_{\overline{X}}]$ is $\alpha$.

unknown variance $\sigma_{\overline{X}}$. First, one can rewrite

$$
\begin{aligned}
1 - \alpha &= P(\mu - z\sigma_{\overline{X}} \le \overline{X} \le \mu + z\sigma_{\overline{X}}) \\
&= P(-z\sigma_{\overline{X}} \le \overline{X} - \mu \le z\sigma_{\overline{X}}) \\
&= P(-\overline{X} - z\sigma_{\overline{X}} \le -\mu \le -\overline{X} + z\sigma_{\overline{X}}) \\
&= P(\overline{X} - z\sigma_{\overline{X}} \le \mu \le \overline{X} + z\sigma_{\overline{X}}).
\end{aligned}
$$

This now states the probability that the true value, which is estimated by the sample mean $\overline{x}$, lies within an interval which is symmetric about the estimate $\overline{x}$. Note that the width $2z\sigma_{\overline{X}}$ is basically given by $\sigma_{\overline{X}} = \sqrt{\text{Var}[\overline{X}]}$. This explains why the mean squared error $\text{MSE}(H) = \text{Var}[H]$, as presented in the beginning of this section, is a good measure for the statistical error made by the estimator. This will be used in Sec. 8.3.4.

To finish, we estimate the true variance $\sigma^2$ using $\frac{n}{n-1}s^2$, hence we get $\sigma_{\overline{X}} = \frac{\sigma}{\sqrt{n}} \approx \frac{S}{\sqrt{n-1}}$. To summarize we get:

$$
P\left(\overline{X} - z\frac{S}{\sqrt{n-1}} \le \mu \le \overline{X} + z\frac{S}{\sqrt{n-1}}\right) \approx 1 - \alpha \qquad (8.62)
$$

Note that this confidence interval, with $l_\alpha = \overline{x} - z(\alpha)S/\sqrt{n-1}$ and $u_\alpha =$

$\bar{x} + z(\alpha)S/\sqrt{n-1}$, is symmetric about $\bar{x}$, which is not necessarily the case for other confidence intervals. Very often in scientific publications, to state the estimate for $\mu$ including the confidence interval, one gives the range where the true mean is located in 68% of all cases ($z = 1$) i.e. $\bar{x} \pm \frac{S}{\sqrt{n-1}}$, this is called the *standard Gaussian error bar* or *one $\sigma$ error bar*. Thus, the sample variance and the sample size determine the error bar/confidence interval.

For the variance, the situation is more complicated, because it is not simply a sum of statistically independent sample points $\{x_0, x_1, \ldots, x_{n-1}\}$. Without going into the details, here only the result from the corresponding statistics literature [Dekking et al (2005); Lefebvre (2006)] is cited: The confidence interval where with probability $1 - \alpha$ the true variance is located is given by $[\sigma_l^2, \sigma_u^2]$ where

$$
\sigma_l^2 = \frac{ns^2}{\chi^2(1 - \alpha/2, n - 1)}
$$
$$
\sigma_u^2 = \frac{ns^2}{\chi^2(\alpha/2, n - 1)}. \tag{8.63}
$$

Here, $\chi^2(\beta, \nu)$ is the inverse of the cumulative chi-squared distribution with $\nu$ degrees of freedom. It states the value where $F(\chi^2, \nu) = \beta$, see page 261. This chi-squared function is implemented in the *GNU scientific library* (see Sec. 7.3) in the function `gsl_cdf_chisq_Pinv()`.

Note that as one alternative, you could regard $y_i \equiv (x_i - \bar{x})^2$ approximately as independent data points and use the above standard error estimate described for the mean of the sample $\{y_i\}$. Also, one can use the bootstrap method as explained below (Sec. 8.3.4), which allows to calculate confidence intervals for arbitrary estimators.

### 8.3.3 *Histograms*

Sometimes, you do not only want to estimate moments of an underlying distribution, but you want to get an impression of the full distribution. In this case you can use *histograms*.

**Definition 8.24** A histogram is given by a set of disjoint intervals

$$
B_k = [l_k, u_k), \tag{8.64}
$$

which are called *bins* and a counter $h_k$ for each bin. For a given *sample* of $n$ measured points $\{x_0, x_1, \ldots, x_{n-1}\}$, bin $h_k$ contains the number of sample points $x_i$ which are contained in $B_k$.

**Example 8.4** For the sample

$$\{x_i\} = \{1.2,\, 1.5,\, 1.0,\, 0.7,\, 1.4,\, 2.0,$$
$$1.5,\, 1.1,\, 0.9,\, 1.9,\, 1.2,\, 0.8\}$$

the bins

$$[0, 0.5),\, [0.5, 1.0)\, [1.0, 1.5),\, [1.5, 2.0),\, [2.0, 2.5)\, [2.5, 3.0),$$

are used, resulting in

$$h_1 = 0,\, h_2 = 3,\, h_3 = 5,\, h_4 = 3,\, h_5 = 1,\, h_6 = 0$$

which is depicted in Fig. 8.20.

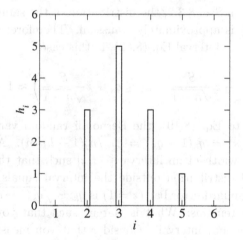

Fig. 8.20 Histogram for the data shown in Ex. 8.4.

In principle, the bins can be chosen arbitrarily. You should take care that the union of all intervals covers all (possible or actual) sample points. Here, it is assumed that the bins are properly chosen. Note also that the width $b_k = u_k - l_k$ of each bin can be different. Nevertheless, often bins with uniform width are used. Furthermore, for many applications, for instance, when assigning different weights to different sample points[6], it is useful to consider the counters as real-valued variables. A simple (fixed-bin width) C implementation of histograms is described in Sec. 5.2. The *GNU scientific library* (see Sec. 7.3) contains data structures and functions which implement histograms allowing for variable bin width.

Formally, for a given random variable $X$, the count $h_k$ in bin $k$ can be seen as a result of a random experiment for the binomial random variable $H_k \sim B(n, p_k)$ with parameters $n$ and $p_k$, where $p_k = P(X \in B_k)$ is the probability that a random experiment for $X$ results in a value which is contained in bin $B_k$. This means that confidence intervals for a histogram bin can be obtained in principle from a binomial distribution. Nevertheless, for each sample the true value for a value $p_k$ is unknown and can only be estimated by $q_k \equiv h_k/n$. Hence, the true binomial distribution is unknown. On the other hand, a binomial random variable is a sum of $n$ Bernoulli random variables with parameter $p_k$. Thus, the estimator $q_k$ is nothing else than a sample mean for a Bernoulli random variable. If the number of sample points $n$ is "large" (see below), from the central limit theorem 8.1 and as discussed in Sec. 8.3.2, the distribution of the sample mean (being binomial in fact) is approximately Gaussian. Therefore, one can use the standard confidence interval Eq. (8.62), in this case

$$P\left(q_k - z\frac{S}{\sqrt{n-1}} \le p_k \le q_k + z\frac{S}{\sqrt{n-1}}\right) \approx 1 - \alpha \qquad (8.65)$$

Here, according to Eq. (8.19), the Bernoulli random variable exhibits a sample variance $s^2 = q_k(1 - q_k) = (h_k/n)(1 - h_k/n)$. Again, $z = z(\alpha)$ denotes the half width of an interval $[-z, z]$ such that the weight of the standard normal distribution outside the interval equals $\alpha$. Hence, the estimate with standard error bar ($z = 1$) is $q_k \pm \sqrt{q_k(1 - q_k)/(n-1)}$.

The question remains: What is "large" such that you can trust this "Gaussian" confidence interval? Consider that you measure for example no point at all for a certain bin $B_k$. This can happen easily in the regions where $p_k$ is smaller than $1/n$ but non-zero, i.e. in regions of the histogram which are used to sample the tails of a probability density function. In this case the estimated fraction can easily be $q_k = 0$ resulting also in a zero-width confidence interval, which is certainly wrong. This means, the number of samples $n$ needed to have a reliable confidence interval for a bin $B_k$ depends on the number of bin entries. A rule of thumb from the

---

[6]This occurs for some advanced simulation techniques.

statistics literature is that $nq_k(1 - q_k) > 9$ should hold. If this condition is not fulfilled, the correct confidence interval $[q_{i,l}, q_{i,u}]$ for $q_k$ has to be obtained from the binomial distribution and it is quite complicated, since it uses the *F distribution* (see Def. 8.21 on page 261)

$$q_{i,l} = \frac{h_k}{h_k + (n - h_k + 1)F_1}$$

$$q_{i,u} = \frac{(h_k + 1)F_2}{(h_k + 1)F_2 + n - h_k},$$ (8.66)

$$\text{where } F_1 = F(1 - \alpha/2; 2n - 2h_k + 2, 2h_k)$$

$$F_2 = F(1 - \alpha/2; 2h_k + 2, 2n - 2h_k)$$

The value $F(\beta; r_1, r_2)$ states the $x$ value such that the distribution function for the F distribution with number of degrees $r_1$ and $r_2$ reaches the value $\beta$. This inverse distribution function is implemented in the *GNU scientific library* (see Sec. 7.3). If you always use these confidence intervals, which are usually *not* symmetric about $q_k$, then you cannot go wrong. Nevertheless, for most applications the standard Gaussian error bars are fine.

Finally, in case you want to use a histogram to represent a sample from a continuous random variable, you can easily interpret a histogram as a sample for a probability density function, which can be represented as a set of points $\{(\tilde{x}_k, p(\tilde{x}_k))\}$. This is called the *histogram pdf* or the *sample pdf*. For simplicity, it is assumed that the interval mid points of the intervals are used as $x$-coordinate. For the normalization, we have to divide by the total number of counts, as for $q_k = h_k/n$ and to divide by the bin width. This ensures that the integral of the sample pdf, approximated by a sum, gives just unity. Therefore, we get

$$\tilde{x}_k \equiv (l_k + u_k)/2$$

$$p(\tilde{x}_k) \equiv h_k/(nb_k).$$ (8.67)

The confidence interval, whatever type you choose, has to be normalized in the same way. A function which prints a histogram as pdf, with simple Gaussian error bars, is shown in Sec. 5.2.

For discrete random variables, the histogram can be used to estimate the pmf.[7] In this case the choice of the bins, in particular the bin widths, is easy, since usually all possible outcomes of the random experiments are known. For a histogram pdf, which is used to describe approximately a

---

[7] For discrete random variables, the $q_k$ values are already suitably normalized.

continuous random variable, the choice of the bin width is important. Basically, you have to adjust the width manually, such that the sample data is represented "best". Thus, the bin width should not be too small nor too large. Sometimes a non-uniform bin width is the best choice. In this case no general advice can be given, except that the bin width should be large where few data points have been sampled. This means that each bin should contain roughly the same number of sample points. Several different rules of thumb exist for uniform bin widths. For example $b = 3.49Sn^{-1/3}$ [Scott (1979)], which comes from minimizing the mean integrated squared difference between a Gaussian pdf and a sample drawn from this Gaussian distribution. Hence, the larger the variance $S$ of the sample, the larger the bin width, while increasing the number of sample points enables the bin width to be reduced.

In any case, you should be aware that the histogram pdf can be only an approximation of the real pdf, due to the finite number of data points and due to the underlying discrete nature resulting from the bins. The latter problem has been addressed in recent years by so-called *kernel estimators* [Dekking et al (2005)]. Here, each sample point $x_i$ is represented by a so-called *kernel function*. A kernel function $k(x)$ is a peaked function, formally exhibiting the following properties:

- It has a maximum at 0.
- It falls off to zero over some distance $h$.
- Its integral $\int k(x)\, dx$ is normalized to one.

Often used kernel functions are, for example, a triangle, a cut upside-down parabola or a Gaussian function. Each sample point $x_i$ is represented such that a kernel function is shifted having the maximum at $x_i$. The estimator $\hat{p}(x)$ for the pdf is the suitably normalized sum (factor $1/n$) of all these kernel functions, one for each sample point:

$$\hat{p}(x) = \frac{1}{n} \sum_i k(x - x_i) \tag{8.68}$$

The advantages of these kernel estimators are that they result usually in a smooth function $\hat{p}$ and that for a value $\hat{p}(x)$ also sample points more distant from $x$ may contribute, with decreasing weight for increasing distance. The most important parameter is the width $h$. A too small value of $h$ will result in many distinguishable peaks, one for each sample point. A too large value of $h$ leads to a loss of important details. This is of similar importance as the choice of the bin width for histograms. The choice of the kernel function (e.g. a triangle, an upside-down parabola or a Gaussian function) seems to be less important.

### 8.3.4 *Resampling using Bootstrap*

As pointed out, an estimator for some parameter $\theta$, given by a function $h(x_0, x_1, \ldots, x_{n-1})$, is in fact a random variable $H = h(X_0, X_1, \ldots, X_{n-1})$. Consequently, to get an impression of how much an estimate differs from the true value of the parameter, one needs in principle to know the distribution of the estimator, e.g. via the pdf $p_H$ or the distribution function $F_H$. In the previous chapter, the distribution was known for few estimators, in particular if the sample size $n$ is large. For instance, the distribution of the sample mean converges to a Gaussian distribution, irrespectively of the distribution function $F_X$ describing the sample points $\{x_i\}$.

For the case of a general estimator $H$, in particular if $F_X$ is not known, one may not know anything about the distribution of $H$. In this case one can approximate $F_X$ by the sample distribution function:

**Definition 8.25** For a sample $\{x_0, x_1, \ldots, x_{n-1}\}$, the *sample distribution function* (also called *empirical distribution function*) is

$$F_{\hat{X}}(x) \equiv \frac{\text{number of sample points } x_i \text{ smaller than or equal to } x}{n} \quad (8.69)$$

Note that this distribution function describes in fact a discrete random variable (called $\hat{X}$ here), but is usually (but not always) used to approximate a continuous distribution function.

The *bootstrap principle* is to use $F_{\hat{X}}$ instead of $F_X$. The name of this principle was made popular by B. Efron [Efron (1979); Efron and Tibshirani (1994)] and comes from the fairy tale of Baron Münchhausen, who dragged himself out of a swamp by pulling on the strap of his boot.[8] Since the distribution function $F_X$ is replaced by the empirical sample distribution function, the approach is also called *empirical bootstrap*, for a variant called parametric bootstrap see below.

Now, having $F_{\hat{X}}$ one could in principle calculate the distribution function $F_{\hat{H}}$ for the random variable $\hat{H} = h(\hat{X}_0, \hat{X}_1, \ldots, \hat{X}_{n-1})$ exactly, which then is an approximation of $F_H$. Usually, this is to cumbersome and one uses a second approximation: One draws so-called *bootstrap samples* $\{\hat{x}_0, \hat{x}_1, \ldots, \hat{x}_{n-1}\}$ from the random variable $\hat{X}$. This is called *resampling*. This can be done quite simply by $n$ times selecting (*with replacement*) one

---

[8]In the European version, he dragged himself out by pulling his hair.

of the data points of the original sample $\{x_i\}$, each one with the same probability $1/n$. This means that some sample points from $\{x_i\}$ may appear several times in $\{\hat{x}_i\}$, some may not appear at all.[9] Now, one can calculate the estimator value $h^* = h(\hat{x}_0, \hat{x}_1, \ldots, \hat{x}_{n-1})$ for each bootstrap sample. This is repeated $K$ times for different bootstrap samples resulting in $K$ values $h_k^*$ ($k = 1, \ldots, K$) of the estimator. The sample distribution function $F_{H^*}$ of this sample $\{h_k^*\}$ is the final result, which is an approximation of the desired distribution function $F_H$. Note that the second approximation, replacing $F_{\hat{H}}$ by $F_{H^*}$ can be made arbitrarily accurate by making $K$ as large as desired, which is computationally cheap.

You may ask: Does this work at all, i.e., is $F_{H^*}$ a good approximation of $F_H$? For the general case, there is no answer. But for some cases there are mathematical proofs. For example for the mean $H = \overline{X}$ the distribution function $F_{\overline{X}^*}$ in fact converges to $F_{\overline{X}}$. Here, only the subtlety arises that one has to consider in fact the *normalized* distributions of $\overline{X} - \mu$ ($\mu = \mathrm{E}[X]$) and $\hat{X} - \overline{x}$ ($\overline{x} = \sum_{i=0}^{n-1} x_i/n$). Thus, the random variables are just shifted by constant values. For other cases, like for estimating the median or the variance, one has to normalize in a different way, i.e., by subtracting the (empirical) median or by dividing by the (empirical) variance. Nevertheless, for the characteristics of $F_H$ we are interested in, in particular in the variance, see below, normalizations like shifting and stretching are not relevant, hence they are ignored in the following. Note that indeed some estimators exist, like the maximum of a distribution, for which one can prove conversely that $F_{H^*}$ does *not* converge to $F_H$, even after some normalization. On the other hand, for the purpose of getting a not too bad estimate of the error bar, for example, bootstrapping is a very convenient and suitable approach which has received high acceptance during recent years.

Now one can use $F_{H*}$ to calculate any desired quantity. Most important is the case of a confidence interval $[h_l, h_u]$ such that the total probability outside the interval is $\alpha$, for given significance level $\alpha$, i.e. $F_{H^*}(h_u) - F_{H^*}(h_l) = 1 - \alpha$. In particular, one can distribute the weight $\alpha$ equally below and above the interval, which allows to determine $h_l, h_u$

$$F_{H^*}(h_u) = F_{H^*}(h_l) = 1 - \alpha/2. \qquad (8.70)$$

Similar to the confidence intervals presented in Sec. 8.3.2, $[h_l, h_u]$ also represents a confidence interval for the unknown parameter $\theta$ which is to

---

[9]The probability for a sample point not to be selected is $(1 - 1/n)^n = \exp(n \log(1 - 1/n)) \to \exp(n(-1/n)) = \exp(-1) \approx 0.367$ for $n \to \infty$.

be estimated from the estimator (if it is unbiased). Note that $[h_l, h_u]$ can be non-symmetric about the actual estimate $h(x_0, x_1, \ldots, x_{n-1})$. This will happen if the distribution $F_{H*}$ is skewed.

For simplicity, as we have seen in Sec. 8.3.2, one can use the variance $\mathrm{Var}[H]$ to describe the statistical uncertainty of the estimator. As mentioned on page 289, this corresponds basically to a $\alpha = 0.32$ uncertainty. The quantity corresponding to the standard error bar is $\sqrt{\mathrm{Var}[H]}$.

The following C function calculates $\mathrm{Var}[H^*]$, as approximation of the unknown $\mathrm{Var}[H]$. One has to pass as arguments the number $n$ of sample points, an array containing the sample points, the number $K$ of bootstrap iterations,

| GET SOURCE CODE |
| --- |
| DIR: **randomness**<br>FILE(S): **bootstrap.c**<br>**bootstrap_test.c** |

and a pointer to the function f which represents the estimator. f has to take two arguments: the number of sample points and an array containing a sample. For an explanation of function pointers, see page 59. The function bootstrap_variance() returns $\mathrm{Var}[H^*]$.

```c
double bootstrap_variance(int n, double *x, int n_resample,
                          double (*f)(int, double *))
{
  double *xb;                              /* bootstrap sample */
  double *h;                          /* results from resampling */
  int sample, i;                           /* loop counters */
  int k;                                  /* sample point id */
  double var;                          /* result to be returned */

  h = (double *) malloc(n_resample * sizeof(double));
  xb = (double *) malloc(n * sizeof(double));
  for(sample=0; sample<n_resample; sample++)
  {
    for(i=0; i<n; i++)                          /* resample */
    {
      k = (int) floor(drand48()*n);      /* select random point */
      xb[i] = x[k];
    }
    h[sample] = f(n, xb);                 /* calculate estimator */
  }

  var = variance(n_resample, h);     /* obtain bootstrap variance */
  free(h);
  free(xb);
  return(var);
}
```

The bootstrap samples $\{\hat{x}_i\}$ are stored in the array xb, while the sampled estimator values $\{h_k^*\}$ are stored in the array h. These arrays are allocated in lines 10–11. In the main loop (lines 12–20) the bootstrap samples are calculated, each time the estimator is obtained and the result is stored in h. Finally, the variance of the sample $\{h_k^*\}$ is calculated (line 22). Here, the function variance() is used, which works similarly to the function mean(), see exercise (5). Your are asked to implement a bootstrap function for general confidence interval in exercise (6).

The most obvious way is to call bootstrap_variance() with the estimator mean as forth argument. For a distribution which is "well behaved" (i.e., where a sum of few random variables resembles the Gaussian distribution), you will get a variance that is, at least if n_resample is reasonably large, very close to the standard Gaussian ($\alpha = 0.32$) error bar.

For calculating properties of the sample mean, the bootstrap approach works fine, but in this case one could also be satisfied with the standard Gaussian confidence interval. The bootstrap approach is more interesting for non-standard estimators. One prominent example from the field of statistical physics is the so-called *Binder cumulant* [Binder (1981)], which is given by:

$$b(x_0, x_1, \ldots, x_{n-1}) = 0.5 \left( 3 - \frac{\overline{x^4}}{[\overline{x^2}]^2} \right) \tag{8.71}$$

where $\overline{\ldots}$ is again the sample mean, for example $\overline{x^2} = \frac{1}{n} \sum_{i=0}^{n-1} x_i^2$. The Binder cumulant is often used to determine phase transitions via simulations, where only systems consisting of a finite number of particles can be studied. For example, consider a ferromagnetic system held at some temperature $T$. At low temperature,

| GET SOURCE CODE |
| --- |
| DIR: randomness |
| FILE(S): |
| binder_L8.dat |
| binder_L10.dat |
| binder_L16.dat |
| binder_L30.dat |

below the *Curie temperature* $T_c$, the system will exhibit a macroscopic magnetization $m$. On the other hand, for temperatures above $T_c$, $m$ will on average converge to zero when increasing the system size. This transition is fuzzy, if the system sizes are small. Nevertheless, when evaluating the Binder cumulant for different sets of sample points $\{m(T, L)_i\}$ which are obtained at several temperatures $T$ and for different system sizes $L$, the $b_L(T)$ curves for different $L$ will all cross [Landau and Binder (2000)] (almost) at $T_c$, which allows for a very precise determination of $T_c$. A sample result for a two-dimensional (i.e. layered) model ferromagnet exhibiting $L \times L$ particles is shown in Fig. 8.21. The Binder cumulant has been useful for the simulation of many other systems like disordered materials, gases, optimization problems, liquids, and graphs describing social systems.

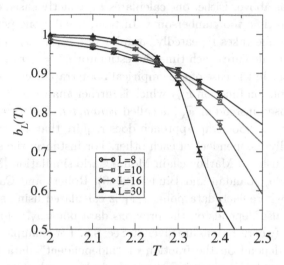

Fig. 8.21   Plot of Binder cumulant of two-dimensional model ferromagnet as function of temperature $T$ (dimensionless units). Each system consists of $L \times L$ particles. The curves for different system sizes $L$ cross very close to the phase transition temperature $T_c = 2.269$ (known from analytical calculations of this model). The error bars shown can be obtained using a bootstrap approach.

A confidence interval for the Binder cumulant is very difficult (or even impossible) to obtain using standard error analysis. Using bootstrapping, it is straightforward. You can use simply the function `bootstrap_variance()` shown above while providing as argument a function which evaluates the Binder cumulant for a given set of data points.

So far, it was assumed that the empirical distribution function $F_{\hat{X}}$ was used to determine an approximation of $F_H$. Alternatively, one can use some additional knowledge which might be available. Or one can make additional assumptions, via using a distribution function $F_{\underline{\lambda}}$ which is parametrized by a vector of parameters $\underline{\lambda}$. For an exponential distribution, the vector would just consist of one parameter, the expectation value, while for a Gaussian distribution, $\underline{\lambda}$ would consist of the expectation value and the variance. In principle, arbitrary complex distributions with many parameters are possible. To make $F_{\hat{\lambda}}$ a "good" approximation of $F_X$, one has to adjust the parameters such that the distribution function represents the sample $\{x_i\}$ "best", resulting in a vector $\hat{\lambda}$ of parameters. Methods and tools to perform this *fitting* of parameters are presented in Sec. 8.6.2. Using $F_{\hat{\lambda}}$ one can proceed as above: Either one calculates $F_{\hat{H}}$ exactly based on $F_{\hat{\lambda}}$, which is most of the time too cumbersome. Instead, usually one performs simulations where one takes repeatedly samples $\{\hat{x}_0, \hat{x}_1, \ldots, \hat{x}_{n-1}\}$ from simulating $F_{\hat{\lambda}}$ and calculates each time the estimator $h^* = h(\hat{x}_0, \hat{x}_1, \ldots, \hat{x}_{n-1})$. This results, as in the case of the empirical bootstrap discussed above, in a sample distribution function $F_{H^*}$ which is further analyzed. This approach, where $F_{\hat{\lambda}}$ is used instead of $F_{\hat{X}}$, is called *parametric bootstrap*.

Note that the bootstrap approach does require that the sample points are statistically independent of each other. For instance, the sample could be generated using a Markov chain Monte Carlo simulation [Newman and Barkema (1999); Landau and Binder (2000); Robert and Casella (2004); Liu (2008)], where each data point $x_{i+1}$ is calculated using some random process, but also depends on the previous data point $x_i$. More details on how to quantify correlations are given in Sec. 8.5. For example, a confidence interval will depend on the fraction of "independent" data points. One can see this easily by assuming that you replace each data point in the original sample $\{x_i\}$ by 10 copies, hence making the sample 10 times larger without adding any information. This will affect the following bootstrap calculations, since the fluctuations of the bootstrap sample will be reduced. Assume, e.g., your original sample has only two independent sample points $x_1 \neq x_2$. Thus, bootstrap samples may exhibit two times $x_1$ or two times $x_2$, or each one once. Thus, the bootstrap average will fluctuate strongly. On the other hand, if you had 10 times data point $x_1$ and 10 times $x_2$ in the sample, the bootstrap sample will contain most of the times about 10 times the point $x_1$ and about 10 times $x_2$, only very rarely you see only (i.e., 20 times) the value $x_2$. Thus, the bootstrap mean will fluctuate less. With this respect, bootstrapping is similar to the classical calculation of confidence intervals explained in Sec. 8.3.2, where also independence of data is assumed and the number of independent data points enters formulas like Eq. (8.62). Hence, to correct for this, one can increase the obtained bootstrap error bars by a factor $\sqrt{\lambda_c}$ where $\lambda_c$ is the typical number of sequentially correlated sample points.

It should be mentioned that bootstrapping is only one of several resampling techniques. Another well known approach is the *jackknife approach*, where one does not sample randomly using $F_{\hat{X}}$ or a fitted $F_{\hat{\lambda}}$. Instead the sample $\{x_i\}$ is divided into $B$ blocks of equal size $n_b = n/B$ (assuming that $n$ is a multiple of $B$). Note that choosing $B = n$ is possible and not uncommon. Next, a number $B$ of so-called *jackknife samples* $b = 1, \ldots, B$ are formed from the original sample $\{x_i\}$ by omitting exactly the sample points from the $b$'th block and including all other points of the original sample. Therefore, each of these jackknife samples consists of $n - n_b$ sample points. For each jackknife sample, again the estimator is calculated, resulting in a sample $\{h_k^{(j)}\}$ of size $B$. Note that the sample distribution function $F^{(j)}$ of this sample is *not* an approximation of the estimator distribution function $F_H$! Nevertheless, it is useful. For instance, the variance $\mathrm{Var}[H]$ can be estimated from $(B - 1)S_h^2$, where $S_h^2$ is the sample variance of $\{h_k^{(j)}\}$. No proof of this is presented here. It is just noted that when increasing the number $B$ of blocks, i.e., making the different jackknife samples more alike, because fewer points are excluded, the sample of estimators values $\{h_k^{(j)}\}$ will fluctuate less. Consequently, this dependence on the number of blocks is exactly compensated via the factor $(B - 1)$. Note that for the jackknife method, in contrast to the boostrap approach, the statistical independence of the original sample is required. If there are correlations between the data points, the jackknife approach can be combined with the so-called blocking method [Flyvbjerg (1998)]. More details on the jackknife approach can be found in [Efron and Tibshirani (1994)].

Finally, you should be aware that there are cases where resampling approaches clearly fail. The most obvious example is the calculation of confidence intervals for histograms, see Sec. 8.3.3. A bin which exhibits no sample points, for example, where the probability is very small, will never get a sample point during resampling either. Hence, the error bar will be of zero width. This is in contrast to the confidence interval shown in Eq. 8.66, where also bins with zero entries exhibit a finite-size confidence interval. Consequently, you have to think carefully before deciding which approach you will use to determine the reliability of your results.

## 8.4  Data plotting

So far, you have learned many methods for analyzing data. Since you do not just want to look at tables filled with numbers, you should visualize the data in viewgraphs. Those viewgraphs which contain the essential results of your work can be used in presentations or publications. To analyze and plot data, several commercial and non-commercial programs are available. Here, two free programs are discussed, *gnuplot*, and *xmgrace*. *Gnuplot* is small, fast, allows two- and three-dimensional curves to be generated and transformed, as well as arbitrary functions to be fitted to the data (see Sec. 8.6.2). In the interactive mode, *gnuplot* uses a command line interface. On the other hand, with a bit of additional effort, also publication-ready plots can be generated, optimally using *gnuplot* scripts. For people who prefer to use menues, buttons and the mouse to generate a plot, *xmgrace* is the better choice. It is very flexible and produces easily nice publication-ready plots.

### 8.4.1  *gnuplot*

The program *gnuplot* is invoked by entering `gnuplot` in a shell, or from a menu of the graphical user interface of your operating system. For a complete manual see [Texinfo].

As always, our examples refer to a UNIX window system like X11, but the program is available for almost all operating systems. After startup, in the window of your shell or the window which pops up for `gnuplot` the prompt (e.g. `gnuplot>`) appears and the user can enter commands in textual form, results are shown in additional windows or are written into files. For a general introduction you can type just `help`.

Before giving an example, it should be pointed out that gnuplot *scripts* can be generated by simply writing the commands into a file, e.g. `command.gp`, and calling `gnuplot command.gp`.

The typical case is that you have available a data file of $x - y$ data or with $x - y - dy$ data (where $dy$ is the error bar of the $y$ data points). Your file might look like this, where

| GET SOURCE CODE |
|---|
| DIR: **randomness** |
| FILE(S): **sg_e0_L.dat** |

the "energy" $e_0$ of a system[10] is stored as a function of the "system size" $L$. The filename is `sg_e0_L.dat`. The first column contains the $L$ values, the second the energy values and the third the standard error of the energy. Please note that lines starting with "#" are comment lines which are ignored on reading:

```
# ground state energy of +-J spin glasses
# L    e_0   error
  3 -1.6710 0.0037
  4 -1.7341 0.0019
  5 -1.7603 0.0008
```

---

[10]It is the ground-state energy of a three-dimensional $\pm J$ spin glass , a protypical system in statistical physics. These spin glasses model the magnetic behavior of alloys like iron-gold.

```
 6 -1.7726 0.0009
 8 -1.7809 0.0008
10 -1.7823 0.0015
12 -1.7852 0.0004
14 -1.7866 0.0007
```

To plot the data enter

```
gnuplot> plot "sg_e0_L.dat" with yerrorbars
```

which can be abbreviated as **p "sg_e0_L.dat" w e**. Please do not forget the quotation marks around the file name. Next, a window pops up, showing the result, see Fig. 8.22.

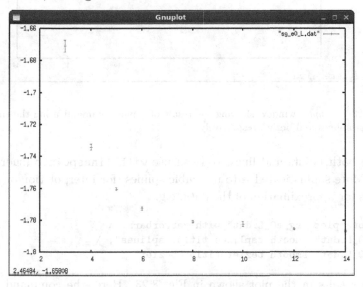

Fig. 8.22  *Gnuplot* window showing the result of a plot command.

For the **plot** command many options and styles are available, e.g. **with lines** produces lines instead of symbols. It is not explained here how to set line styles or symbol sizes and colors, because this is usually not necessary for a quick look at the data. For "nice" plots used for presentations, we recommend *xmgrace*, see next section. Anyway, **help plot** will tell you all you have to know about the **plot** command.

Sometimes you want to add lines to the data points, as a guide to the eyes. The most simple option is to use the option **with lines** (shortcut **w l**), which shows lines connecting the points, but not the points. If you want

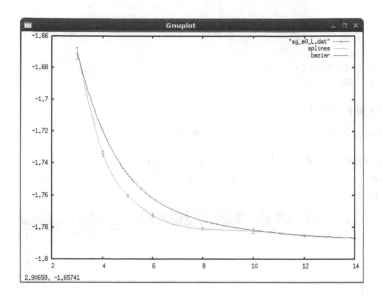

Fig. 8.23 *Gnuplot* window showing the result of a plot command using the smooting options csplines and bezier, respectively.

to see both points and lines, you can use **with linespoints** (shortcut **w lp**). More sophisticated is to use cubic splines for interpolation or Bézier curves for approximation of the data, e.g.

```
gnuplot> plot "sg_e0_L.dat" with yerrorbars, \
"sg_e0_L.dat" smooth csplines title "splines", \
"sg_e0_L.dat" smooth bezier title "bezier"
```

which results in the plot shown in Fig. 8.23. Here, the command is distributed over several lines: The "\" indicates that the command is continued on the next line. Using the **title** option, different names are given to the different lines plotted. Note that other options for smoothing the data are available see **help smooth**. Alternatively, one can fit a function to the data and plot the fitted function, see Sec. 8.6.2.

Among the important options of the **plot** command is that one can specify ranges. This can be done by specifying the range directly after the command, e.g.

```
gnuplot> plot [7:20]  "sg_e0_L.dat" with yerrorbars
```

will only show the data for $x \in [7, 20]$. Also an additional $y$ range can be specified like in

```
plot [7:20][-1.79:-1.77]  "sg_e0_L.dat" with yerrorbars
```

If you just want to set the $y$ range, you have to specify [ ] for the $x$-range. You can also fix the ranges via the set xrange and the set yrange commands, such that you do not have to give them each time with the plot command, see help set xrange or help unset xrange for unsetting a range.

Gnuplot knows a lot of built-in functions like $\sin(x)$, $\log(x)$, powers, roots, Bessel functions, error function,[11] and many more. For a complete list type help functions. These function can be also plotted. Furthermore, using these functions and standard arithmetic expressions, you can also define your own functions, e.g. you can define a function ft(x) for the Fischer-Tippett pdf (see Eq. (8.43)) for parameter $\lambda$ (called lambda here) and show the function via

```
gnuplot> ft(x)=lambda*exp(-lambda*x)*exp(-exp(-lambda*x))
gnuplot> lambda=1.0
gnuplot> plot ft(x)
```

You can also include arithmetic expressions in the plot command. To plot a shifted and scaled Fischer-Tippett pdf you can type:

```
gnuplot> plot [0:20] 0.5*ft(0.5*(x-5))
```

The Fischer-Tippett pdf has a tail which drops off exponentially. This can be better seen by a logarithmic scaling of the $y$ axis.

```
gnuplot> set logscale y
gnuplot> plot [0:20] 0.5*ft(0.5*(x-5))
```

will produce the plot shown in Fig. 8.24.

Furthermore, it is also possible to plot several functions in one plot, via separating them via commas, e.g. to compare a Fischer-Tippett pdf to the standard Gaussian pdf, here the predefined constant pi is used:

```
gnuplot> plot ft(x), exp(-x*x/2)/sqrt(2*pi)
```

It is possible to read files with multi columns via the using *data modifier*, e.g.

```
gnuplot> plot "test.dat" using 1:4:5 w e
```

---

[11]The error function is $\mathrm{erf}(x) = (2/\sqrt{\pi}) \int_0^x dx' \exp(-x'^2)$.

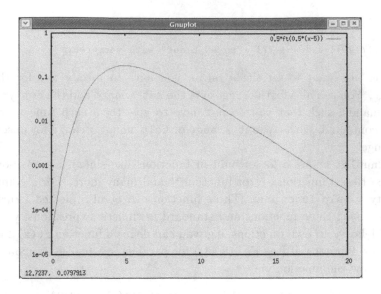

Fig. 8.24   *Gnuplot* window showing the result of plotting a shifted and rescaled Fischer-Tippett pdf with logarithmically scaled $y$-axis.

displays the fourth column as a function of the first, with error bars given by the 5th column. The elements behind the **using** are called entries. Within the **using** data modifier you can also perform transformations and calculations. Each entry, where some calculations should be performed have to be embraced in ( ) brackets. Inside the brackets you can refer to the different columns of the input file via **$1** for the first column, **$2** for the second, etc. You can generate arbitrary expressions inside the brackets, i.e. use data from different columns (also combine several columns in one entry), operators, variables, predefined and self-defined functions and so on. For example, in Sec. 8.6.2, you will see that the data from the **sg_e0_L.dat** file follows approximately a power law behavior $e_0(L) = e_\infty + aL^b$ with $e_\infty \approx -1.788$, $a \approx 2.54$ and $b \approx -2.8$. To visualize this, we want to show $e_0(L) - e_\infty$ as a function of $L^b$. This is accomplished via:

```
gnuplot> einf=-1.788
gnuplot> b=-2.8
gnuplot> plot "sg_e0_L.dat" u ($1**b):($2-einf)
```

Now the *gnuplot* window will show the data as a straight line (not shown, but see Fig. 8.33).

It is also possible to display vector fields using **gnuplot**. They have to be given, for planar fields, in four column format in the form $x$ position, $y$ position, $\Delta x$, $\Delta y$. When plotting a vector field, the **plot** command has to be used with the style **vectors**, e.g.

| GET SOURCE CODE |
| --- |
| DIR: **randomness** <br> FILE(S): <br> **vector_field.dat** |

```
gnuplot> plot "vector_field.dat" u 1:2:($3/10):($4/10) with vectors
```

Note that the vector lengths (columns three and four) have been rescaled by a factor $1/10$ to prevent that the vectors are intersecting. The resulting plot is shown in Fig. 8.25.

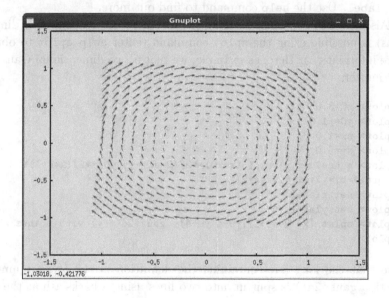

Fig. 8.25 *Gnuplot* window showing the result of plotting a two-dimensional function using **splot**.

Vectors can also specified individually on the *gnuplot* command line or in a gnuplot script. Here one uses the **set arrow** command, the arrow will appear the next time a plot (or splot) command is issued. One has to specifiy a tag, which identifies the arrow, a starting position using **from** and a destination using **to**, e.g.

```
gnuplot> set arrow 1 from 0,0 to 1,0.5
```

More options, like relative coordinates, line styles, etc., can be obtained by typing `help set arrow`.

So far, all output has appeared on the screen. It is possible to redirect the output, for example, to an encapsulated postscript file (by setting `set terminal postscript` and redirecting the output `set output "test.eps"`). When you now enter a plot command, the corresponding postscript file will be generated.

Note that not only several functions but also several data files or a mixture of both can be combined into one figure. To remember what a plot exported to files means, you can set axis labels of the figure by typing `set xlabel "L"`, which becomes active when the next `plot` command is executed. Also you can use `set title` or place arbitrary labels via `set label`. Use the `help` command to find out more.

Also three-dimensional plotting (in fact a projection into two dimensions) is possible using the `splot` command (enter `help splot` to obtain more information). Here, as example, we plot a two-dimensional Gaussian distribution:

```
gnuplot> x0=3.0
gnuplot> y0=-1.0
gnuplot> sx=1.0
gnuplot> sy=5.0
gnuplot> gauss2d(x,y)=exp(-(x-x0)**2/(2*sx)-(y-y0)**2/(2*sy))\
> /sqrt(4*pi**2*sx**2*sy**2)
gnuplot> set xlabel "x"
gnuplot> set ylabel "y"
gnuplot> splot [x0-2:x0+2] [y0-4:y0+4]  gauss2d(x,y) with points
gnuplot>
```

Note that the long line containing the definition of the (two-argument) function `gauss2d()` is split up into two lines using a backslash at the end of the first line. Furthermore, some of the variables are used inside the interval specifications at the beginning of the `splot` command. Clearly, you also can plot data files with three-dimensional data. The resulting plot appearing in the output window is shown in Fig. 8.26. You can drag the mouse inside the window showing the plot, which will alter the view.

Finally, to stop the interactive execution of *gnuplot*, enter the command `exit`. These examples should give you already a good impression of what can be done with *gnuplot*. More can be found in the documentation or the online help. How to fit functions to data using *gnuplot* is explained in Sec. 8.6.2. So far, the plots were quickly made to get an impression

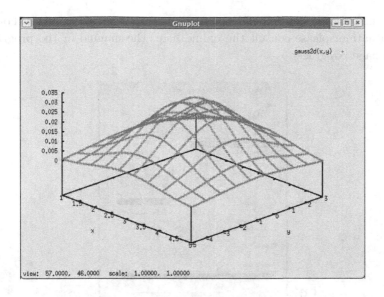

Fig. 8.26  *Gnuplot* window showing the result of plotting a two-dimensional function using `splot`.

of the data, but they did not look nice. It is also possible to make, with some effort, publication-suitable figures using *gnuplot*, see Sec. 9.2.1. On the other hand, if you prefer a direct and interactive tool, you can achieve this with *xmgrace*, which is presented in the following section.

### 8.4.2  xmgrace

The *xmgrace* (X Motiv GRaphing, Advanced Computation and Exploration of data) program is much more powerful than *gnuplot* and produces nicer output, commands are issued by clicking on menus and buttons and it offers WYSIWYG. The *xmgrace* program offers almost every feature you can imagine for two-dimensional data plots, including multiple plots (insets), fits, fast Fourier transform, interpolation. The look of the plots may be altered in any kind of way you can imagine like choosing fonts, sizes, colors, symbols, styles for lines, bar charts etc. Also, you can create manifold types of labels / legends and it is possible to add elements like texts, labels, lines or other geometrical objects in the plot. The plots can be exported to various format, in particular encapsulated postscript (`.eps`) Advanced users also can program it or use it for real-time visualization of simulations. On the other hand, its handling is a little bit slower compared to *gnuplot* and the program has the tendency to fill your screen with windows. For full information, please consult the online help, the manual or the program's web page [xmgrace].

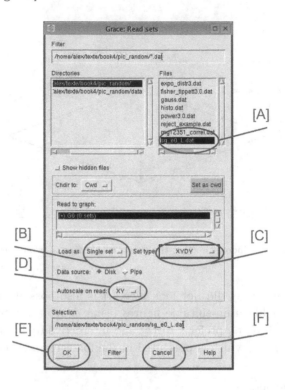

Fig. 8.27 The Grace:Read Set window of the *xmgrace* program. Among others, you can select a file [A], choose the type of the input file [B], choose the format of the data [C], what axes should be rescaled automatically on input [D]. You can actually load the data by hitting on the OK button [E] and closing the window by hitting on the Cancel button [F].

Here, just the main steps to produce a simple but nice plot are shown and some further directions are mentioned. You will be given here the most important steps to create a similar plot to the first example, shown for the *gnuplot* program, but ready for publication. First you have to start the program by typing **xmgrace** into a shell (or to start it from some window/operating system menu). Then you choose the Data menu[12], next the

---

[12]The underlined character appears also in the menu name and refers to the key one has to hit together with Alt button, if one wants to open the menu via key strokes.

Import sub menu and finally the ASCII.. sub sub menu. Then a "Grace:Read Set" window will pop up (see Fig. 8.27) and you can choose the data file to be loaded (here `sg_e0_L.dat`) [A], the type of the input file (Single Set) [B], the format of the data (XYDY) [C]. This means you have three columns, and the third one is an error bar for the second. Then you can hit on the OK button [E]. The data will be loaded and shown in the main window (see Fig. 8.28). The axis ranges have been adjusted to the data, because the "Autoscale on read" is set by default to "XY" [D]. You can quickly change the part of the data shown by the buttons (magnifier, AS, Z, z, ←, →, ↓, ↑) on the left of the main window just below the Draw button.

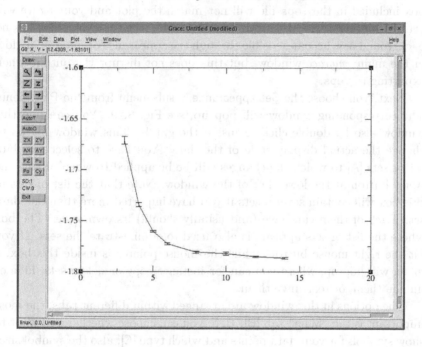

Fig. 8.28 The main *xmgrace* window after the data set has been loaded (with auto scale).

Note that another important input file type is "Block data" where the files consist of many columns of which you only want to show some. When you hit the OK button [E], another window (Grace:Edit block data) will pop up, where you have to select the columns which you actually want to display. For the data format (also when loading block data), some other important choices are XY (no error bars) and XYDYDY (full confidence interval, maybe non-symmetric). Finally, you can close the file selection window, by hitting on the Cancel button [F]. The OK and Cancel buttons are common to all *xmgrace* windows and will not be mentioned explicitly in the following.

In the form the loaded data is shown by default, it is not suitable for publication purposes, because the symbols, lines and fonts are usually too small/ too thin. To adjust many details of your graph, you should go to the Plot menu. First, you choose the Plot appearance... sub menu. A corresponding window will pop up. Here, you should just unselect the "Fill" toggle box (upper right corner), because otherwise the bounding box included in the .eps file will not match the plot and your figure will overwrite other parts of your manuscript. The fact that your plot has no background now becomes visible through the appearance of some small dots in the main *xmgrace* window, but this does not disrupt the output when exporting to .eps.

Next, you choose the Set appearance... sub menu from the Plot menu. The corresponding window will pop up, see Fig. 8.29. You can pop this window also by double-clicking inside the graph. This window allows to change the actual display style of the data. You have to select the data set or sets [A] to which the changes will be be applied to when hitting the Apply button at the lower left of the window. Note that the list of sets in this box will contain several sets if you have imported more than one data set. Each of them can have (and usually should) its own style. The box where the list of sets appears is also used to administrate the sets. If you hit the right mouse button, while the mouse pointer is inside this box, a menu will pop up, where you can for instance copy or delete sets, hide or unhide them, or rearrange them.

The options in this window are arranged within different tabs, the most important is the "Main" tab [B]. Here you can choose whether you want to show symbols for your data points and which type [C], also the symbol sizes and colors. If you want to show lines as well (Line properties area at the right), you can choose the style like "straight" and others, but also "none" is no lines should be displayed. The style can be full, dotted, dashed, and different dotted-dashed styles. For presentations and publications it is important that lines are well visible, in this example a line width of 2 is chosen [D] and a black color [E]. For presentations you can distinguish different data sets also by different colors, but for publications in scientific journals you should keep in mind that the figures are usually printed in

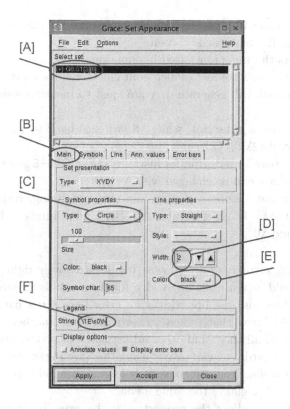

Fig. 8.29 The Grace:Set Appearance window of the *xmgrace* program. First you have to select the set or sets which should be addressed by the changes [A]. Due to the large amount of adjustable parameters, the window is organized into different tabs. The most import one is "Main" [B], which is shown here. Among others, you can select a symbol type [C] (below: symbol size, symbol color), choose the width of the lines [D] (also: line type, style) and the color [E]. Furthermore, the label for this data appearing in the legends can be states [F].

black and white, hence light colors are not visible.[13]

Each data set can have a legend (see below how to activate it). Here, the legend string can be stated. You can enter it directly, with the help of some formatting commands which are characters preceded by a backslash \. The most important ones are

- \\ prints a backslash.

---

[13] Acting as referee reading scientific papers submitted to journals, I experienced many times that I could not recognize or distinguish some data because they were obviously printed in a light color, or with a thin line width, or with tiny symbols ....

- \0 selects the Roman font, which is also the default font. A font is active until a new one is chosen.
- \1 selects the *italic* font, used in equations.
- \x selects a symbol font, which contains Greek characters. For example \xabchqL will generate $\alpha\beta\chi\eta\theta\Lambda$, just to mention some important symbols.
- \s generates a subscript, while \N switches back to normal. For example \xb\s2\N\1(x) generates $\beta_2(x)$.
- \S generates a superscript, for instance \1A\S3x\N-5 generates $A^{3x} - 5$.
- The font size can be changed with \+ and \-.
- With \o and \0 one can start and stop overlining, respectively, for instance \1A\oBC\OD generates $A\overline{BC}D$. Underlining can be controlled via \u and \U.

By default, error bars are shown (toggle box lower right corner). At least you should increase the line width for the symbols (Symbols tab) and increase the base and rise line widths for error bars (Error bars tab).

You should know that, when you are creating another plot, you do not have to redo all these and other adjustments of styles. Once you have found your standard, you can save it using the Save Parameters... sub menu from the Plot menu. You can conversely load a parameter set via the Load Parameters... sub menu of the same menu.

Next, you can adjust the properties of the axes, by choosing the Set appearance... sub menu from the Plot menu or by double-clicking on an axis. The corresponding window will pop up, see Fig. 8.30. You have to select the axis where the current changes apply to [A]. For the $x$ axis you should set the range in the fields **Start** [B] and **Stop** [C], here to the values 1 and 15. Below these two fields you find the important Scale field, where you can choose linear scaling (default), logarithmic or reciprocal, to mention the important ones.

The most important adjustments you can perform within the Main tab [D]. Here you enter the label shown below the axis in the Label string field [E]. The format of the string is the same as for the data set legends. Here you enter just \1L, which will show as $L$. The major spacing of the major (with labels) and minor ticks can be chosen in the corresponding fields [F,G]. Below there is a Format field, where you can choose how the tick labels are printed. Among the many formats, the most common are General $(1, 5, 10, \ldots)$, Exponential $(1.0\text{e}+00, 5.0\text{e}+00, 1.0\text{e}+01, \ldots)$, and Power, which is useful for logarithmic scaled axes $(10^1, 10^2, 10^3, \ldots)$. For the tick

[A]

[B]

[C]

[D]

[E]

[F]

[G]

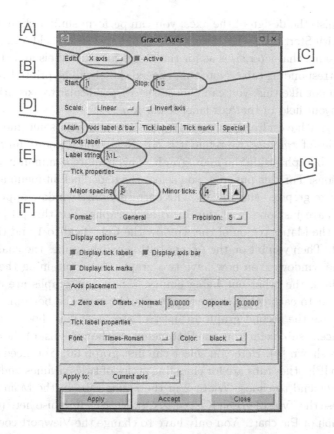

Fig. 8.30 The **Grace:** Axes window of the *xmgrace* program. First you have to select the axis which should be addressed by the changes [A]. Among others, you can change the range in the **Start** [B] and **Stop** [C] fields. Here the **Main** tab [D] is shown. You can enter an axis label in the **Label string** field [E] and select the spacing of the major and minor ticks [F,G]

labels, you can also choose a Precision. This and other fields of this tab you can leave at their standard values here. Nevertheless, you should also adjust the Char size of the axis labels (tab Axis label & bar) and of the tick labels (tab Tick labels). For publications, character sizes above 150% are usually well readable. Note that in the Axis label & bar tab, there is a field Axis transform where you can enter formulas to transform the axis more or less arbitrarily, see the manual for details. All tabs have many other fields, which are useful as well, but here we stay with the standard choices. Note that sometimes the Special tab is useful, where you can enter all major and minor ticks individually.

To finish the design of the axes, you can perform similar changes to the $y$ axis, with Start field −1.8, Stop field −1.6, Label string field \1E\s0\N(L) and the same character sizes as for the $x$ axis for axis labels and tick labels in the corresponding tabs. Note that the axis label will be printed vertically. If you do not like this, you can choose the Perpendicular to axis orientation in the Layout field of the Axis label & bar tab.

Now you have already a nice graph. To show you some more of the capabilities of *xmgrace*, we refine it a bit. Next, you generate an inset, i.e. a small subgraph inside the main graph. This is quite common in scientific publications. For this purpose, you select the underlineEdit menu and there the Arrange graph... sub menu. The corresponding window appears. We want to have just one inset, i.e. in total 2 graphs. For this purpose, you select in the Matrix region of the window the Cols: field to 1 and the Rows: field to 2. Then you hit on the Accept button which applies the changes and closes the window. You now have two graphs, one containing the already loaded data, the other one being empty. These two graphs are currently shown next to each other, one at the top and one at the bottom.

To make the second graph an inset of the first, you choose the Graph appearance... sub menu from the Plot menu. At the top a list of the available graphs is shown [A]. Here you select the first graph G0. You need only the Main tab [B], other tabs are for changing styles of titles, frames and legends. We recommend to choose Width 2 in the Frame tab. In the Main tab, you can choose the Type of graph [C], e.g. XY graph, which we use here (default), Polar graph or Pie chart. You only have to change the Viewport coordinates [D] here. These coordinates are relative coordinates, i.e. the standard full viewport including axes, labels and titles is $[0, 1] \times [0, 1]$. For the main graph G0, you choose Xmin and Ymin 0.15 and Xmax and Ymax 0.85. Note that below there is a toggle box Display legend [E], where you can control whether a legend is displayed. If you want to have a legend, you can control its position in the Leg. box tab. Now the different graphs overlap. This does not bother you, because next you select graph G1 in the list at the top of the window. We want to have the inset in the free area of the plot, in the upper right region. Thus, you enter the viewport coordinates Xmin 0.38, Ymin 0.5, Xmax 0.8 and Ymax 0.8.

Now the second graph is well placed, but empty. We want to show a scaled version of the data in the inset. Hence, you import the data again in the same way as explained above, while choosing Read to graph G1 in the Grace: Read sets window. In Sec. 8.6.2, you will see that the data follows

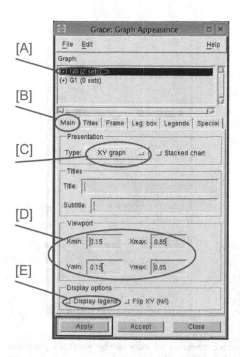

Fig. 8.31 The Grace: Graph Appearance window of the *xmgrace* program. At the top one can select to which graph changes should apply [A]. The window is divided into different tabs [B], here the Main tab is shown. The Type of the graph can be selected [C], also Title and Subtitle (empty here). The extensions of the graph can be selected in the Viewport area [D]. This allows to make one graph an inset of another. Using the Display legend toggle [E] the legend can be switched on and off.

approximately a power law behavior $e_0(L) = e_\infty + aL^b$ with $e_\infty \approx -1.788$, $a \approx 2.54$ and $b \approx -2.8$. To visualize this, we want to show $e_0(L) - e_\infty$ as a function of $L^b$. Hence, we want to *transform* the data. You choose from the Data menu the Transformations sub menu and there the Evaluate expression sub sub menu. Note that here you can also find many other transformations, e.g. Fourier transform, interpolation and curve fitting. Please consult the manual for details. In this case, the evaluateExpression window pops up, see Fig. 8.32 (if you did not close the windows you have used before, your screen will be already pretty populated). A transformation always takes the data points from one *source* set, applies a formula to all data points (or to a subset o points) and stores the result in a *destination* set. These sets can be selected at the top of the window in the Source [A] and Destination [B] fields for graph and set separately. Note that the data in the destination

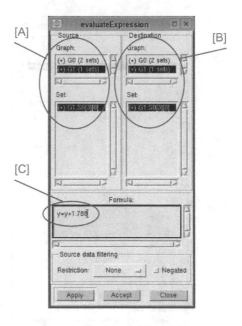

Fig. 8.32 The evaluateExpression window of the *xmgrace* program. At the top you can select Source [A] and Destination [B] sets of the transformation. The actual transformation is entered at the bottom [C].

set is overwritten. If you want to write the transformed data to a new set, you can first copy an existing set (click on the right mouse button in the Destination Set window and choose Duplicate). In our case, we want to replace the data, hence you select for source and destination the data set from graph G1. The transformation is entered below [C], here you first enter y=y+1.788 to shift the data. The you hit the Apply button at the bottom. Next you change the transformation to x=x^(-2.8) and hit the Apply button again. When you now select the second graph by clicking into it, and hit the AS (auto scale) button on the left of the main window, you will see that the data points follow a nice straight line in the inset, which confirms the behavior of the data.

Again you should select symbols, line stiles, and axis labels for the inset. Usually smaller font sizes are used here. Note that all operations always apply to the *current* graph, which can be selected for example by clicking near the corners of the boundary boxes of the graph (which does not always work, depending on which other windows are open) or by double clicking on the corresponding graph in the graph list in the Grace: Graph Appearance

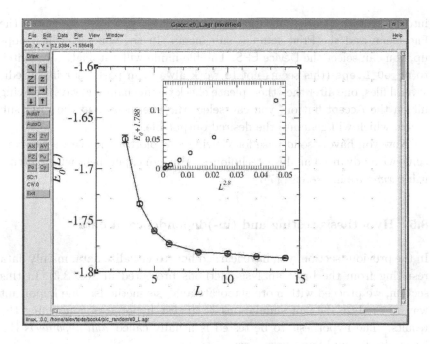

Fig. 8.33  The main *xmgrace* window after all adjustments have been made.

window. The final main window is shown in Fig. 8.33. Note that the left axis label is not fully visible. This is no problem when exporting the file as encapsulated postscript; everything will be shown. But if you do not like it, you can adjust the Xmin value of graph G0.

Finally, if you choose the menu Window and the sub menu Drawing objects a window will pop up which enable many graphical elements like texts, lines, boxes and ellipses (again with a variety of choices for colors, styles, sizes etc.) tobe added/changed and deleted in your plot. We do not go into details here.

Now you should save your plot using the File menu and the S<u>a</u>ve as... sub menu, e.g. with file name sg_e0_L.agr, where .agr is the typical postfix of *xmgrace* source files. When

| GET SOURCE CODE |
| --- |
| DIR: randomness<br>FILE(S): sg_e0_L.dat |

you want to create another plot with similar layout later, it is convenient to start from this saved file by copying it to a new file and subsequently using again *xmgrace* to modify the new file.

To export your file as encapsulated postscript, suitable for including it into presentations or publications (see Sec. 9.3), you have to choose the File menu and the Print setup... sub menu. In the window, which pops up, you can select the Device EPS. The file name will automatically switch to sg_e0_L.eps (this seems not to work always, in particular if you edit several files, one after the other, please check the file names always). Having hit on the Accept button, you can select the File menu and the Print sub menu, which will generate the desired output file.[14]

Now you have a solid base for viewing and plotting, hence we can continue with advanced analysis techniques. You can experiment with plotting using *xmgrace* in exercise (7).

## 8.5  Hypothesis testing and (in-)dependence of data

In the previous section, you have learned how to visualize data, mainly data resulting from the basic analysis methods presented in Sec. 8.3. In this section, we proceed with more elaborate analysis methods. One important way to analyze data of simulations is to test hypotheses concerning the results. The hypothesis to be tested is usually called *null hypothesis* $H_0$. Examples for null hypotheses are:

(A) In a traffic system, opening a new track will decrease the mean value of the travel time $\bar{t}_{A\to B}$ for a connection A→B below a target threshold $t_{target}$.

(B) Within an acquaintance network, a change of the rules describing how people meet will change the distribution of the number of people each person knows.

(C) The distribution of ground-states energies in disordered magnets follows a Fisher-Tippett distribution.

(D) Within a model of an ecological system, the population size of foxes is dependent on the population size of beetles.

(E) For a protein dissolved in water at room temperature, adding a certain salt to the water changes the structure of the protein.

One now can model these situations and use simulations to address the above questions. The aim is to find methods which tell us whether or not, depending on the results of the simulations, we should *accept* a null hypothesis. There is no general approach. The way we can test $H_0$ depends

---

[14]Using the tool epstopdf you can convert the postcript file also to a *pdf* file. With other tools like *convert* or *gimp* you can convert to many other styles.

on the formulation of the null hypothesis. In any case, our result will again be based on a set of measurements, such as a sample of independent data points $\{x_0, x_1, \ldots, x_{n-1}\}$, formally obtained by sampling from random variables $\{X_0, X_1, \ldots, X_{n-1}\}$ (here again, all described by the same distribution function $F_X$). To get a solid statistical interpretation, we use a *test statistics*, which is a function of the sample $t = t(x_0, x_1, \ldots, x_{n-1})$. Its distribution describes a corresponding random variable $T$. This means, you can use any estimator (see page 284), which is also a function of the sample, as test statistics. Nevertheless, there are many test statistics, which usually are not used as estimators.

To get an idea of what a test statistics $t$ may look like, we discuss now test statistics for the above list of examples. For (A), one can use obviously the sample mean. This has to be compared to the threshold value. This will be performed within a statistical interpretation, enabling a null hypothesis to be accepted or rejected, see below. For (B) one needs to compare the distributions of the number of acquaintances before and after the change, respectively. Comparing two distributions can be done in many ways. One can just compare some moments, or define a distance between them based on the difference in area between the distribution function, just to mention two possibilities. For discrete random variables, the mean-squared difference is particularly suitable, leading to the so-called chi-squared test, see Sec. 8.5.1. For the example (C), the task is similar to (B), only that the empirical results are compared to a given distribution and that the corresponding random variables are continuous. Here, a method based on the maximum distance between two distribution functions is used widely, called Kolmogorov-Smirnov (KS) test (see Sec. 8.5.2). To test hypothesis (D), which means to check for statistical independence, one can record a two-dimensional histogram of the population size of foxes and beetles. This is compared with the distribution where both populations are assumed to be independent, i.e. with the product of the two single-population distribution functions. Here, a variant of the chi-squared test is applied, see Sec. 8.5.4. In the case (E), the sample is not a set of just one-dimensional numbers, instead the simulation results are conformations of proteins given by $3N$–dimensional vectors of the positions $\underline{r}_i$ ($i = 1, \ldots, N$) of $N$ particles. Here, one could introduce a method to compare two protein conformations $\{\underline{r}_i^A\}$, $\{\underline{r}_i^B\}$ in the following way: First, one "moves" the second protein towards the first one such that the positions of the center of masses agree. Second, one considers the axes through the center of masses and through the first atoms, respectively. One rotates the second protein around its center of mass such that these axes become parallel. Third, the second protein is rotated around the above axis such that the distances between the last atoms of the two proteins are minimized. Finally, for these normalized positions $\{\underline{r}_i^{B\star}\}$, one calculates the squared difference of all pairs of atom positions $d = \sum_i (\underline{r}_i^A - \underline{r}_i^{B\star})^2$ which serves as test function. For a statistical analysis, the distribution of $d$ for one thermally fluctuating protein can be determined via a simulation and then compared to the average value observed when changing the conditions. We do not go into further details here.

The general idea to test a null hypothesis using a test statistics in a statistical meaningful way is as follows:

(1) You have to know, at least to an approximate level, the probability distribution function $F_T$ of the test statistics *under the assumption that the null hypothesis is true*. This is the main step and will be covered in detail below.
(2) You select a certain significance level $\alpha$. Then you calculate an interval $[a_l, a_u]$ such that the cumulative probability of $T$ outside the interval equals to $\alpha$, for instance by distributing the weight equally outside the interval via $F(a_l) = \alpha/2$, $F(a_u) = 1 - \alpha/2$. Sometimes one-sided intervals are more suitable, e.g. $[\infty, a_u]$ with $F(a_u) = 1 - \alpha$, see below concerning example (A).
(3) You calculate the actual value $t$ of the test statistics from your simulation. If $t \in [a_l, a_u]$ then you *accept* the hypothesis, otherwise you reject it. Correspondingly, the interval $[a_l, a_u]$ is called *acceptance interval*.

Since this is a probabilistic interpretation, there is a small probability $\alpha$ that you do not accept the null hypothesis, although it is true. This is called a *type I error* (also called *false negative*), but this error is under control, because $\alpha$ is known.

On the other hand, it is important to realize that in general the fact that the value of the test statistics falls inside the acceptance interval does *not* prove that the null hypothesis is true! A different hypothesis $H_1$ could indeed hold, just your test statistics is not able to discriminate between the two hypotheses. Or, with a small probability $\beta$, you might obtain some value for the test statistics which is unlikely for $H_1$, but likely for $H_0$. Accepting the null hypothesis, although it is not true, is called a *type II error* (also called *false positive*). Usually, $H_1$ is not known, hence $\beta$ cannot be calculated explicitly. The different cases and the corresponding possibilities are summarized in Fig. 8.34. To conclude: If you want to prove a hypothesis H (up to some confidence level $1 - \alpha$), it is better to use the opposite of H as null hypothesis, if this is possible.

| test decision \ reality | $H_0$ is true | $H_1$ is true |
|---|---|---|
| accept $H_0$ | correct decision $1-\alpha$ | type II error $\beta$ |
| reject $H_0$ | type I error $\alpha$ | correct decision $1-\beta$ |

Fig. 8.34 The null hypothesis $H_0$ might be true, or the alternative (usually unknown) hypothesis $H_1$. The test of the null hypothesis might result in an acceptance or in a rejection. This leads to the four possible scenarios which appear with the stated probabilities.

Indeed, in general the null hypothesis must be suitably formulated, such that it can be tested, i.e. such that the distribution function $F_T$ describing $T$ can be obtained, at least in principle. For example (A), since the test statistics $T$ is a sample mean, it is safe to assume a Gaussian distribution for $T$: One can perform enough simulations rather easily, such that the central limit theorem applies. We use as null hypothesis the opposite of the formulated hypothesis (A). Nevertheless, it is impossible to calculate an acceptance interval for the Gaussian distribution based on the assumption that the mean is *larger* than a given value. Here, one can change the null hypothesis, such that instead an expectation value *equal to* $t_{\text{target}}$ is assumed. Hence, the null hypothesis assumes that the test statistics has a Gaussian distribution with expectation value $t_{\text{target}}$. The variance of $T$ is unknown, but one can use, as for the calculation of error bars, the sample variance $s^2$ divided by $n - 1$. Now one calculates on this basis an interval $[a_l, \infty]$ with $F_T(a_l) = \alpha$. Therefore, one rejects the null hypothesis if $t < a_l$, which happens with probability $\alpha$. On the other hand, if the true expectation value is even larger than $t_{\text{target}}$, then the probability of finding a mean with $t < a_l$ becomes even smaller than $\alpha$, i.e. less likely. Hence, the hypothesis (A) can be accepted or rejected on the basis of a fixed expectation value.

For a general hypothesis test, to evaluate the distribution of the test statistics $T$, one can perform a *Monte Carlo simulation*. This means one draws repeatedly samples of size $n$ according to a distribution $F_X$ determined by the null hypothesis. Each time one calculates the test statistics $t$ and records a histogram of these values (or a sample distribution function

$F_{\hat{T}}$) which is an approximation of $F_T$. In this way, the corresponding statistical information can be obtained. To save computing time, in most cases no Monte Carlo simulations are performed, but some knowledge is used to calculate or approximate $F_T$.

In the following sections, the cases corresponding to examples (B), (C), (D) are discussed in detail. This means, it is explained how one can test for equality of discrete distributions via the chi-squared test and for equality of continuous distributions via the KS test. Finally, some methods for testing concerning (in-)dependence of data and for quantifying the degree of dependence are stated.

### 8.5.1   Chi-squared test

The chi-squared test is a method to compare histograms and discrete probability distributions. The test works also for discretized (also called *binned*) continuous probability distributions, where the probabilities are obtained by integrating the pdf over the different bins. The test comes in two variants:

- Either you want to compare the histogram $\{h_k\}$ for bins $B_k$ (see Sec. 8.3.3) describing the sample $\{x_0, x_1, \ldots, x_{n-1}\}$ to a given discrete or discretized probability mass function with probabilities $\{p_k\} = P(x \in B_k)$. The null hypothesis $H_0$ is: "*the sample follows a distribution given by $\{p_k\}$*".
  Note that the probabilities are fixed and independent of the data sample. If the probabilities are parametrized and the parameter is determined by the sample (e.g. by the mean of the data) such that the probabilities fit data best, related methods as described in Sec. 8.6.2 have to be applied.
- Alternatively, you want to compare two histograms $\{h_k\}$, $\{\hat{h}_k\}$ obtained from two different samples $\{x_0, x_1, \ldots, x_{n-1}\}$ and $\{\hat{x}_0, \hat{x}_1, \ldots, \hat{x}_{n-1}\}$ defined for the same bins $B_k$. The null hypothesis $H_0$ is: "*the two samples follow the same distribution*".[15]

In case the test is used to compare intrinsically discrete data, the intervals $B_k$ can conveniently be chosen such that each possible outcome

---

[15]Note that here we assume that the two samples have the same size, which is usually easy to achieve in simulations. A different case occurs when also the number of sample points is a random variable, hence a difference in the number of sample points makes the acceptance of $H_0$ less likely, see [Press et al. (1995)].

corresponds to one interval. Note that due to the binning process, the test can be applied to high-dimensional data as well, where the sample is a set of vectors. Also non-numerical data can be binned. In these cases each bin represents either a subset of the high-dimensional space or, in general, a subset of the possible outcomes. For simplicity, we restrict ourselves here to one-dimensional numerical samples.

Fig. 8.35   Chi-squared statistics: A histogram (solid line) is compared to a discrete probability distribution (dashed line). For each bin, the sum of the squared differences of the bin counter $h_k$ to the expected number of counts $np_k$ is calculated (dotted vertical lines), see Eq. (8.72). In this case, the differences are quite notable, thus the probability that the histogram was obtained via random experiments from a random variable described by the probabilities $\{p_k\}$ (null hypothesis) will be quite small.

We start with the first case, where a sample histogram is compared to a probability distribution, corresponding to example (C) on page 320. The test statistics, called $\chi^2$, is defined as:

$$\chi^2 = \sum_k{}' \frac{(h_k - np_k)^2}{np_k} \qquad (8.72)$$

with $np_k$ being the expected number of sample points in bin $B_k$. The prime at the sum symbol indicates that bins with $h_k = np_k = 0$ are omitted. The number of contributing bins is denoted by $K'$. If the pmf $p_k$ is nonzero for an infinite number of bins, the sum is truncated for terms $np_k \ll 1$. This means that the number of contributing bins will be always finite. Note that bins exhibiting $h_k > 0$ but $p_k = 0$ are not omitted. This results in an infinite value of $\chi^2$, which is reasonable, because for data with $h_k > 0$ but $p_k = 0$, the data cannot be described by the probabilities $p_k$.

The chi-squared distribution with $\nu = K' - 1$ degrees of freedom (see Eq. (8.45)) describes the chi-squared test statistics, if the number of bins and the number of bin entries is large. The term $-1$ in the number of degrees of freedom comes from the fact that the total number of data points $n$ is equal to the total number of expected data points $\sum_k n_k p_k = n \sum_k p_k = n$, hence the $K'$ different summands are not statistically independent. The probability density of the chi-squared distribution is given in Eq. (8.45). To perform the actual test, it is recommended to use the implementation in the *GNU scientific library* (GSL) (see Sec. 7.3).

Next, a C function `chi2_hd()` is shown which calculates the cumulative probability (*p-value*) that a value of $\chi^2$ or larger is obtained, given the null hypothesis that the

| GET SOURCE CODE |
| --- |
| DIR: `randomness` |
| FILE(S): `chi2.c` |

sample was generated using the probabilities $p_k$. Arguments of `chi2_hd()` are the number of bins, and two arrays `h[]` and `p[]` containing the histogram $h_k$ and the probabilities $p_k$, respectively:

```
1  double chi2_hd(int n_bins, int *h, double *p)
2  {
3    int n;                        /* total number of sample points */
4    double chi2;                            /* chi^2 value */
5    int K_prime;                  /* number of contributing bins */
6    int i;                                   /* counter */
7
8    n = 0;
9    for(i=0; i<n_bins; i++)
10     n += h[i];        /* calculate total number of sample_points */
11
12   chi2 = 0.0; K_prime = 0;
13   for(i=0; i<n_bins; i++)                    /* calculate chi^2 */
14   {
15     if(p[i] > 0)
16     {
17       chi2 += (h[i]-n*p[i])*(h[i]/(n*p[i])-1.0);
18       K_prime ++;
19     }
20     else if(h[i] >0)        /* bin entry for zero probability ? */
21     {
22       chi2 = 1e60;
23       K_prime ++;
24     }
25   }
26   return(gsl_cdf_chisq_Q(chi2, K_prime-1));
27 }
```

First, in lines 8–10, the total number of sample points is obtained from summing up all histogram entries. In the main loop, lines 12–25, the value of $\chi^2$ is calculated. In parallel, the number of contributing bins is determined. Finally (line 26) the p-value is obtained using the GSL function `gsl_cdf_chisq_Q()`. This p-value can be compared with the significance level $\alpha$. If the p-value is larger, the null hypothesis is accepted, otherwise rejected.

Note that the result for the p-value clearly depends on the number of bins, and, if applicable, on the actual choice of bins. Nevertheless, all reasonable choices, although maybe leading to somehow different numerical results, will lead to the same decisions concerning the null hypothesis in most cases.

Next, we consider the case, where we want to compare two histograms $\{h_k\}, \{\hat{h}_k\}$ corresponding to example (B) on page 320. In this case the $\chi^2$ statistics reads

$$\chi^2 = {\sum_k}' \frac{(h_k - \hat{h}_k)^2}{h_k + \hat{h}_k} \tag{8.73}$$

The sum runs over all bins where $h_k \neq 0$ or $\hat{h}_k \neq 0$, and $K'$ being the corresponding number of contributing bins. Consequently, the bins which should be included are uniquely defined, in contrast to the case where a histogram is compared to a distribution defined for infinitely many outcomes. Note that in the denominator the sum of the bin entries occurs, not the average. The reason is that the chi-squared distribution is a sum of standard Gaussian distributed numbers (variance 1) and here, where the differences of two (approximately) Gaussian quantities are taken, the resulting variance is the sum of the individual variances, approximated roughly by the histogram entries. To calculate the p-value, again the chi-squared distribution with $\nu = K' - 1$ degrees of freedom is to be applied. Here, no C implementation is shown, rather we refer the reader to exercise (8). In case the two sample sizes are different, e.g, $n$ and $\hat{n}$, respectively, Eq. (7.69) must be changed to [Gagunashvili (2009)]

$$\chi^2 = \frac{1}{n\hat{n}} {\sum_k}' \frac{(\hat{n} h_k - n \hat{h}_k)^2}{h_k + \hat{h}_k}$$

### 8.5.2 *Kolmogorov-Smirnov test*

Next, we consider the case where the statistical properties of a sample $\{x_0, x_1, \ldots, x_{n-1}\}$, obtained from a repeated experiment using a continuous random variable, is to be compared to a given distribution function $F_X$. One could, in principle, compare a histogram and a correspondingly binned probability distribution using the chi-squared test explained in the previous section. Unfortunately, the binning is artificial and has an influence on the results (imagine few very large bins). Consequently, the method presented in this section is usually preferred, since it requires no binning. Note that if the distribution function is parametrized and if the parameter is determined by the sample (e.g. by the mean of the data) such that the $F_X$ fits the data best, the methods from Sec. 8.6.2 have to be applied.

The basic idea of the *Kolmogorov-Smirnov* test is to compare the distribution function to the empirical sample distribution function $F_{\hat{X}}$ defined in Eq. (8.69). Note that $F_{\hat{X}}(x)$ is piecewise constant with jumps of size $1/n$ at the positions $x_i$ (assuming that each data point is contained uniquely in the sample).

Here again, one has several choices for the test statistics. For instance, one could calculate the area between $F_X$ and $F_{\hat{X}}$. Instead, usually just the maximum difference between the two functions is used:

$$d_{\max} \equiv \max_x \left| F_X(x) - F_{\hat{X}}(x) \right| \tag{8.74}$$

Since the sample distribution function changes only at the sample points, one has to perform the comparison just before and just after the jumps. Thus, Eq. (8.74) is equivalent to

$$d_{\max} \equiv \max_{x_i} \left\{ \left| F_X(x_i) - 1/n - F_{\hat{X}}(x_i) \right|, \left| F_X(x_i) - F_{\hat{X}}(x_i) \right| \right\}$$

This sample statistics is visualized in Fig. 8.36.

The p-value, i.e. the probability of a value of $d_{\max}$ as measured ($d_{\max}^{\text{measured}}$) or worse, given the null hypothesis that the sample is drawn from $F_X$, is approximately given by (see [Press et al. (1995)] and references therein):

$$P(d_{\max} \geq d_{\max}^{\text{measured}}) = Q_{\text{KS}} \left( [\sqrt{n} + 0.12 + 0.11/\sqrt{n}] d_{\max}^{\text{measured}} \right) \tag{8.75}$$

This approximation is already quite good for $n \geq 8$. Here, the following

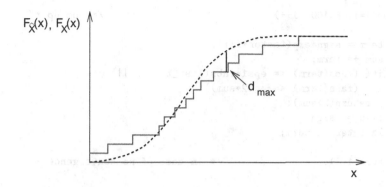

Fig. 8.36  Kolmogorov-Smirnov test: A sample distribution function (solid line) is compared to a given probability distribution function (dashed line). The sample statistics $d_{max}$ is the maximum difference between the two functions.

auxiliary probability function is used:

$$Q_{KS}(\lambda) = 2 \sum_{i=1}^{\infty} (-1)^{i+1} e^{-2i^2 \lambda^2} \qquad (8.76)$$

with $Q_{KS}(0) = 1$ and $Q_{KS}(\infty) = 0$. This function can be implemented most easily by a direct summation [Press et al. (1995)]. The function Q_ks() receives the value of $\lambda$ as argument and returns $Q_{KS}(\lambda)$:

> GET SOURCE CODE
>
> DIR: **randomness**
> FILE(S): **ks.c**

```
1  double Q_ks(double lambda)
2  {
3    const double eps1 = 0.0001;  /* relative margin for stop */
4    const double eps2 = 1e-10;   /* relative margin for stop */
5    int i;                              /* loop counter */
6    double sum;                         /* final value */
7    double factor;             /* constant factor in exponent */
8    double sign;                        /* of summand */
9    double term, last_term;    /* summands, last summand */
10
11   sum = 0.0; last_term = 0.0; sign = 1.0;    /* initialize */
12   factor = -2.0*lambda*lambda;
```

```
13    for(i=1; i<100; i++)                          /* sum up */
14    {
15      term = sign*exp(factor*i*i);
16      sum += term;
17      if( (fabs(term) <= eps1*fabs(last_term)) ||
18          (fabs(term) <= eps2*sum))
19        return(2*sum);
20      sign =- sign;
21      last_term = term;
22    }
23    return(1.0);                  /* in case of no convergence */
24  }
```

The summation (lines 13–22) is performed for at most 100 iterations. If the current term is small compared to the previous one or very small compared to the sum obtained so far, the summation is stopped (line 17–18). If this does not happen within 100 iterations, the sum has not converged (which means $\lambda$ is very small) and $Q(0) = 1$ is returned.

This leads to the following C implementation for the KS test. The function ks() expects as arguments the number of sample points n, the sample x[] and a pointer F to the distribution function:

```
1   double ks(int n, double *x, double (*F)(double))
2   {
3     double d, d_max;                    /* (maximum) distance */
4     int i;                              /* loop counter */
5     double F_X;              /* empirical distribution function */
6
7     qsort(x, n, sizeof(double), compare_double);
8
9     F_X = 0; d_max = 0.0;
10    for(i=0; i<n; i++)                      /* scan through F_X */
11    {
12      d = fabs(F_X-F(x[i]));    /* distance before jump of F_X */
13      if( d> d_max)
14        d_max = d;
15      F_X += 1.0/n;
16      d = fabs(F_X-F(x[i]));    /* distance after jump of F_X */
17      if( d> d_max)
18        d_max = d;
19    }
20    return(Q_ks( d_max*(sqrt(n)+0.12+0.11/sqrt(n))));
21  }
```

First the sample is sorted (line 7). This allows for a simple implementation of the sample distribution function, because at each sample data point, in the order of occurrence, the value of $F_{\hat{X}}$ is increased by $1/n$. When obtaining the maximum distance (lines 10–19), one has to compare $F_{\hat{X}}$ to the distribution function $F_X$ just before (lines 12–14) and after (lines 15–18) the jumps. Note that this implementation works also for samples, where some data points occur multiple times.

For the actual test, one calculates the p-value for the given sample using `ks()`. If the p-value exceeds the indented significance level $\alpha$, the null hypothesis is accepted, i.e. the data is compatible with the distribution with high probability. Usually quite small significances are used, e.g. $\alpha = 0.05$. This means that even substantial values of $d_{max}$ are accepted. Thus, one rejects the null hypothesis only, as usual, in case the probability for an error of type I is quite small.

It is also possible to compare two samples of sizes $n_1, n_2$ via the KS test. The test statistics for the two sample distribution functions is again the maximum distance. The probability to find a value of $d_{max}$ as obtained or worse, given the null hypothesis that the samples are drawn from the same distribution, is as above in Eq. (8.75), only one has to replace $n$ by the "effective" sample size $n_{eff} = n_1 n_2/(n_1 + n_2)$, for details see [Press et al. (1995)] and references therein. It is straightforward to implement this test when using the C function `ks()` shown above as template.

### 8.5.3 *ROC analysis*

The *Receiver-operator characteristics* (ROC) is a method to evaluate a classification test. We assume that there are two types A and B of objects and we measure some quantity $Q$ in order to establish whether an object is of type A or B (null hypothesis $H_0$ in Fig. 8.34). As an experimental example, we measure the weight of an animal in order to find out whether the animal is grown up or not. Or, in the field of computer simulations, one could simulate the evolution of proteins, compare different proteins via so called *alignment* algorithms, and use the resulting *alignment score* to find out whether two proteins picked from the population are evolutionary related or not [Wolfsheimer et al. (2012)].

Usually, the distributions of the measured quantity $Q$ follow for type A and B different distributions. These distributions might differ by shape or by the values of the parameters, e.g., two Gaussians centered at different values $\mu_A$ and $\mu_B$ (and exhibiting variances $\sigma_A^2$ and $\sigma_B^2$). Now, when measuring $Q$, to decide whether the object belongs to A or B, a threshold $\theta$ can

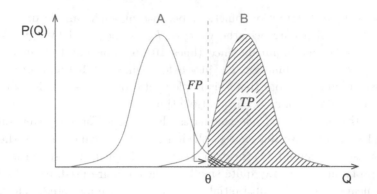

Fig. 8.37  Distributions of quantity Q for objects of type A and B, respectively. A test using a threshold $\theta$ is used to classify the objects. All objects where $Q > \theta$ is measured are assumed to be of type B. For a good choice of $\theta$ a large fraction $TP$ ("true positive") is correctly classified as type B, while a small fraction $FP$ ("false positive") of A objects is also classified by the test as B.

be used, see Fig. 8.37. If $Q > \theta$ holds, it is assumed that the object belongs to type B ("positive"), otherwise to type A ("negative"). This type of test, with a threshold value $\theta$, is the same as for example (A) from page 320. In case the two distributions overlap, there will not only be objects of type B which exhibit $Q > \theta$ (null-hypothesis $H_0$ true, "true positive") but also objects of type A where $Q$ is above the threshold ("false positive"). Clearly both true and false positive will increase when decreasing the threshold $\theta$. To evaluate the test, within an ROC analysis one draws the fraction $TP$ of true positive as a function of false positive $FP$ while varying the threshold. Therefore, the curve is given by $(1 - F_B(\theta), 1 - F_A(\theta))$, while varying $\theta$, where $F_A$ and $F_B$ are the cumulative distribution functions for type A and B, respectively. Such curves look typically like as shown in Fig. 8.38, where the simple case of two Gaussian distributions ($\mu_A = 0$, $\sigma_A^2 = 1$, $\sigma_B^2 = 1$) is shown for three selected values of $\mu_B$.

A good choice of the threshold corresponds to that part of the curve, where the true positives are large while the false positive are still low, hence, where the curve is close to its upper left corner. As visible from Fig. 8.38, quite intuitively, the quality of the test increases when the two distributions are better separated, i.e., when $\mu_B$ increases. For this example, since it was assumed the distributions are Gaussian, the ROC curve could be obtained exactly. To evaluate such tests for general cases, simulations are the most natural choice to obtain the ROC curve.

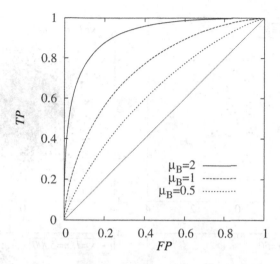

Fig. 8.38 Receiver-operator characteristic for the case where a random population consists of two types A,B. Some quantity $Q$ is measured. For elements of type A, $Q$ is N(0,1) distributed while for type B, the quantity $Q$ is N($\mu_B$,1) distributed. A test is used, where elements are classified as type B (Null hypothesis) if $Q > \theta$, $\theta$ being a adjustable threshold. The ROC plot displays the true positive *(TP)* as a function of the false positive *(FP)* when varying $\theta$. Here, three curves for different value $\mu_B$ are shown. Also the diagonal *TP=FP* is shown, which represents the case where no meaningful classification is possible, corresponding to $\mu_B = 0 = \mu_A$.

### 8.5.4 *Statistical (in-)dependence*

Here, we consider samples, which consist of pairs $(x_i, y_i)$ $(i = 0, 1, \ldots, n-1)$ of data points. Generalizations to higher-dimensional data is straightforward. The question is, whether the $y_i$ values depend on the $x_i$ values (or vice versa). In this case, one also says that they are *statistically related*. If yes, this means that if we know one of the two values, we can predict the other one with higher accuracy. The formal definition of statistical (in-) dependence was given in Sec. 8.1. An example of statistical dependence occurs in weather simulations: The amount of snowfall is statistically related to the temperature: If it is too warm or too cold, it will not snow. This also shows, that the dependence of two variables it not necessarily monotonous. In case one is interested in monotonous and even linear dependence, one usually says that the variables are *correlated*, see below.

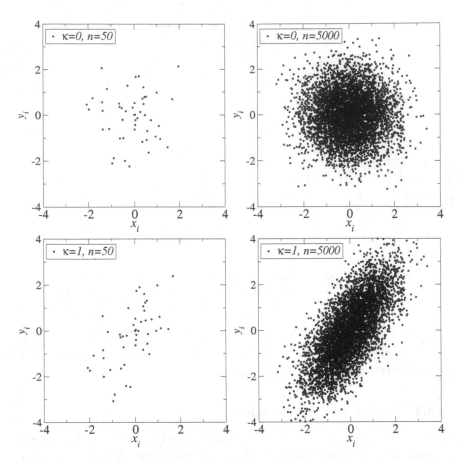

Fig. 8.39　Scatter plots for $n$ data points $(x_i, y_i)$ where the $x_i$ numbers are generated from a standard Gaussian distribution (expectation value 0, variance 1), while each $y_i$ number is drawn from a Gaussian distribution with expectation value $\kappa x_i$ (variance 1).

It is important to realize that we have to distinguish between statistical *significance* of a statistical dependence and the *strength* of the dependence. Say that our test tells us that the $x$ values are statistically related with high probability. This usually just means that we have a large sample. On the other hand, the strength of the statistical dependence can be still small. It could be, for example, that a given value for $x$ will influence the probability distribution for $y$ only

```
GET SOURCE CODE

DIR: randomness
FILE(S): points0A.dat
points0B.dat
points1A.dat
points1B.dat
```

slightly. One the other hand, the strength can be large, which means, for example, knowing $x$ almost determines $y$. But if we have only few sample points, we cannot be very sure whether the data points are related or not. Nevertheless, there is some connection: the larger the strength, the easier it is to show that the dependence is significant. For illustration consider a sample where the $x_i$ numbers are generated from a standard Gaussian distribution (expectation value 0, variance 1), while each $y_i$ number is drawn from a Gaussian distribution with expectation value $\kappa x_i$ (variance 1).[16] Hence, if $\kappa = 0$, the data points are independent. Scatter plots, where each sample point $(x_i, y_i)$ is shown as dot in the $x - y$ plane are exposed in Fig. 8.39. Four possibilities are presented, $\kappa = 0/1$ combined with $n = 50/5000$. Below, we will also present what the methods we use here will tell us about these data sets.

In this section, first a variant of the chi-squared test is presented, which enables us to check whether data is independent. Next, the *linear correlation coefficient* is given, which states the strength of linear correlation. Finally, it is discussed how one can quantify the dependence *within* a sample, for example between sample points $x_i, x_i + \tau$.

To test statistical dependence for a sample $\{(x_0, y_0), (x_1, y_1), \ldots, (x_{n-1}, y_{n-1})\}$, one considers usually the null hypothesis: $H_0 = $ "The $x$ sample points and the $y$ sample points are independent." To test $H_0$ one puts the pairs of sample points into two-dimensional histograms $\{h_{kl}\}$. The counter $h_{kl}$ receives a count, if for data point $(x_i, y_i)$ we have $x_i \in B_k^{(x)}$ and $y_i \in B_l^{(y)}$, for suitably determined bins $\{B_k^{(x)}\}$ and $\{B_l^{(y)}\}$. Let $k_x$ and $k_y$ be the number of bins in $x$ and $y$ direction, respectively. Next, one calculates single-value (or one-dimensional) histograms $\{\hat{h}_k^{(x)}\}$ and $\{\hat{h}_l^{(y)}\}$ defined by

$$\hat{h}_k^{(x)} = \sum_l h_{kl}$$

$$\hat{h}_l^{(y)} = \sum_k h_{kl} \tag{8.77}$$

These one-dimensional histograms describe how many counts in a certain bin arise for one variable, regardless of the value of the other variable. It is assumed that all entries of these histograms are not empty. If not, the bins should be adjusted accordingly. Note that $n = \sum_k \hat{h}_k^{(x)} = \sum_l \hat{h}_l^{(y)} = \sum_{kl} h_{kl}$ holds.

---

[16]This is an example, where the random variables $Y_i$ which described the sample are not identical.

Relative frequencies, which are estimates of probabilities, are obtained by normalizing with $n$, i.e. $\hat{h}_k^{(x)}/n$ and $\hat{h}_l^{(y)}/n$. If the two variables $x_i, y_i$ are independent, then the relative frequency to obtain a pair of values $(x, y)$ in bins $\{B_k^{(x)}\}$ and $\{B_l^{(y)}\}$ should be the product of the single-value relative frequencies. Consequently, by multiplying with $n$ one obtains the corresponding expected number $n_{kl}$ of counts, under the assumption that $H_0$ holds:

$$n_{kl} = n\frac{\hat{h}_k^{(x)}}{n}\frac{\hat{h}_l^{(y)}}{n} = \frac{\hat{h}_k^{(x)}\hat{h}_l^{(y)}}{n} \tag{8.78}$$

These expected numbers are compared to the actual numbers in the two-dimensional histogram $\{h_{kl}\}$ via the $\chi^2$ test statistics, comparable to Eq. (8.72):

$$\chi^2 = \sum_{kl}\frac{(h_{kl} - n_{kl})^2}{n_{kl}} \tag{8.79}$$

The statistical interpretation of $\chi^2$ is again provided by the chi-squared distribution. The number of degrees of freedom is determined by the number of bins $(k_x k_y)$ in the two-dimensional histogram minus the number of constraints. The constraints are given by Eq. (8.77), except that the total number of counts being $n$ is contained twice, resulting in $k_x + k_y - 1$. Consequently, the number of degrees of freedom is

$$\nu = k_x k_y - k_x - k_y + 1. \tag{8.80}$$

Therefore, under the assumption that the $x$ and $y$ sample points are independent, $p = 1 - F(\chi^2, \nu)$ gives the probability (p-value) of observing a test statistics of $\chi^2$ or larger. $F$ is here the distribution function of the chi-squared distribution, see Eq. (8.45). This p-value has to be compared to the significance level $\alpha$. If $p < \alpha$, the null hypothesis is rejected.

The following C function implements the chi-squared independence test `chi2_indep()`. It receives the number of bins in $x$ and $y$ direction as arguments, as well as a two-dimensional array, which carries the histogram:

| GET SOURCE CODE |
|---|
| DIR: **randomness** |
| FILE(S): **chi2indep.c** |

```
1   double chi2_indep(int n_x, int n_y, int **h)
2   {
3     int n;                          /* total number of sample points */
4     double chi2;                                      /* chi^2 value */
5     int k_x, k_y;                     /* number of contributing bins */
6     int k, l;                                           /* counters */
7     int *hx, *hy;                   /* one-dimensional histograms */
8
9     hx = (int *) malloc(n_x*sizeof(int));             /* allocate */
10    hy = (int *) malloc(n_y*sizeof(int));
11
12    n = 0;              /* calculate total number of sample_points */
13    for(k=0; k<n_x; k++)
14      for(l=0; l<n_y; l++)
15        n += h[k][l];
16
17    k_x = 0;                  /* calculate 1-dim histogram for x */
18    for(k=0; k<n_x; k++)
19    {
20      hx[k] = 0;
21      for(l=0; l<n_y; l++)
22        hx[k] += h[k][l];
23      if(hx[k] > 0)                       /* does x bin contribute ? */
24        k_x++;
25    }
26
27    k_y = 0;                  /* calculate 1-dim histogram for y */
28    for(l=0; l<n_y; l++)
29    {
30      hy[l] = 0;
31      for(k=0; k<n_x; k++)
32        hy[l] += h[k][l];
33      if(hy[l] > 0)                       /* does y bin contribute ? */
34        k_y++;
35    }
36
37    chi2 = 0.0;
38    for(k=0; k<n_x; k++)                            /* calculate chi^2 */
39      for(l=0; l<n_y; l++)
40        if( (hx[k] != 0)&&(hy[l] != 0) )
41          chi2 += pow(h[k][l]-(double) hx[k]*hy[l]/n, 2.0)/
42            ((double) hx[k]*hy[l]/n);
43    free(hx);
44    free(hy);
45    return(gsl_cdf_chisq_Q(chi2, k_x*k_y - k_x -k_y + 1));
46  }
```

First, the one-dimensional histograms are allocated (lines 9–10). Then the total number of counts, i.e. the sample size, is calculated (lines 12–15). In lines 17–26, the one-dimensional histogram for the $x$ direction is obtained. Also the effective number of bins in that direction is calculated. In lines 27–35, the same happens for the $y$ direction. The actual value of the $\chi^2$ test statistics is determined in lines 37–42. After being used, the allocated memory is freed (lines 43–44). Finally, the p-value is calculated (line 45), again the GSL function `gsl_cdf_chisq_Q()` is used for this purpose.

The p-values for the sample sets shown in Fig. 8.39 are as follows: $p(\kappa = 0, n = 50) = 0.077$, $p(\kappa = 0, n = 5000) = 0.457$, $p(\kappa = 1, n = 50) = 0.140$, $p(\kappa = 1, n = 5000) < 10^{-100}$. Hence, the null hypothesis of independence would not be rejected (say $\alpha = 0.05$) for the case $\kappa = 1, n = 50$, which is actually correlated. On the other hand, if the number of samples is large enough, there is no doubt.

Once it is established that a sample contains dependent data, one can try to measure the strength of dependence. A standard way is to use the *linear correlation coefficient* (also called *Pearson's r*) given by

$$r \equiv \frac{\sum_i (x_i - \overline{x})(y_i - \overline{y})}{\sqrt{\sum_i (x_i - \overline{x})^2}\sqrt{\sum_i (y_i - \overline{y})^2}} \,. \tag{8.81}$$

This coefficient assumes, as indicated by the name, that a linear correlation exists within the data. The implementation using a C function is straight forward, see exercise (9). For the data shown in Fig. 8.39, the following correlation coefficients are obtained: $r(\kappa = 0, n = 50) = 0.009$, $r(\kappa = 0, n = 5000) = 0.009$, $r(\kappa = 1, n = 50) = 0.653$, $r(\kappa = 1, n = 5000) = 0.701$. Here, also in the two cases, where the statistics is low, the value of $r$ reflects whether or not the data is correlated. Nevertheless, this is only the case because we compare strongly correlated data to uncorrelated data. If we compare weakly but significantly correlated data, we will still get a small value of $r$. Hence, to test for significance, it is better to use the hypothesis test based on the $\chi^2$ test statistics.

Finally, note that a different type of correlation may arise: So far it was always assumed that the different sample points $x_i, x_j$ (or sample vectors) are statistically independent of each other. Nevertheless, it could be the case, for instance, that the sample is generated using a Markov chain Monte Carlo simulation [Newman and Barkema (1999); Landau and Binder (2000); Robert and Casella (2004); Liu (2008)], where each data point $x_{i+1}$ is calculated using some random process, but also depends on the previous data point $x_i$, hence $i$ is a kind of artificial sample time of the simulation. This dependence decreases with growing time distance between sample points. One way to see how quickly this dependence decreases is to use a variation of the correlation coefficient Eq. (8.81), i.e. a *correlation function*:

$$\tilde{C}(\tau) = \frac{1}{n-\tau} \sum_{i=0}^{n-1-\tau} x_i x_{i+\tau}$$

$$- \left( \frac{1}{n-\tau} \sum_{i=0}^{n-1-\tau} x_i \right) \times \left( \frac{1}{n-\tau} \sum_{i=0}^{n-1-\tau} x_{i+\tau} \right) \qquad (8.82)$$

The term $\frac{1}{n-\tau} \sum_{i=0}^{n-1-\tau} x_i \times \frac{1}{n-\tau} \sum_{i=0}^{n-1-\tau} x_{i+\tau}$ will converge to $\overline{x}^2$ for $n \to \infty$ if it can be assumed that the distribution of the sample points is stationary, i.e. does not depend on the sample time. Therefore, $\tilde{C}(\tau)$ is approximately $\frac{1}{n-\tau} \sum_{i=0}^{n-1-\tau} (x_i - \overline{x})(x_{i+\tau} - \overline{x})$, comparable to the nominator of the linear correlation coefficient Eq. (8.81). Usually one normalizes the correlation function by $\tilde{C}(0)$, which is just the sample variance in the stationary case, see Eq. (8.55):

$$C(\tau) = \tilde{C}(\tau)/C(0) . \qquad (8.83)$$

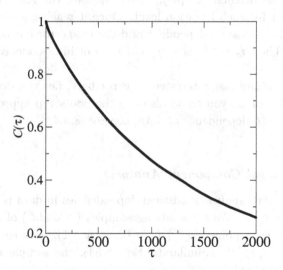

Fig. 8.40 Correlation function $C(\tau)$ for a simulation of a ferromagnetic system, $x_i$ being the magnetization at time step $i$. (For experts: Ising system of size $16 \times 16$ spins simulated with single-spin flip Metropolis Monte Carlo at a (reduced) temperature $T = 2.269$ close to the phase transition temperature, where correlation times $\tau_c$ are large).

Consequently, for any data, for example obtained from a Markov chain Monte Carlo simulation, $C(0) = 1$ will always hold, Then $C(\tau)$ decreases with increasing difference $\tau$, see for example Fig. 8.40. Very often the functional form is similar to an exponential $\sim \exp(-\tau/\tau_c)$. In theory, $C(\tau)$ should converge to zero for $\tau \to \infty$, but due to the finite size of the sample, usually strong fluctuations appear for $\tau$ approaching $n$. A typical time $\tau_c$ which measures how fast the dependence of the sample points decreases is given by $C(\tau_c) = 1/e$, which is consistent with the above expression, if the correlation function decreases exponentially. At twice this distance, the correlation is already substantially decreases (to $1/e^2$). Consequently, if you want to obtain error bars for samples obtained from dependent data, you could include for instance only points $x_0, x_{2\tau_c}, x_{4\tau_c}, x_{6\tau_c}, \dots$ in a sample, or just use $n/(2\tau_c)$ instead of $n$ in any calculation of error bars. Although these error bars are different from those if the sample was really independent, it gives a fairly good impression of the statistical error.

Alternatively, to obtain a typical time $\tau_c$ without calculating a correlation function, you can also use the *blocking method* [Flyvbjerg (1998)]. Within this approach, you iteratively merge neighboring data points via $x_i^{(z+1)} = (x_{2i}^{(z)} + x_{2i+1}^{(z)})/2$ and $n^{(z+1)} = n^{(z)}/2$ (iteration level $z = 0$ corresponds to the original sample). You calculate the standard error bar $\sigma^{(z)}/\sqrt{n^{(z)} - 1}$ for each iteration level. Once it reaches a plateau at level $z_c$, the data is (almost) independent and the true error bar is given by the level value. Then $\tau_c = 2^{z_c}$ is a typical time of independence of the data points.

If you are really just interested in error bars, i.e. you do not need to know the value of $\tau_c$, you could also use the bootstrap approach which is not susceptible to dependence of data, see Sec. 8.3.4.

### 8.5.5 *Principal Component Analysis*

A different way to analyse statistical dependencies in data is the *principal component analysis*. We are studying samples ("clouds") of $n$ real-valued $d$-dimensional data points $\{\underline{\tilde{x}}^{(i)}\}$ ($i = 0, \dots, n - 1$), with each data point $\underline{\tilde{x}}^{(i)} = (\tilde{x}_1^{(i)}, \dots, \tilde{x}_d^{(i)})^T$. Similar to Eq. (8.54), the sample mean, or the "center" of the cloud, is

$$\underline{\bar{x}} = \frac{1}{n} \sum_i \underline{\tilde{x}}^{(i)},$$

where the sum $\sum_i$ here and in the following runs from 0 to $n-1$. Since we will be interested in the spread of the cloud, i.e., in the variance of the data, it is sufficient to consider the center of the cloud shifted to the origin. Therefore, we normalize the data points by the mean:

$$\underline{x}^{(i)} \equiv \underline{\tilde{x}}^{(i)} - \underline{\overline{x}}. \tag{8.84}$$

An example of such a sample for $d = 3$ dimensions, shifted to the origin, is shown in Fig. 8.41.

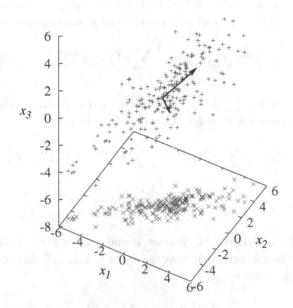

Fig. 8.41 A sample of random three-dimensional vectors $(x_1, x_2, x_3)^T$ ("+" symbols). The "×" symbols show the same sample projected on the $x_1, x_2$ plane. The two arrows indicate to the most important principal components, i.e., the two eigenvectors of the covariance matrix corresponding to the two largest eigenvalues. The length of the vectors are the square roots of the corresponding eigenvalues.

The basic idea is to find directions $\underline{r}$ (with length $|\underline{r}| = 1$) in the cloud which characterize the data sample most, i.e., along which the variance of the data is largest. These are the *principal components*. For the sample data, the two most important directions are indicated in Fig. 8.41 by arrows.

Now we will see step by step that obtaining the principal components can be actually performed via calculating eigenvectors and eigenvalues of the (empirical) covariance matrix of the sample of data points. For this purpose we project each normalized data point on the direction $\underline{r}$, i.e., calculate the scalar products $q^{(i)} \equiv \underline{r} \cdot \underline{x}^{(i)}$. Thus, for the sample variance in direction $\underline{r}$ we obtain using Eq. (8.55):

$$\text{Var}_{\underline{r}}(\{\underline{x}^{(i)}\}) = \frac{1}{n}\sum_i (q^{(i)} - \bar{q})^2 \overset{\bar{q}=0}{=} \frac{1}{n}\sum_i (q^{(i)})^2 = \frac{1}{n}\sum_i (\underline{r} \cdot \underline{x}^{(i)})^2 \,,$$

where the average $\bar{q} = \frac{1}{n}\sum_i q^{(i)}$ vanishes because the data points have been normalized by their mean. Using the transpose $\ldots^T$ a scalar product can be written as $\underline{a} \cdot \underline{b} = \underline{a}^T \underline{b}$. When additionally using the symmetry of the scalar product $\underline{a} \cdot \underline{b} = \underline{b} \cdot \underline{a}$ we can rewrite the above expression as

$$\text{Var}_{\underline{r}}(\{\underline{x}^{(i)}\}) = \frac{1}{n}\sum_i \underline{r}^T \underline{x}^{(i)} (\underline{x}^{(i)})^T \underline{r} \,. \tag{8.85}$$

The expression $\frac{1}{n}\sum_i \underline{x}^{(i)} (\underline{x}^{(i)})^T$ denotes a $d \times d$ matrix which is called the (sample) *covariance matrix* $\mathbf{C}$, which has the entries

$$C_{kl} \equiv \frac{1}{n}\sum_i x_k^{(i)} x_l^{(i)} = \frac{1}{n}\sum_i (x_k^{(i)} - \bar{x}_k)(x_l^{(i)} - \bar{x}_l)$$

$$= \frac{1}{n}\sum_i x_k^{(i)} x_l^{(i)} - \bar{x}_k \bar{x}_l \,. \tag{8.86}$$

The covariance matrix is real, symmetric and positive semi-definite,[17] i.e., it exhibits only non-negative eigenvalues $\lambda_z$ ($z = 1, \ldots, d$) and corresponding normalized eigenvectors $\underline{e}_z$, i.e.,

$$\mathbf{C}\underline{e}_z = \lambda_z \underline{e}_z \quad (\lambda_z \geq 0)\,. \tag{8.87}$$

Without loss of generality, we can assume that the eigenvalues are ordered in descending order $\lambda_1 > \lambda_2 > \ldots > \lambda_d$ and that the eigenvectors form an orthonormal ($\underline{e}_z \cdot \underline{e}_{z'} = \delta_{z,z'}$) basis of the $d$-dimensional vector space.[18] This means, any vector can be represented as linear combination of the eigenvectors. Thus, we can write $\underline{r} = \sum_z r_z \underline{e}_z$, where here and in the following sums $\sum_z$ run from 1 to $d$, and $r_z \equiv \underline{r} \cdot \underline{e}_z$. This we can insert into Eq. (8.85), resulting in

---

[17]Any matrix of the form $\mathbf{A}\mathbf{A}^T$ is semi-definite.

[18]In case of degeneracy, i.e., if some eigenvalues appear multiple times, one can turn the set of eigenvectors into an orthonormal basis.

$$\text{Var}_{\underline{r}}(\{\underline{x}^{(i)}\}) = \underline{r}^T \mathbf{C} \underline{r} = \left(\sum_z r_z \underline{e}_z\right)^T \mathbf{C} \left(\sum_{z'} r_{z'} \underline{e}_{z'}\right)$$

$$\overset{(8.87)}{=} \left(\sum_z r_z \underline{e}_z\right)^T \left(\sum_{z'} r_{z'} \lambda_{z'} \underline{e}_{z'}\right)$$

$$= \sum_z r_z^2 \lambda_z . \tag{8.88}$$

where for the last equality, we have used the orthonormality of the basis of eigenvectors. As mentioned in the beginning, we are interested in the direction which maximizes the variance of the projection of the data sample onto the $\underline{r}$ direction, hence in the maximum of (8.88). Due to the normalization of $\underline{r}$ we have $1 = \underline{r}^2 = (\sum_z r_z \underline{e}_z)^2 = \sum_z r_z^2$. Since the eigenvalues are sorted in descending order, we obtain the maximum for $\text{Var}_{\underline{r}}(\{\underline{x}^{(i)}\})$ simply by setting $r_1 = 1$ and $r_z = 0$ for $z = 2, \ldots, d$, thus $\underline{r} = \underline{e}_1$. The direction of the maximum spread of the data, i.e., the first principal component, is simply the direction of the eigenvector corresponding to the largest eigenvalue of the (sample) covariance matrix, and the variance in that direction is the largest eigenvalue itself.

Thus, it is straightforward to see that the direction of the second largest spread of the data (orthogonal to the direction of the largest spread), i.e., the second principal component, is the eigenvector corresponding to eigenvalue $\lambda_2$ and so on. Finally, the total spread (variance) of the data is simply the sum $\lambda = \sum_z \lambda_z$ of all eigenvalues and the amount $\lambda_z/\lambda$ denotes the relative amount of the spread corresponding to the $z$th eigenvector, i.e., corresponding to the $z$th most important principal component. Thus, using the eigenvalues one can somehow estimate how many really important independent degrees of freedom the data exhibits. For example, if in a sample of $d = 10$ dimensional data points, the first three eigenvalues sum up to 99% of the total variance $\lambda$ of the sample, then in fact the data can be represented by three more or less independent variables. The dependencies inherent in the variables are described by the eigenvectors, which are often, unfortunately, hard to interpret.

The C function `principal_components()` calculates the principal components of a data sample. Here we again use some data types and functions from the GNU scientific library,

| GET SOURCE CODE |
| --- |
| DIR: `randomness` |
| FILE(S): `princ_comp.c` |

in particular for the calculation of the eigenvalues and eigenvectors, which keeps the code compact. Thus, we use the GSL data types gsl_vector and gsl_matrix, see page 239 in Sec. 7.3.

The function receives a matrix data which contains the data points, $\{\tilde{\underline{x}}^{(i)}\}$, each column of the matrix contains one data point. To return the eigenvalues, a vector eval is used. The corresponding eigenvectors are returned in the matrix evec, again one vector per column.

```
1   void principal_components(gsl_vector *eval, gsl_matrix *evec,
2                             gsl_matrix *data)
3   {
4     int t,d1, d2;                          /* loop counters */
5     int dim;           /* number of components of data point vectors */
6     int num_points;                     /* number of data points */
7     gsl_matrix *cov;                      /* covariance matrix */
8     gsl_vector *avg;                             /* averages */
9     gsl_eigen_symmv_workspace *w;     /* memory for eigenvalues etc */
```

Locally, we need a vector avg to hold the averages of the different components of the data points, and a matrix cov which stores the covariance matrix. These are allocated next, using the dimension $d$ which is read off from the number of rows of data and stored in variable dim. The number $n$ of data points is stored in variable num_points:

```
10    num_points = data->size2;                     /* initialize */
11    dim = data->size1;
12    avg = gsl_vector_alloc(dim);
13    cov = gsl_matrix_alloc(dim, dim);
```

Next, the averages are calculated. Here, the following two GSL functions are used: gsl_vector_set(), which sets an entry in the vector (taking the vector, the index of the entry and the value as arguments) and gsl_vector_get(), which returns the value of an entry (taking the vector and the index of the entry as arguments). The GSL function gsl_vector_scale() (line 21) multiplies every entry of a given vector with some number.

```
14    for(d1=0; d1<dim; d1++)                   /* calculate averages */
15    {
16      gsl_vector_set(avg, d1, 0);
17      for(t=0; t<num_points; t++)
18        gsl_vector_set(avg, d1, gsl_vector_get(avg, d1)
19              +gsl_matrix_get(data, d1, t));
20    }
```

```
21   gsl_vector_scale(avg, 1.0/ num_points);
```

Using these averages, the covariance matrix can be calculated according to Eq. (8.86). Similar to the corresponding vector functions, the GSL matrix access functions `gsl_matrix_get()` and `gsl_matrix_set()` are used, where in both cases a row (2nd argument) and a column index (3rd argument) have to be supplied.

```
22   for(d1=0; d1<dim; d1++)              /* calculate covariance matrix */
23     for(d2=d1; d2<dim; d2++)
24     {
25       gsl_matrix_set(cov, d1, d2,
26                      -gsl_vector_get(avg, d1)*gsl_vector_get(avg, d2));
27       for(t=0; t<num_points; t++)
28         gsl_matrix_set(cov, d1, d2, gsl_matrix_get(cov, d1, d2)+
29                                     gsl_matrix_get(data, d1, t)
30                              *gsl_matrix_get(data, d2, t)/num_points);
31       gsl_matrix_set(cov, d2, d1, gsl_matrix_get(cov, d1, d2));
32     }
```

The actual calculation of the eigenvalues and eigenvectors is now very simple using the GSL build-in functions. First, one has to provide some "work space" `w` (line 33) and then the actual calculation is performed and the results are stored in the eigenvalue vector `eval` and in the matrix `evec` of eigenvectors:

```
33   w = gsl_eigen_symmv_alloc(dim); /* calculate eigenvectors/values */
34   gsl_eigen_symmv(cov, eval, evec, w);
```

Finally, the eigenvalues and the corresponding eigenvectors are sorted in descending eigenvalue order. For brevity, a simple neighbor exchange algorithm is used here. Since the dimension $d$ is typically small, there is no benefit in using a more complicated algorithm which pays off only asymptotically. Finally, all locally used memory is freed.

```
35   for(t=0; t<dim; t++) /* order eigenvalues (and corresp. vectors) */
36     for(d1=0; d1<dim-1; d1++)
37       if(gsl_vector_get(eval, d1) < gsl_vector_get(eval, d1+1))
38       {
39         gsl_matrix_swap_columns(evec, d1, d1+1);
40         gsl_vector_swap_elements(eval, d1, d1+1);
41       }
42
43   gsl_eigen_symmv_free(w);
```

```
44    gsl_vector_free(avg);
45    gsl_matrix_free(cov);
46  }
```

### 8.5.6   *Clustering Data*

Often one wants to find similarities in a set of $n$ objects, characterized by "feature vectors". As in the case of principal component analysis (Sec. 8.5.5), we assume that the data is given by $n$ $d$-dimensional real-valued data points $\{\underline{x}^{(i)}\}$ $(i = 0, \ldots, n-1)$, with each data point $\underline{x}^{(i)} = (x_1^{(i)}, \ldots, x_d^{(i)})^T$. Furthermore, let the data exhibit some substructure, i.e., one can organize the data into groups, called *clusters*, such that the objects within the groups are more similar to each other compared to objects belonging to different groups. Note that this is not a precise definition. In fact, a good definition does not exist. Thus, what is a good clustering always depends on the application and on the data. This is already illustrated by the two sample data sets A and B, which are shown in Fig. 8.42. For a detailed discussion of clustering, see Ref. [Jain and Dubes (1988)]. Here, we discuss three approaches, the *k-means* algorithm, *neighbor-based clustering*, and an *agglomerative clustering* method.

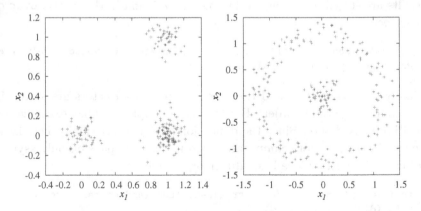

Fig. 8.42   Two sample sets A ($n = 200$, left) and B ($n = 200$, right) for sets of two-dimensional data points, which will subsequently be used to test clustering algorithms, aiming at identifying subsets of similar data points.

Following the *k-means* approach, one wants to partition the data set into $k$ clusters, $k$ being a somehow given parameter. Below we discuss shortly the influence of the choice of $k$ on the resulting clustering. The $k$-means approach is based on a geometric point of view. Each cluster $c = 0 \ldots, k-1$ shall be represented by a *center* vector $\underline{\xi}^{(c)}$. Let us assume that each data point with index $i \in \{0, \ldots, n-1\}$ is assigned to some (initially possibly randomly chosen) cluster $c(i) \in \{0, \ldots, k-1\}$.

We calculate the mean-squared difference (or "spread") $\chi^2$ of all data points to the center of its cluster:

$$\chi^2 = \frac{1}{n} \sum_{i=0}^{n-1} \left( \underline{\xi}^{(c(i))} - \underline{x}^{(i)} \right)^2 .$$

We assume that the best choice of the center vectors and of the assignment to the clusters is the one which minimizes the spread. Thus, for a fixed assignment of data points to clusters and any cluster $c \in \{0, \ldots, k-1\}$ we have for each direction $a \in \{1, \ldots, d\}$ the condition that the partial derivative of the spread with respect to the $a$'th component of the center vector $\underline{\xi}^{(c)}$ vanishes:

$$0 \stackrel{!}{=} \frac{\partial \chi^2}{\partial \xi_a^c} = \frac{2}{n} \sum_{i=0}^{n-1} \delta_{c,c(i)} \left( \xi_a^{(c(i))} - x_a^{(i)} \right) = 2 \frac{n_c}{n} \xi_a^{(c)} - \frac{2}{n} \sum_{i=0}^{n-1} \delta_{c,c(i)} x_a^{(i)} ,$$

where $n_c = \sum_{i=0}^{n-1} \delta_{c,c(i)}$ is the size of cluster $c$. Thus, each center vector $\underline{\xi}^{(c)}$ is, as the name suggest, the geometric center of the data points assigned to cluster $c$:

$$\underline{\xi}^{(c)} = \frac{1}{n_c} \sum_{i=0}^{n-1} \delta_{c,c(i)} \underline{x}^{(i)} . \tag{8.89}$$

On the other hand, for fixed centers $\underline{\xi}^{(c)}$, minimizing $\chi^2$ can be achieved by assigning each data point to its closest cluster:

$$c(i) = \operatorname{argmin}_{c=0,\ldots,k-1} \left\{ \left( \underline{\xi}^{(c)} - \underline{x}^{(i)} \right)^2 \right\} . \tag{8.90}$$

Thus, a very simple algorithm can be obtained by starting with a random assignments of the data points to clusters and then iterating Eqs. (8.89) and (8.90) until convergence, e.g. until the relative change of the center vectors is less than a small given threshold $\epsilon$. Note that this approach does *not* guarantee a convergence to a solution where the spread $\chi^2$ assumes its *global* minimum. See below for an example.

Next, we discuss a short C implementation of the *k*-means approach. Note that
the file cluster.c also contains auxiliary and
test functions, like cluster_test_data1() and

| GET SOURCE CODE |
| --- |
| DIR: randomness |
| FILE(S): cluster.c |

cluster_test_data2() which generate the test sets A and B, respectively.
You can just use the code it as it is, or use it as a starting point for a
more refined approach, e.g. by introducing additional weights signifying
the importance of the data points. The function cluster_k_means() receives a matrix data, which contains the data points as column vectors,
and the number *k* of clusters. For convenience, we use the GSL data types
gsl_vector and gsl_matrix, see page 239 in Sec. 7.3. Also we use a GSL
random number generator rng for the initial assignment of the data points
to the clusters, see page 236 in Sec. 7.3. The function returns an array,
which contains for each data point an integer specifying its cluster. The
array is created inside the function. Furthermore, the function returns the
final spread, via a pointer spread_p which is passed as argument.

```
1   int *cluster_k_means(gsl_matrix *data, int k, gsl_rng *rng,
2                        double *spread_p)
3   {
4     int *cluster;          /* holds for each point its cluster ID */
5     gsl_matrix *center;       /* holds for each cluster its center */
6     int *cluster_size;        /* holds for each cluster its #points */
7     int dim;        /* number of components of data point vectors */
8     int num_points;                   /* number of data points */
9     int t, d, c;                          /* loop counters */
10    double spread, spread_old;        /* total distance to centers */
11    double dist, dist_min;    /* (minimum) dist. between point/center */
12    double diff;        /* lateral distance between point/center */
13    int c_min;              /* center which is closest to a point */
14    int do_print = 0;                      /* for debugging */
```

For initializing, the number num_points of data points and the number dim
of entries are take from the GSL matrix data structure (lines 15 and 16).
Using this, the array cluster, which is returned, the array cluster_size,
which holds for each cluster the number of assigned data points, and a GSL
matrix for the centers are allocated (lines 17–19). Also, each data point is
assigned initially to a randomly chosen cluster (lines 21,22):

```
15    num_points = data->size2;                      /* initialize */
16    dim = data->size1;
17    cluster = (int *) malloc(num_points*sizeof(int));
```

```
18    cluster_size = (int *) malloc(k*sizeof(int));
19    center = gsl_matrix_alloc(dim, k);
20
21    for(t=0; t<num_points; t++)    /* intial assignments to clusters */
22      cluster[t] = (int) k*gsl_rng_uniform(rng);
```

The main loop (lines 25–66) is performed until the spread changes by less than one percent (line 25). In each iteration, for the given assignments of the data points to clusters, the cluster sizes and the cluster centers are updated (lines 27–42) according to Eq. (8.89). This is achieved by first initializing centers and cluster sizes to zero (lines 27–29), by next iterating over all data points (lines 30–37), and by finally normalizing the centers by the cluster sizes $n_c$ (lines 38–42). Note that in C, the entries $1, \ldots, d$ of the data points run from 0 to dim$-1$.

For each iteration, second, for each data point its closest cluster is determined and the spread is recalculated (lines 44–65). This involves in particular iterating for each data point over all cluster centers (lines 49–62), determining the distance between the data point and a center (lines 51–56) and determining the closest center (lines 57–61).

```
23    spread = 1e100;
24    spread_old = 2e100;
25    while ( (spread_old-spread)>0.01*spread_old)        /* main loop */
26    {
27      gsl_matrix_set_all(center, 0.0);
28      for(c=0; c<k; c++)
29        cluster_size[c] = 0;
30      for(t=0; t<num_points; t++)                   /* determine centers */
31      {
32        cluster_size[cluster[t]]++;
33        for(d=0; d<dim; d++)
34          gsl_matrix_set(center, d, cluster[t],
35                         gsl_matrix_get(center, d, cluster[t])+
36                         gsl_matrix_get(data, d, t));
37      }
38      for(c=0; c<k; c++)
39        if(cluster_size[c] > 0)
40        for(d=0; d<dim; d++)
41          gsl_matrix_set(center, d, c,
42                         gsl_matrix_get(center, d, c)/cluster_size[c]);
43
44      spread_old = spread;
```

```
45      spread = 0;
46      for(t=0; t<num_points; t++)          /* determine closest center */
47      {
48        c_min = -1;
49        for(c=0; c<k; c++)                     /* test with all centers */
50        {
51          dist = 0;                    /* calculate distance point/center */
52          for(d=0; d<dim; d++)
53          {
54            diff = gsl_matrix_get(center,d,c)-gsl_matrix_get(data,d,t);
55            dist += diff*diff;
56          }
57          if( (c_min == -1)||(dist_min > dist))  /* closest center ? */
58          {
59            c_min = c;
60            dist_min = dist;
61          }
62        }
63        cluster[t] = c_min;
64        spread += dist_min;
65      }
66    }
```

At the end of the function, the current spread is stored in the external variable which is given by the pointer **spread_p**. Also the memory for the center vectors and the cluster sizes is freed and finally the **cluster** array containing the result is returned:

```
67    *spread_p = spread;
68    gsl_matrix_free(center);
69    free(cluster_size);
70    return(cluster);
71  }
```

In the upper left of Fig. 8.43 the result for the cluster analysis of data set A is shown for the choice $k = 3$. Also shown are the "paths" the centers have taken during the iteration of the algorithm. Obviously, the clustering represents the structure of the data well. This changes in case the value of $k$ does not represent the data well, see upper right of Fig. 8.43, where the result for $k = 5$ is shown. Since the algorithm is forced to have five clusters, it subdivides the cluster around $(1, 0)^T$ into three clusters. This case where $k$ is not well adapted serves also as an example to show that the simple iterative algorithm does not necessarily converge to the global minimum spread. When repeating the clustering for $k = 5$ with different

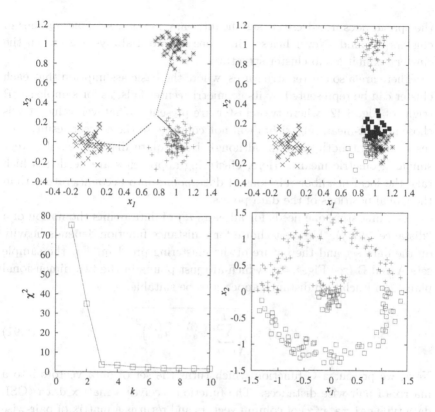

Fig. 8.43 Upper left: Result of the clustering of sample set A with the $k$-means algorithm for $k = 3$. Different symbols correspond to different clusters. The lines show how the centers have moved during the iterations of the algorithm. Upper right: result of the $k$-means algorithm for sample set B and $k = 5$. Here, the algorithm mistakenly subdivides the cluster around $(1, 0)^T$ into three sub clusters. Lower left: spread $\chi^2$ as function of the number of clusters $k$. Above the most suitable number $k = 3$ the spread decreases only slightly when increasing the number of clusters. Lower right: For sample set B, $k$-means fails even if the most suitable number $k = 2$ is chosen.

seeds for the random number generator, different spreads and thus different cluster assignments will occur. Such a non-unique convergence, observed after restarting the `cluster_k_means()` function, may also be used as an indicator that $k$ is not well chosen.

Often, the most suitable number $k$ of clusters is in fact not known in advance. In this case, it helps sometimes to perform the clustering for several values of $k$ and observe the spread $\chi^2$ as a function of $k$, see lower left of Fig. 8.43. The spread shrinks monotonously when increasing $k$. When the spread does not decrease significantly any more, a suitable number of clusters is found. Nevertheless, this does not work always, e.g. when the clusters exhibit a sub-cluster structure.

There are also cluster structures, where the basic assumption that each cluster can be represented by its geometric center fails, as for sample set B (right of Fig. 8.42) where two cluster are present. Whatever value of $k$ is chosen, the $k$-means algorithm will not converge to the correct result. The reason is that both clusters, although being quite distinct, exhibit very similar geometric means. Here, clustering approaches are needed, which take the local neighbor relations of data points into account, rather than the global positions of the data points.

As a first step, one needs for all pairs $i, j$ of data points the notion of a "distance" $d(i, j)$. The best choice for a distance function depends heavily on the data set and the nature of the clustering problem. For the sample sets A and B (see Fig. 8.42), which are just points in the two-dimensional plane, the Euclidean distance appears to be suitable:

$$d(i, j) \equiv \sqrt{\sum_{a=1}^{d} \left( x_a^{(i)} - x_a^{(j)} \right)^2} \tag{8.91}$$

Next, we present a C function which turns the set of data vectors into a matrix of pair-wise distances. The function receives a matrix **data** (GSL data type **gsl_matrix**) of column vectors and returns a matrix of pair-wise distances. Note that the number of data points and the dimensions, i.e., the number of entries, can be taken from the matrix **data** (lines 9 and 10). The main loop over all pairs of data points is performed in lines 13–24. The calculation of the distance is done in lines 16–21. As usually in C, the elements $1, \ldots, d$ of the data points are stored in entries 0 through **dim**$-1$.

```
1   gsl_matrix *cluster_distances(gsl_matrix *data)
2   {
3     gsl_matrix *dist;                    /* matrix containing distances  */
4     int dim;            /* number of components of data point vectors */
5     int num_points;                        /* number of data points */
6     int t1, t2, d;                            /* loop counters */
7     double distance, diff;            /* auxiliary distance variables */
8
9     dim = data->size1;
10    num_points = data->size2;                        /* initialize */
11    dist = gsl_matrix_alloc(num_points, num_points);
12
```

```
13    for(t1=0; t1<num_points; t1++)              /* iterate over all pairs */
14      for(t2=0; t2<=t1; t2++)
15      {
16        distance = 0;
17        for(d=0; d<dim; d++)                    /* calculate distance */
18        {
19          diff = gsl_matrix_get(data,d,t1) - gsl_matrix_get(data,d,t2);
20          distance += diff*diff;
21        }
22        gsl_matrix_set(dist, t1, t2, sqrt(distance));        /* set */
23        gsl_matrix_set(dist, t2, t1, gsl_matrix_get(dist, t1, t2));
24      }
25
26    return(dist);
27  }
```

The basic idea of the *neighbor-based clustering* is to translate the data set into a graph, see Sec. 6.8. For each data point $x^{(i)}$, there is a node $i$ in the graph. Furthermore, all pairs $i, j$ of nodes are connected by an (undirected) edge $\{i, j\}$, if the distance between the corresponding data points is smaller than some given threshold $\theta$, i.e., if $d(i, j) < \theta$. This is achieved by the following function, which uses the graph data structures as previously introduced in Sec. 6.8.2. The function receives the matrix of distances and the threshold value $\theta$. The code is rather concise, because one needs only to determine the number of nodes (line 8), set up the nodes of the graph (line 9) and iterate over all pairs of nodes to set an edge whenever the distance is below the threshold (lines 10–13):

```
1   gs_graph_t *cluster_threshold_graph(gsl_matrix *distance,
2                                       double threshold)
3   {
4     gs_graph_t *g;
5     int num_nodes;
6     int n1, n2;                                 /* node counter */
7
8     num_nodes = distance->size1;
9     g = gs_create_graph(num_nodes);
10    for(n1=0; n1<num_nodes; n1++)      /* loop over all pairs of nodes */
11      for(n2=n1+1; n2<num_nodes; n2++)
12        if(gsl_matrix_get(distance, n1, n2) < threshold)   /* edge ? */
13          gs_insert_edge(g, n1, n2);
14
15    return(g);
16  }
```

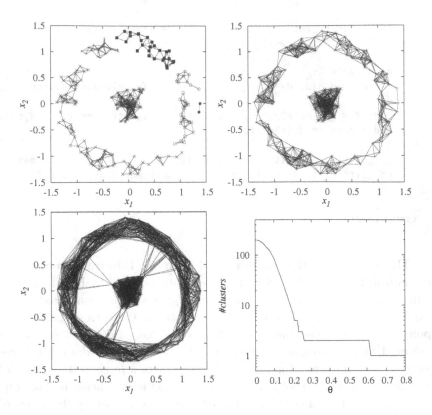

Fig. 8.44  Result of the clustering of sample set B with the neighbor-based clustering. Upper left: result for threshold $\theta = 0.2$. Upper right: result for threshold $\theta = 0.3$. Lower left: result for threshold $\theta = 0.7$. Lower right: Number of clusters as a function of the threshold $\theta$.

Finally, the actual clustering is fairly simple: one just determines the connected components using the function gs_components() as presented in Sec. 6.8.4. Each connected component corresponds to one cluster!

As example, the neighbor-based clustering algorithm is applied to sample set B, where the $k$-means approach failed. As visible from Fig. 8.44, the result depends on the choice of the threshold $\theta$: If the threshold is too small, too many clusters will be detected, while for a threshold being too large, just one cluster is found. For intermediate values of the threshold, the most suitable result of two clusters is found. If the correct threshold is not known in advance, on can, e.g., study the number of clusters as a function of the threshold $\theta$. As visible from the lower right of Fig. 8.44, the number of clusters does not change for a large range of thresholds $\theta \in [0.26, 0.6]$, indicating that the most natural number of clusters for sample set B is two.

However, be aware that also neighbor-based clustering might fail. Imagine that for sample set B there is a small "bridge" of data points between the two clusters. In this case, neighbor-based clustering will also not be able to distinguish the two clusters. In this case, more advanced techniques are needed, which are based on the idea that a group of several close-by points should influence the outcome of the clustering as a group (similar to the $k$ means clustering) but in terms of distances to other points or groups of points (unlike $k$-means clustering where only absolute positions are relevant). This is the fundamental notion underlying *hierarchical clustering* methods. These methods are also often able to detect substructures, like clusters inside clusters etc. Here, we will focus on an *agglomerative* clustering approach, namely the *average-linkage* approach.

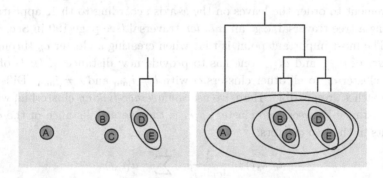

Fig. 8.45 Example for agglomerative clustering: Initially one has a set of $n = 5$ data points corresponding to $n$ clusters A, B, C, D, and E (bottom part). Iteratively the closest clusters are merged (illustrated by ellipses). For each merger, a branch in a *dendrogram* (tree) is generated (top part). Left: Situation after the first two single-point clusters D,E have been merged into a two-point cluster DE. Right: Final situation, after the merger of B with C, followed by the merger of BC with DE and finally the merger of A with BCDE. The dendrogram represents the hierarchical cluster structure.

The basic idea of agglomerative clustering is that one considers the initial set of $n$ data points as a set of $n$ clusters $C = \{c_1, \ldots, c_n\}$ with $c_i = \{\underline{x}^{(i)}\}$. One defines cluster distances $d_c(i,j)$ between pairs of the initial clusters $c_i$ and $c_j$ as given by the selected point-to-point distance function $d(i,j)$, like the Euclidean distance Eq. (8.91) or any other suitable distance function. Within agglomerative clustering iteratively the two closest clusters $c_{i_{\min}}$ and $c_{j_{\min}}$, i.e., where

$$i_{\min}, j_{\min} = \operatorname{argmin}_{i,j} d_c(i,j) \,,$$

are merged into one new single cluster $k = c_{i_{\min}} \cup c_{j_{\min}}$. Thus, within the first step, two clusters containing a single data point each will be merged. During the next steps, single-data point clusters or multiple-data point clusters will be merged. This is illustrated in Fig. 8.45. During each iteration the number of clusters will be decreased by one, hence, this process stops after $n - 1$ iterations when all data points are collected in one single cluster. The merging process can be represented by a tree, called *dendrogram*: The leaves of the tree are given by the initial data points, i.e., the clusters $c_1, \ldots, c_n$. Whenever two cluster are merged, a new (non-leaf) node is created, which has the two clusters as descendants. Therefore, the root of the tree is the node which has those two clusters as descendants, which were joined during the last iteration. Note when drawing the tree, it is convenient to order the leaves on the $x$-axis according to their appearance during a tree traversal, e.g. an *inorder* traversal (see page 190 in Sec. 6.7).

The most important point is that when creating a cluster $c_k$ through a merger of $c_{i_{\min}}$ and $c_{j_{\min}}$, one has to provide new distances $d_c(k, l)$ of the new cluster $c_k$ to all other clusters $c_l$ with $l \neq i_{\min}$ and $l \neq j_{\min}$. Different approaches are possible. Here, we use the *average-linking* clustering, where the distance between two cluster $c_k, c_l$ is the average distance of the data points in the two clusters:

$$d_c(k, l) = \frac{1}{|c_k||c_l|} \sum_{i \in c_k, j \in c_l} d(i, j) \,,$$

where $|c_k|$ and $|c_l|$ represent the number of data points in the clusters $c_k$ and $c_l$, respectively. Thus, when cluster $c_k$ is created by merging $c_{i_{\min}}$ and $c_{j_{\min}}$, the distance of new cluster $c_k$ to all other clusters $c_l$ can be conveniently calculated via

$$d_c(k, l) = \frac{1}{|c_k|} \left\{ |c_{i_{\min}}| d(i_{\min}, l) + |c_{j_{\min}}| d(j_{\min}, l) \right\} \,.$$

Many other choices for calculating cluster distances exists, basically they only have to have the property that the distances between clusters are monotonically increasing when merging. Common examples are taking the minimum or the maximum of the point-wise distances between the nodes of the cluster, specifying *single-linkage* and *complete-linkage* clustering. Another widely used method is *Ward's* approach, where the geometric centers of the clusters are also taken into account. For details about many clustering algorithms, see Ref. [Jain and Dubes (1988)].

Once the clustering procedure is completed and the dendrogram calculated, the full clustering information is contained in the dendrogram, in particular the hierarchical structure, i.e., if clusters contain sub clusters that in turn contain sub clusters etc. To obtain a single set of clusters, a common approach is to use a threshold $\theta$ such that all inter-cluster distances are larger than $\theta$ and all intra-cluster distances are smaller or equal to $\theta$. This is similar to the neighbor-based clustering presented before, only that the intra-cluster distances for agglomerative clustering represent joint properties of sub clusters instead of single pairs of nodes. When drawing the dendrogram, one usually uses the $\delta = 0$ (height) position for the leaves. For all other nodes, representing mergers of two clusters $i_{min}, j_{min}$ , one uses a height $\delta \sim d_c(i_{min}, j_{min})$, i.e., the distance of the two clusters which are merged. Thus, using a threshold $\theta$ corresponds to drawing a horizontal line at $\delta = \theta$ and cutting off all nodes above this line, c.f. Fig. 8.47. The remaining trees located below the line represent the clusters. Often a meaningful choice of $\theta$ is to cut the tree at a height value inside the largest interval where no node has its height in. This correspond to the iteration where the difference between the distances of the last and the current mergers is largest.

In the following, we discuss the C implementation of the single-linkage agglomerative clustering. First, we need a data structure for the nodes of the dendrogram. Each node stores the

| GET SOURCE CODE |
| --- |
| DIR: `randomness` |
| FILE(S): `cluster.c` |

ID of the corresponding cluster and the size of the cluster. If the cluster was merged from two clusters, the node stores pointers (`left` and `right`) to nodes corresponding to these clusters as well as the distance of these two clusters, otherwise the corresponding entries are NULL (or 0). For this structure a new type name `cluster_node_t` is introduced:

```
typedef struct cluster_node
{
    int ID;                     /* ID of cluster */
    int size;                   /* number of members */
    double dist;          /* distance of sub clusters */
    struct cluster_node *left;      /* sub cluster */
    struct cluster_node *right;     /* sub cluster */
} cluster_node_t;
```

The function `cluster_agglomerative()` performs the actual clustering.

It receives a matrix `distance` (GSL type `gsl_matrix`) of point-to-point distances, as calculated, e.g. by the function `cluster_distances()`. The function returns a pointer to the root of the dendrogram, which represents the clustering.

```
1   cluster_node_t *cluster_agglomerative(gsl_matrix *distance)
2   {
3     cluster_node_t *tree;                    /* root of dendrogram */
4     cluster_node_t *node;                    /* nodes of dendrogram */
5     int num_points;            /* number of points to be clustered */
6     int num_clusters;            /* current number of clusters */
7     int next_ID;                        /* ID of next cluster */
8     int ID_curr;                     /* ID of current cluster */
9     int ID_min1, ID_min2;    /* IDs of clusters having min distance */
10    int last_ID;              /* ID of cluster in last row/column */
11    int entry_min1, entry_min2;       /* entry having min distance */
12    int c1, c2;                           /* loop counters */
13    int *pos;              /* position of cluster in distance matrix */
14    int *cluster;  /* ID of cluster in each row/colum, inv. of 'pos' */
15    double delta;                       /* auxiliary distance */
```

The distances among the data points as well as all cluster created during the process will be stored in the matrix `distance`. Since there are at most $n$ clusters existing at any time, the matrix `distance` is large enough. When two clusters are merged, the entries of one cluster will be used to store the distances of the merged cluster, while the entries of the other cluster will be disregarded; they will be exchanged with the distances stored in the last column and row. Thus, after a merger, the last column and row will not be used any more. In this way, the matrix `distance` is overwritten. The current number of used columns and rows, equal to the current number of clusters is stored in the variable `num_clusters`. Note that the cluster IDs are allocated in increasing manner, i.e., the IDs 0 to $n-1$ are for the single-data point clusters, the ID $n$ is for the first cluster created by a merger, the ID $n+1$ for the second, and so forth. Since the rows and columns of `distance` contain entries for all clusters, also for those which are created by mergers, i.e., with IDs larger than $n-1$, two additional arrays are used: The array `pos` stores for each cluster in which row and column the corresponding distances are stored currently. Inversely to `pos`, the array `cluster` stores for each row and column, which cluster is currently represented there. Thus, we have always `pos[cluster[i]]==i` and `cluster[pos[i]]==i`. The data arrangement is illustrated in Fig. 8.46.

In the C code, the number of data points is determined from the size

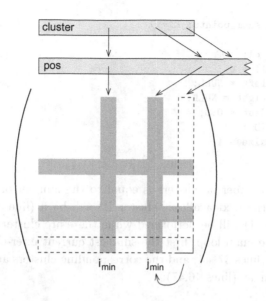

Fig. 8.46  When merging clusters with IDs $i_{min}$ and $j_{min}$ ($i_{min} < j_{min}$), the distances of the new merged cluster are stored in the row and column where the distances of $i_{min}$ were stored, while the entries corresponding to cluster with ID $j_{min}$ are swapped with the last row and column. Top part: for each cluster, the current column and row is stored in the array pos, while for each column and row the current cluster is stored in the array cluster.

of the matrix distance (line 16). Next, memory is allocated for the arrays cluster, pos and nodes (line 18–21). The latter two have $2n - 1$ entries since this is the total number of clusters considered during the construction procedure. The initialization is completed by setting up the entries of pos and cluster and the nodes for the original data points (lines 23–32):

```
16    num_points = distance->size1;
17
18    cluster = (int *) malloc(num_points*sizeof(int));
19    pos = (int *) malloc( (2*num_points-1)*sizeof(int));
20    node = (cluster_node_t *)
21            malloc( (2*num_points-1)*sizeof(cluster_node_t));
22
```

```
23   for(c1=0; c1<num_points; c1++)                    /* initialize  */
24   {
25     pos[c1] = c1;
26     cluster[c1] = c1;
27     node[c1].left = NULL;
28     node[c1].right = NULL;
29     node[c1].dist = 0.0;
30     node[c1].ID = c1;
31     node[c1].size = 1;
32   }
```

Initially, the number of clusters is equal to the number of data points $n$ (line 33) and the next available cluster ID will be $n$ (line 34). The main loop (lines 35–81) will be performed while there are clusters left for being merged. In the main loop, first the smallest current inter-cluster distance is determined (lines 37–44) and the corresponding clusters are obtained via the cluster array (lines 46,47):

```
33   num_clusters = num_points;
34   next_ID = num_clusters;
35   while(num_clusters > 1)           /* until all clusters are merged */
36   {
37     entry_min1=0; entry_min2=1;   /* search min. off-diag distance */
38     for(c1=0; c1<num_clusters; c1++)
39       for(c2=c1+1; c2<num_clusters; c2++)
40         if(gsl_matrix_get(distance, c1, c2) <
41            gsl_matrix_get(distance, entry_min1, entry_min2))
42         {
43           entry_min1=c1, entry_min2=c2;
44         }
45
46     ID_min1 = cluster[entry_min1];       /* determine cluster IDs */
47     ID_min2 = cluster[entry_min2];
```

Now, a new node can be set up. It contains pointers to its two sub clusters, its ID, its size which is the sum of the sizes of the two sub clusters, and the distance of the two sub clusters:

```
48   node[next_ID].left = &(node[ID_min1]);       /* merge clusters */
49   node[next_ID].right = &(node[ID_min2]);
50   node[next_ID].ID = next_ID;
51   node[next_ID].size = node[ID_min1].size + node[ID_min2].size;
52   node[next_ID].dist =
53     gsl_matrix_get(distance, entry_min1, entry_min2);
```

Next, the distances of the remaining clusters to the new clusters are calculated. These distances are stored in the entries of the first of the two merged clusters:

```
54    for(c1=0; c1<num_clusters; c1++)   /* distances to new cluster */
55      if(c1 == entry_min1)
56        gsl_matrix_set(distance, entry_min1, c1, 0);
57      else if(c1 != entry_min2)
58      {
59        ID_curr = cluster[c1];
60        delta = node[ID_min1].size*
61          gsl_matrix_get(distance, entry_min1, c1)+
62          node[ID_min2].size*
63          gsl_matrix_get(distance, entry_min2, c1);
64        delta /= node[next_ID].size;
65        gsl_matrix_set(distance, entry_min1, c1, delta);
66        gsl_matrix_set(distance, c1, entry_min1, delta);
67      }
```

Finally, the current number of clusters is reduced by one (line 68), the root of the dendrogram is set if necessary (lines 69 and 70) the entries of the last current row and column are put to the row and column where previously the distances of the second cluster were stored (lines 71–75), the entries of pos and cluster for the new cluster are set (lines 77 and 78), and the counter for the next available cluster ID is increased by one (line 79). After the main loop has finished, the memory which is associated to those data structures which are not used any more is freed (lines 83 and 84):

```
68    num_clusters--;
69    if(num_clusters == 1)
70      tree = &(node[next_ID]);                    /* set root of tree */
71    last_ID = cluster[num_clusters];/* last cluster -> entry_min2 */
72    pos[last_ID] = entry_min2;
73    cluster[entry_min2] = last_ID;
74    gsl_matrix_swap_rows(distance, num_clusters, entry_min2);
75    gsl_matrix_swap_columns(distance, num_clusters, entry_min2);
76
77    cluster[entry_min1] = next_ID;
78    pos[next_ID] = entry_min1;
79    next_ID++;
80
81  }
82
83  free(pos);                                          /* clean up */
```

```
84    free(cluster);

85

86    return(tree);

87  }
```

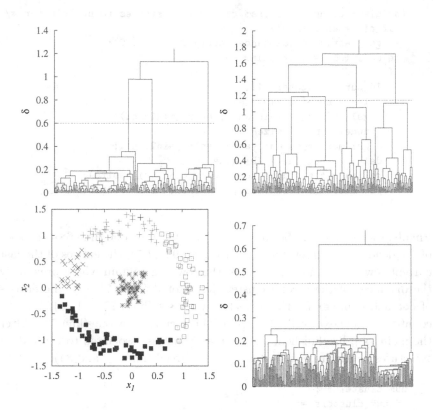

Fig. 8.47   Results of the clustering of sample set B with agglomerative clustering. Upper left: dendrogram using average linkage clustering for sample set A. Upper right: dendrogram using average linkage clustering for sample set B. Lower left: clusters for sample set B obtained when cutting the dendrogram at height $\delta = 1.12$. Lower right: dendrogram using single-linkage clustering for sample set B.

In Fig. 8.47 the resulting dendrograms for sample sets A and B are shown. When cutting the dendrogram for sample set A at the most obvious height, indeed three clusters emerge. On the other hand, one has to cut the dendrogram for sample set B at a lower height to obtain a clustering where the cluster in the middle is separate from the "ring", resulting in five clusters. When considering a height where four clusters emerge, the "central" cluster will be merged with the cluster to the left indicated by the symbol ×, thus "ring" and "central" part are not separated. More successful is the single-linkage agglomerative approach (not shown here), but this is essentially equivalent to the neighbor-based clustering. The difference (and improvement) is that also a dendrogram is obtained which allows to obtain the most natural threshold and to analyze hierarchical sub structures.

Note that the source code `cluster.c` also contains the function `cluster_list_tree()` which prints for a given dendrogram and a given threshold $\theta$ the positions of the data points ordered by the clusters, i.e., between every cluster there will be printed two empty lines.[19] This function can be easily extended that, e.g. cluster IDs are assigned to the initial data points.

## 8.6 General estimators

In Sec. 8.3, different methods are presented of how to estimate parameters which can be obtained directly and simply from the given sample $\{x_0, x_1, \ldots, x_{n-1}\}$. In this section, a general method is considered which enables estimators to be obtained for arbitrary parameters of probability distributions. The method is based on the *maximum-likelihood principle*, which is exposed in Sec. 8.6.1. This principle can be extended to the *modeling of data*, where often a sample of triplets $\{(x_0, y_0, \sigma_0), (x_1, y_1, \sigma_1), \ldots, (x_{n-1}, y_{n-1}, \sigma_{n-1})\}$ is given. Typically the $x_i$ data points represent some control parameter, which can be chosen in the simulation, such as the temperature of a gas. It is assumed that all $x_i$ values are different. Consequently, the simulation has been carried out at $n$ different values of the control parameter. The $y_i$ data points are averages of measurements (e.g. the density of the gas) obtained in the simulations for the fixed value $x_i$ of the control parameter. The $\sigma_i$ values are the corresponding error bars.[20] Modeling the data means that one wants to determine a relationship $y = y(x)$. Usually some assumptions or knowledge about the relationship are available, which means one has available one parametrized test function $y_\theta(x)$. Consequently, the set of parameters $\theta$ has to be adjusted such that the function $y_\theta(x)$ *fits* the sample "best". This is called *data fitting* and will be explained in Sec. 8.6.2. This approach can also be used

---

[19]This can be used in `gnuplot` using the `index` plot keyword to plot the data points of different clusters using different symbols.

[20]Sometimes also the $x_i$ data points are measured quantities which are also characterized by error bars. The generalization of the methods to this case is straightforward.

to compare several fitted test functions to determined which represents the most suitable model.

### 8.6.1　*Maximum likelihood*

Here, we consider the following task: For a given sample $\{x_0, x_1, \ldots, x_{n-1}\}$ and a probability distribution represented by a pmf $p_{\underline{\theta}}(x)$ or a pdf $f_{\underline{\theta}}(x)$, we want to determine the parameters $\underline{\theta} = (\theta_1, \ldots, \theta_{n_\mathrm{p}})$ such that the pmf or pdf represents the data "best". This is written in parentheses, because there is not unique definition what "best" means, or even a mathematical way to derive a suitable criterion. If one assumes no prior knowledge about the parameters, one can use the following principle:

**Definition 8.26**　The *maximum-likelihood principle* states that the parameters $\underline{\theta}$ should be chosen such that the likelihood of the data set, given the parameters, is maximal.

In case of a discrete random variable, if it can be assumed that the different data points are independent, the likelihood of the data is just given by the product of the single data point probabilities. This defines the *likelihood function*

$$L(\underline{\theta}) \equiv p_{\underline{\theta}}(x_1) p_{\underline{\theta}}(x_2) \ldots p_{\underline{\theta}}(x_{n-1}) = \prod_{i=0}^{n-1} p_{\underline{\theta}}(x_i) \tag{8.92}$$

For the continuous case, the probability is zero that one obtains during a random experiment a certain sample exactly. Nevertheless, for a small uncertainty parameter $\epsilon$, the probability to obtain a value in the interval $[\tilde{x} - \epsilon, \tilde{x} + \epsilon]$ is $P(\tilde{x} - \epsilon \leq X < \tilde{x} + \epsilon) = \int_{\tilde{x}-\epsilon}^{\tilde{x}+\epsilon} f_{\underline{\theta}}(x)\,dx \approx f_{\underline{\theta}}(\tilde{x})2\epsilon$. Since $2\epsilon$ enters just as a factor, it is not relevant to determining the maximum. Consequently, for the continuous case, one considers the following likelihood function

$$L(\underline{\theta}) \equiv f_{\underline{\theta}}(x_1) f_{\underline{\theta}}(x_2) \ldots f_{\underline{\theta}}(x_{n-1}) = \prod_{i=0}^{n-1} f_{\underline{\theta}}(x_i) \tag{8.93}$$

To find the maximum of a likelihood function $L(\underline{\theta})$ analytically, one has to calculate the first derivatives with respect to all parameters, respectively, and requires them to be zero. Since calculating the derivative of a product involves the application of the product rule, it is usually more convenient to consider the *log-likelihood function*

$$l(\underline{\theta}) \equiv \log L(\underline{\theta}).  \tag{8.94}$$

This turns the product of single-data-points pmfs or pdfs into a sum, where the derivatives are easier to obtain. Furthermore, since the logarithm is a monotonous function, the maximum of the likelihood function is the same as the maximum of the log-likelihood function. Hence, the parameters which suit "best" are determined within the maximum-likelihood approach by the set of equations

$$\frac{\partial l(\underline{\theta})}{\partial \theta_k} \stackrel{!}{=} 0 \quad (k = 1, \dots, n_{\mathrm{p}})  \tag{8.95}$$

Note that the fact that the first derivatives are zero only assures that an extremal point is obtained. Furthermore, these equations often have several solutions. Therefore, one has to check explicitly which solutions are indeed maxima, and which is the largest one. Note that maximum-likelihood estimators, since they are functions of the samples, are also random variables $\mathrm{ML}_{\theta_k,n}(X_0, \dots, X_{n-1})$.

As a toy example, we consider the exponential distribution with the pdf given by Eq. (8.39). It has one parameter $\mu$. The log-likelihood function for a sample $\{x_0, x_1, \dots, x_{n-1}\}$ is in this case

$$\begin{aligned}
l(\mu) &= \log \prod_{i=0}^{n-1} f_\mu(x_i) \\
&= \sum_{i=0}^{n-1} \log \left\{ \frac{1}{\mu} \exp\left(-\frac{x_i}{\mu}\right) \right\} \\
&= \sum_{i=0}^{n-1} \left( \log\left\{ \frac{1}{\mu} \right\} - \frac{x_i}{\mu} \right) \\
&= n \log \left\{ \frac{1}{\mu} \right\} - \frac{n}{\mu} \bar{x}
\end{aligned}$$

Taking the derivative with respect to $\mu$ we obtain:

$$0 \stackrel{!}{=} \frac{\partial L(\underline{\theta})}{\partial \mu} = n \frac{-1}{\mu^2} \mu - \frac{-n}{\mu^2} \bar{x} = \frac{-n}{\mu^2}(\mu - \bar{x})$$

This implies $\mu = \bar{x}$. It is easy to verify that this corresponds to a maximum. Since the expectation value for the exponential distribution is just $\mathrm{E}[X] =$

$\mu$, this is compatible with the result from Sec. 8.3, where it was shown that the sample mean is an unbiased estimator of the expectation value.

If one applies the maximum-likelihood principle to a Gaussian distribution with parameters $\mu$ and $\sigma^2$, one obtains (not shown here, see for example [Dekking et al (2005)]) as maximum-likelihood estimators the sample mean $\bar{x}$ (for $\mu$) and the sample variance $s^2$ (for $\sigma^2$), respectively. This means (see Eq. (8.58)) that the maximum-likelihood estimator for $\sigma^2$ is biased. Fortunately, we know that the bias disappears asymptotically for $n \to \infty$. Indeed, it can be shown, under rather mild conditions on the underlying distributions, that all maximum-likelihood estimators $\mathrm{ML}_{\theta_k,n}(X_0,\ldots,X_{n-1})$ for a parameter $\theta_k$ are asymptotically unbiased, i.e.

$$\lim_{n \to \infty} \mathrm{E}[\mathrm{ML}_{\theta_k,n}] = \theta_k \tag{8.96}$$

In contrast to the exponential and Gaussian cases, for many applications the maximum-likelihood parameter is not directly related to a standard sample estimator. Furthermore, $\mathrm{ML}_{\theta_k,n}$ can often even not be determined analytically. In this case, one has to optimize the log-likelihood function numerically, for example, using the corresponding methods from the *GNU scientific library* (GSL) (see Sec. 7.3).

As example, we consider the Fisher-Tippett distribution, see Eq. (8.43), shifted to exhibit the maximum at $x_0$ instead of at 0. Hence, we have two parameters $\lambda$ and $x_0$ to adjust. The

---
| GET SOURCE CODE |
| :-- |
| DIR: `randomness` |
| FILE(S): `max_likely.c` |
---

function to be optimized (the *target function*), i.e. the log-likelihood function here, must be of a special format when using the minimization functions of the GSL. This first argument of the target function contains the pdf parameters to be adjusted, i.e. the main argument vector of the target function. This argument must be of the type `gsl_vector`, which is a GSL type for vectors. One needs to include `<gsl/gsl_vector.h>` to use this data type. These vectors are created using `gsl_vector_alloc()`, set elements via `gsl_vector_set()`, access elements via `gsl_vector_get()` and delete the vectors via `gsl_vector_free()`. The usage of these functions should be self-explanatory from the examples below, but you may also have a look at the GSL documentation [Galassi et al. (2006)].

The second argument of the target function contains *one* pointer to all additional data needed to calculate the target function, i.e. the sample in this case. Thus, the sample must be stored in *one* chunk of memory. For this purpose, we use the following structure type:

```
typedef struct
{
  int      n;         /* number of sample points; */
  double  *x;                          /* sample */
}
sample_t;
```

Since the GSL package contains actually minimization functions, while we are interested in a maximum, the actual log-likelihood function returns minus the log-likelihood. The log-likelihood function reads as follows:

```
1   double ll_ft(const gsl_vector *par, void *param)
2   {
3       double lambda, x0;                     /* parameters of pdf */
4       sample_t *sample;                             /* sample */
5       double sum;          /* sum of log-likelihood contributions */
6       int i;                                    /* loop counter */
7
8       lambda = gsl_vector_get(par, 0);            /* get data */
9       x0 = gsl_vector_get(par, 1);
10      sample = (sample_t *) param;
11
12      sum = sample->n*log(lambda);    /* calculate log likelihood */
13      for(i=0; i<sample->n; i++)
14        sum -= lambda*(sample->x[i]-x0) +
15               exp(-lambda*(sample->x[i]-x0));
16
17      return(-sum);                    /* return - log likelihood */
18  }
```

First, we convert the pointers passed as arguments to the data format that we find useful (lines 8–10). Next, the actual log likelihood

$$l(\lambda, x_o) = n \log \lambda - \lambda \sum_{i=0}^{n-1}(x_i - x_0) - \sum_{i=0}^{n-1} \exp(-\lambda(x_i - x_0))$$

is calculated in lines 12–15 and finally returned with inverted sign (line 17).

The GSL has built in several minimization algorithms. They are all put under one of two frameworks. One framework is for algorithms which require the target function and its first derivatives. The other framework contains algorithms where just the target function is sufficient. Here we use the *simplex algorithm*, which belongs to the latter form. It works by spanning a simplex,[21] evaluating the target functions at the corners of the

---

[21] A simplex is a convex set in an $n$-dimensional space generated by $n+1$ corner points.

simplex, and iteratively changing the simplex until it is very small and contains the solution. Note that the algorithm is only able to find local minima, and only one of them. If several minima exist, the choice of the initial parameters strongly influence the final results; Here, one maybe has to try several parameters. For details see [Galassi et al. (2006)]. Here we only show how to use the minimizer. The minimizer itself is stored in a special data structure of type `gsl_multimin_fminimizer`. The target function has to be put into a "surrounding" variable of type `gsl_multimin_function`. Furthermore, one needs two `gsl_vector` variables to store the current estimate for the optimum (specifying the position of the simplex) and to store the size of the simplex. Also, `par` is used here to state the dimension of the target function argument (2) and `sample` to store the sample.

These variables are declared as follows:

```
int num_par;                          /* number of parameters */
sample_t sample;                                    /* sample */

gsl_multimin_fminimizer *s;           /* the full mimimizer */
gsl_vector *simplex_size;         /* (relative) simplex size */
gsl_vector *par; /* params to be optimized = args of target */
gsl_multimin_function f;  /* holds function to be optimized */
```

The actual allocation and initialization of these variables may look as follows:

```
sample.n = 10000;                       /* initilization */
sample.x = (double *) malloc(sample.n*sizeof(double));
num_par = 2;

f.f = &ll_ft;                     /* initialize minimization */
f.n = num_par;
f.params = &sample;
simplex_size = gsl_vector_alloc(num_par);  /* alloc simplex */
gsl_vector_set_all(simplex_size, 1.0);     /* init simplex */
par = gsl_vector_alloc(num_par);  /* alloc + init arguments */
gsl_vector_set(par, 0, 1.0);
gsl_vector_set(par, 1, 1.0);
s =
    gsl_multimin_fminimizer_alloc(gsl_multimin_fminimizer_nmsimplex,
                                num_par);
gsl_multimin_fminimizer_set(s, &f, par, simplex_size);
```

The set-up of the minimizer object comes in two steps, first allocation using `gsl_multimin_fminimizer_alloc()`, then initialization via `gsl_multimin_fminimizer_set()` while passing the target function, the starting point `par` and the (initial) simplex size.[22] The `sample.x[]` array has to be filled with the actual sample (not shown here).

The minimization loop looks as follows:

```
do                                      /* perform minimization */
{
  iter++;
  status = gsl_multimin_fminimizer_iterate(s);   /* one step */
  if(status)     /* error ? */
    break;
  size = gsl_multimin_fminimizer_size(s);     /* converged ? */
  status = gsl_multimin_test_size(size, 1e-4);
}
while( (status == GSL_CONTINUE) && (iter<100) );
```

The main work is done in `gsl_multimin_fminimizer_iterate()`. Then it is checked whether an error has occurred. Next, the size of the simplex is calculated and finally tested whether the size falls below some limit, $10^{-4}$ here.

The actual estimate of the parameters can be obtained via `gsl_vector_get(s->x, 0)` and `gsl_vector_get(s->x, 1)`. Note that finally all allocated memory should be freed:

```
gsl_vector_free(par);                    /* free everything */
gsl_vector_free(simplex_size);
gsl_multimin_fminimizer_free(s);
free(sample.x);
```

As an example, $n = 10000$ data points were generated according to a Fisher-Tippett distribution with parameters $\lambda = 3.0$, $x_0 = 2.0$. With the above starting parameters, the minimization converged to the values $\hat{\lambda} = 2.995$ and $\hat{x}_0 = 2.003$ after 39 iterations.

### 8.6.2 *Data fitting*

In the previous section, the parameters of a probability distribution are chosen such that the distribution describes the data best. Here, we consider a more general case, called *modeling of data*. As explained above, here

---

[22]The simplex is spanned by `par` and the $n$ vectors given by `par` plus $(0, \ldots, 0,$ `simplex_size[i]`$, 0, \ldots, 0)$ for $i = 1, \ldots, n$.

a sample of triplets $\{(x_0, y_0, \sigma_0), (x_1, y_1, \sigma_1), \ldots, (x_{n-1}, y_{n-1}, \sigma_{n-1})\}$ is given. Typically, the $y_i$ are measured values obtained from a simulation with some control parameter (e.g. the temperature) fixed at different values $x_i$; $\sigma_i$ is the corresponding error bar of $y_i$. Here, one wants to determine parameters $\underline{\theta} = (\theta_1, \ldots, \theta_{n_p})$ such that the given parametrized function $y_{\underline{\theta}}(x)$ fits the data "best", one says one wants to *fit* the function to the data. Similar to the case of fitting a pmf or a pdf, there is no general principle of what "best" means.

Let us assume that the $y_i$ are random variables, i.e. comparing different simulations. Thus, the measured values are scattered around their "true" values $y_{\underline{\theta}}(x_i)$. This scattering can be described approximately by a Gaussian distribution with mean $y_{\underline{\theta}}(x_i)$ and variance $\sigma_i^2$:

$$q_{\underline{\theta}}(y_i) \sim \exp\left(-\frac{(y_i - y_{\underline{\theta}}(x_i))^2}{2\sigma_i^2}\right). \tag{8.97}$$

This assumption is often valid, e.g. when each sample point $y_i$ is itself a sample mean obtained from a simulation performed at control parameter value $x_i$, and $\sigma_i$ is the corresponding error bar. The log-likelihood function for the full data sample is

$$l(\underline{\theta}) = \log \prod_{i=0}^{n-1} q_{\underline{\theta}}(y_i)$$

$$\sim -\sum_{i=0}^{n-1} \frac{1}{2}\left(\frac{y_i - y_{\underline{\theta}}(x_i)}{\sigma_i}\right)^2$$

Maximizing $l(\underline{\theta})$ is equivalent to minimizing $-2l(\underline{\theta})$, hence one minimizes the *mean-squared difference*

$$\chi_{\underline{\theta}}^2 = \sum_{i=0}^{n-1}\left(\frac{y_i - y_{\underline{\theta}}(x_i)}{\sigma_i}\right)^2 \tag{8.98}$$

This means the parameters $\underline{\theta}$ are determined such that function $y_{\underline{\theta}}(x)$ follows the data points $\{(x_0, y_0), \ldots (x_{n-1}, y_{n-1})\}$ as close as possible, where the deviations are measured in terms of the error bars $\sigma_i$. Hence, data points with smaller error bar enter with more *weight*. The full procedure is called *least-squares fitting*.

The minimized mean-squared difference is a random variable. Note that the different terms are not statistically independent, since they are related by the $n_p$ parameters $\hat{\underline{\theta}}$ which are determined via minimizing $\chi_{\underline{\theta}}^2$. As a consequence, the distribution of $\chi^2_{\hat{\theta}}$ is approximately given by chi-squared distribution (see Eq. (8.45) for the pdf) with $n - n_{\mathrm{p}}$ degrees of freedom. This distribution can be used to evaluate the statistical significance of a least-squares fit, see below.

In case, one wants to model the underlying distribution function for a sample as in Sec. 8.6.1, say for a continuous distribution, it is possible in principle to use the least-squares approach as well. In this case one would fit the parametrized pdf to a histogram pdf, which has also the above mentioned sample format $\{(x_i, y_i, \sigma_i)\}$. Nevertheless, although the least-squares principle is derived using the maximum-likelihood principle, usually different parameters are obtained if one fits a pdf to a histogram pdf compared to obtaining these parameters from a direct maximum-likelihood approach. Often [Bauke (2007)], the maximum-likelihood method gives more accurate results. Therefore, one should use a least-squares fit mainly for a fit of a non-pmf/non-pdf function to a data set.

Fortunately, to actually perform least-squares fitting, you do not have to write your own fitting functions, because there are very good fitting implementations readily available. Both programs presented in Sec. 8.4, *gnuplot* and *xmgrace*, offer fitting to arbitrary functions. It is advisable to use *gnuplot*, since it offers higher flexibility for that purpose and gives you more information useful to estimate the quality of a fit.

As an example, let us suppose that you want to fit an algebraic function of the form $f(L) = e_\infty + aL^b$ to the data set of the file `sg_e0_L.dat` shown on page 302. First, you have to define the function and supply some rough (non-zero) estimations for the unknown parameters. Note that the exponential operator is denoted by `**` and the standard argument for a function definition is `x`, but this depends only on your choice:

```
gnuplot> f(x)=e+a*x**b
gnuplot> e=-1.8
gnuplot> a=1
gnuplot> b=-1
```

The actual fit is performed via the `fit` command. The program uses the nonlinear least-squares Levenberg-Marquardt algorithm [Press et al. (1995)], which allows a fit data to almost all arbitrary functions. To issue the command, you have to state the fit function, the data set and the parameters which are to be adjusted. First, we consider the case where just two columns of the data are used or available (in this case, *gnuplot* assumes $\sigma_i = 1$). For our example you enter:

```
gnuplot> fit f(x) "sg_e0_L.dat" via e,a,b
```

Then *gnuplot* writes log information to the output describing the fitting process. After the fit has converged it prints for the given example:

```
After 8 iterations the fit converged.
final sum of squares of residuals : 7.55104e-06
rel. change during last iteration : -2.54894e-10

degrees of freedom    (FIT_NDF)                        : 5
rms of residuals      (FIT_STDFIT) = sqrt(WSSR/ndf)    : 0.00122891
variance of residuals (reduced chisquare) = WSSR/ndf   : 1.51021e-06

Final set of parameters           Asymptotic Standard Error
=======================           ==========================

e            = -1.78786           +/- 0.0008548     (0.04781%)
a            = 2.54248            +/- 0.2282        (8.976%)
b            = -2.80103           +/- 0.08264       (2.951%)


correlation matrix of the fit parameters:

              e       a       b
e           1.000
a           0.708   1.000
b          -0.766  -0.991   1.000
```

The most interesting lines are those where the results $\hat{\theta}$ for your parameters along with the standard error bar are printed.[23] Additionally, the quality of the fit can be estimated by the information provided in the three lines beginning with "**degree of freedom**". The first of these lines states the number of degrees of freedom, which is just $n - n_{\mathrm{p}}$. As visible in the brackets, within *gnuplot* this is available in the variable FIT_NDF:

```
gnuplot> print FIT_NDF
5
```

The mean-squared difference $\chi^2_{\hat{\theta}}$ is denoted as WSSR in the *gnuplot* output. The root of this mean-squared difference per degree of freedom is stored in the *gnuplot* variable FIT_STDFIT.

---

[23]These "error bars" are calculated in a way which is in fact correct only when fitting linear functions; hence, they have to be taken with care.

A measure of quality of the fit is the probability $Q$ that the value of the mean-squared difference is equal or larger compared to the value from the current fit, given the assumption that the data points are distributed as in Eq. (8.97) [Press et al. (1995)]. The larger the value of $Q$, the better is the quality of the fit. As mentioned above, $Q$ can be evaluated from a chi-squared distribution with $n - n_p$ degrees of freedom. Since the cumulative chi-squared distribution is related to the normalized incomplete gamma function, which is available in *gnuplot*, one can evaluate Q directly in gnuplot:

```
gnuplot> Q = 1 - igamma(0.5 * FIT_NDF, 0.5 * FIT_NDF*FIT_STDFIT**2)
```

Note that in this case we obtain $Q = 1$, which is so large, because $\sigma_i = 1$ was used, see below.

To watch the result of the fit along with the original data, just enter

```
gnuplot> plot "sg_e0_L.dat" w e, f(x)
```

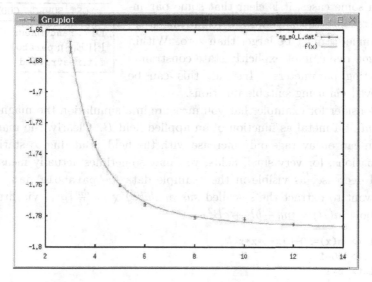

Fig. 8.48   *Gnuplot* window showing the result of a fit command along with the input data.

The result is displayed in Fig. 8.48. Please note that the convergence depends on the initial choice of the parameters. The algorithm may be trapped into a local minimum in case the parameters are too far away from the best values. Try the initial values e=1, a=-3 and b=1! Furthermore, not all function parameters have to be subjected to the fitting. Alternatively, you can set some parameters to fixed values and omit them from the `via` list at the end of the `fit` command. Remember that in the above example all data points enter into the result with the same weight, i.e. $\sigma_i = 1 \, \forall i$ is assumed. You can tell the algorithm to consider the error bars, for example supplied in the third column, by typing

```
gnuplot> fit f(x) "sg_e0_L.dat" using 1:2:3 via e,a,b
```

Then, data points with larger error bars have less influence on the results. In this case a different result whith smaller value of $Q$ will arise (try it !).

You can also restrict the data points which are considered for the fit, which is applicable if only a subset of the sample follows the function law you are considering. This can be done in the same way as restricting the range of plotted values, for instance using

```
gnuplot> fit [5:12] f(x) "sg_e0_L.dat" using 1:2:3 via e,a,b
```

In some cases, it is clear that some parameters to be fitted fall into a certain range, e.g. a parameter must be larger than zero. Within *gnuplot*, one cannot explicitly state constraints to fitting parameters. Instead, this can be achieved via using suitable functions.

| GET SOURCE CODE |
|---|
| DIR: `randomness` |
| FILE(S): `parabola.dat` |
| `fit_restricted.gp` |

Consider for example that you measure in a simulation the magnetization $m$ of a metal as function of an applied field $B$. Clearly, the magnetization can on average only increase with the field. But due to statistical fluctuations, for very small fields, you may sometimes actually measure a small decrease, as visible in the example data file **parabola.dat**. Now, you want to extract the so-called *susceptibility* $\chi = \frac{dm}{dB}|_{B \to 0}$ via fitting a parabola $m(B) = m_0 + bB + cB^2$:

```
gnuplot> m(x)=m0+b*x+c*x*x
gnuplot> m0=1
gnuplot> b=1
gnuplot> c=1
gnuplot> fit m(x) "parabola.dat" u 1:2:3 via m0,b,c
```

This will result in a negative value for the susceptibility $\chi =$ b:

Final set of parameters            Asymptotic Standard Error
=======================            ==========================

```
m0                = 1.20905           +/- 0.03342        (2.765%)
b                 = -0.0576451        +/- 0.25           (433.6%)
c                 = 3.0935            +/- 0.3017         (9.753%)
```

Notice the large error bar given for b which is compatible with zero. One can restrict the coefficient of the linear term via using the exponential function, which can only take positive values:

```
m2(x)=m2+exp(b2)*x+c2*x*x
m2=1
b2=-10
c2=1
fit m2(x) "parabola.dat" u 1:2:3 via m2,b2,c2
```

resulting in

```
Final set of parameters            Asymptotic Standard Error
=======================            ==========================

m2                = 1.20357           +/- 0.03354        (2.786%)
b2                = -10.5721          +/- 9840           (9.308e+04%)
c2                = 3.02815           +/- 0.3027         (9.996%)
```

Thus, the resulting value of b2 will be negative, i.e., the effective value of the linear coefficient is close to zero but positive: $b = \exp(b2) = 2.6 \times 10^{-5}$. Note that here the effect is rather small, the resulting function m2(B) cannot be distinguished from m(B). But for other cases, in particular when the fit converges to something completely off the data, which sometimes happens, restricting the values of fitting parameters as shown above can help a lot. For stronger restrictions, e.g. within an interval $[b_0, b_1]$, the application of other functions may help. For example, the tanh(), which is available in *gnuplot* as well, is limited to the range [-1,1]. Therefore, identifying $b \equiv (b_0 + (b_1 - b_0)(\tanh(\hat{b}) + 1)/2)$ and fitting $\hat{b}$ will give the desired restriction for $b$.

Finally, we discuss *multi-branch fitting*. This means that several function sharing one or more parameters, possibly having also individual parameters in addition, are fitted to several independent data sets.

| GET SOURCE CODE |
|---|
| DIR: randomness |
| FILE(S): exp1.dat |
| multi_fit.gp |

As an example, we consider the case of three exponential functions $f_i(x) = a_i \exp(-x/\mu)$ $(i = 1, 2, 3)$, which have different prefactors but share the same parameter $\mu$. We want to fit the three functions, i.e., determine the values of the parameters $a_0, a_1, a_2$, and $\mu$, to three sets of data generated from exponential distributions with three different values of a lower cutoff, but all exhibiting the parameter $\mu = 1$. To make this multi-branch fitting work within *gnuplot*, all three data sets have to be contained in one single file, expl.dat here, separated by blank lines. First, we define the three functions:

```
gnuplot> f0(x)=a0*exp(-x/mu)
gnuplot> f1(x)=a1*exp(-x/mu)
gnuplot> f2(x)=a2*exp(-x/mu)
```

Next, we have to combine these three functions into a single function, f(x,s) where the first argument is the same argument as to the functions f1(x), f2(x), f3(3). The second argument s is used to select among the three functions. For this purpose, we first define an auxiliary function delta(a,b), which returns 1 is a=b and 0 else. The definition of the function uses the conditional operator (see page 20) as known from the programming language C. This delta(a,b) is used for the actual definition of f(x,s):

```
gnuplot> delta(a,b) = (a==b) ? 1 : 0
gnuplot> f(x,s) = delta(s,0)*f0(x)+delta(s,1)*f1(x)+delta(s,2)*f2(x)
```

From *gnuplot* 4.4 onwards, you can use as an alternative the value command, which takes any string as the name of a variable and returns the value assigned currently to the variable. For our example the command can be used in the following way:

```
gnuplot> f(x,s) = value(sprintf("a%d",s))*exp(-x/mu)
```

Note that all given *gnuplot* commands are collected in multi_fit.gp and can be executed using gnuplot multi_fit.gp. The fit is performed such that the first column is used as $x$ argument. The argument s is the *index* of the current data file expo1.dat. This can be taken automatically from the data file via stating the "column number" -2 in the second column of the using part of the fit command, stating the function value at the third column, and the error bars as fourth column[24]

```
gnuplot> fit f(x,y) "expo1.dat" using 1:-2:2:3 via a0,a1,a2,mu
```

---

[24]Apparently, the multi-branch fit seems not to work if no error bars are included.

which results in a joint fit of all three functions (combined into one) to all three data sets:

```
Final set of parameters          Asymptotic Standard Error
=======================          =========================

a0           = 1.00658           +/- 0.003899      (0.3874%)
a1           = 2.75671           +/- 0.01422       (0.5157%)
a2           = 7.54958           +/- 0.05115       (0.6775%)
mu           = 0.992714          +/- 0.001955      (0.1969%)
```

Note that the error bar to the parameter $\mu$ is smaller compared to fitting any of the functions of just one of the data sets (please try), which does not necessarily mean that the obtained parameter is actually closer to the real value.

Finally, the original data, containing all three sets, is plotted together with the three different exponentials:

```
gnuplot> set logscale y
gnuplot> plot "expo1.dat" u 1:2:3 w e, \
             f(x,0) title sprintf("a0=%f", a0),\
             f(x,1) title sprintf("a1=%f", a1), \
             f(x,2) title sprintf("a2=%f", a2)
```

Note that here also the sprintf() C-like command is used to allow for formatted printing of strings and labels.

More information on how to use the fit command, such as fitting higher-dimensional data, can be obtained when using the *gnuplot* online help via entering help fit.

# Exercises

(solutions: can be downloaded from http://www.worldscientific.com/r/9019-supp)

(1) **Simple sampling from discrete distribution**

Design, implement and test a function, which returns a random number which is distributed according to some discrete distribution function stored in an array F. Use the simple approach as described in Sec. 8.2.2. The function prototype reads as follows:

```
SOLUTION SOURCE CODE
DIR: randomness
FILE(S): poisson.c
```

```
/****************** rand_discrete() ****************/
/** Returns natural random number distributed       **/
/** according a discrete distribution given by the  **/
/** distribution function in array 'F'              **/
/** Uses search in array to generate number         **/
/** PARAMETERS: (*)= return-parameter               **/
/**       n: number of entries in array             **/
/**       F: array with distribution function       **/
/** RETURNS:                                         **/
/**       random number                             **/
/****************************************************/
int rand_discrete(int n, double *F)
```

For simplicity, you can use the **drand48()** function from the standard C library to generate random numbers distributed according to $U(0,1)$.

Furthermore, design, implement and test a function, which allocates and initializes the array F for a Poisson distribution with parameter $\mu$, see Eq. (8.27) for the probability mass function. The function should determine automatically how many entries of F are needed, depending on the paramater $\mu$. The function prototype reads as follows:

```
/****************** init_poisson() ****************/
/** Generates  array with distribution function      **/
/** for Poisson distribution with mean mu:           **/
/** p(k)=mu^k*exp(-mu)/k!                             **/
/** The size of the array is automatically adjusted. **/
/** PARAMETERS: (*)= return-parameter                **/
/**   (*) n_p: p. to number of entries in table      **/
/**       mu: parameter of distribution              **/
/** RETURNS:                                          **/
/**    pointer to array with distribution function   **/
/****************************************************/
double *init_poisson(int *n_p, double mu)
```

Hints: To determine the array sizes, you can first loop over the probabilities and take the first value k_0 where $p(k\_0) = 0$ within the precision of the numerics. This value of k_0 serves as array size. Alternatively, you start with some size and extend the array if needed by doubling its size. For testing purposes, you can generate many numbers, calculate the mean and compare it with $\mu$. Alternatively, you could record a histogram (see Chap. 5) and compare with Eq. (8.27).

(2) **Constant-time sampling from discrete distribution**

Correspondingly to exercise (1), implement the approach for drawing random numbers from a discrete distribution in constant time. You should first setup the table representing the pmf of the Poisson distribution with parameter $\mu$, see Eq. (8.27) with a function double *init_poisson_pmf(int *n_p, double mu), similar to the function double *init_poisson() from exercise (1), which sets up the distribution function.

For representing the table, you can use the following data structure:

```
typedef struct
{
    int num_entries;      /* number of outcomes */
    double *q;       /* splitting probabilities */
    int *a;              /* events for rand<=q */
    int *b;              /* events for rand>q */
} discrete_variate_t;
```

Set up the table according Walker's method as shown in the second part of Sec. 8.2.2. YOu can use the following prototype:

```
/****************** setup_table() *******************/
/** Sets up table to generate discrete random numbers **/
/** in constant time using the Walker's method , as **/
/** implemented in K. Fukui & S. Todo, J. Comp. Phys. **/
/** 228 (2009) 2629-2642                              **/
/** PARAMETERS: (*)= return-paramter                  **/
/**   num_entries: number of possible results         **/
/**            p: original probabilities              **/
/** RETURNS:                                          **/
/**     table with auxiliary variables                **/
/*******************************************************/
discrete_variate_t setup_table(int num_entries, double *p)
```

You should also implement a function for drawing a random number:

```
/****************** draw_number() *******************/
/** Draw discrete number using the Walker's method.  **/
/** PARAMETERS: (*)= return-paramter                 **/
/**    table: for drawing discrete random numbers    **/
/** RETURNS:                                          **/
/**        random number                             **/
/*****************************************************/
int draw_number(discrete_variate_t * table)
```

Test your implementation, as in exercise 1, by drawing many random numbers, calculating the mean and compare it with $\mu$, or, even better, by recording a histogram (see Chap. 5) and comparing it with Eq. (8.27).

(3) **Inversion Method for Fisher-Tippett distribution**

Design, implement and test a function, which returns a random number which is distributed according to the Fisher-Tippett distribution Eq. (8.43) with parameter $\lambda$. Use the inversion method.

| SOLUTION SOURCE CODE |
| --- |
| DIR: **randomness** |
| FILE(S): |
| **fischer_tippett.c** |

The function prototype reads as follows:

```
/****************** rand_fisher_tippett() ***********/
/** Returns random number which is distributed      **/
/** according the Fisher-Tippett distribution        **/
/** PARAMETERS: (*)= return-parameter                **/
/**        lambda: parameter of distribution         **/
/** RETURNS:                                          **/
/**    random number                                 **/
/*****************************************************/
double rand_fisher_tippett(double lambda)
```

Remarks: For simplicity, you can use the **drand48()** function from the standard C library to generate random numbers distributed according to $U(0, 1)$. To test your function, you can calculate the mean of the generated numbers, for instance, and compare it with the expectation value $\sim 0.57721/\lambda$.

(4) **Rejection Method for Gaussian**

Implement a function which returns a Gaussian distributed (pseudo) random number using the rejection approach via bordering the Gaussian by an exponential $\exp(-x)$, as shown in Sec. 8.2.4.

| SOLUTION SOURCE CODE |
| --- |
| DIR: **randomness** |
| FILE(S): |
| **gauss_reject.c** |

The function prototype reads as follows:

```
/*************** reject_gaussian() *****************/
/** Generates Gaussian distributed random number    **/
/** using the rejection method via bordering the     **/
/** positive part of the Gaussian by exp(-x). With   **/
/** probability 1/2 the sign is negated.             **/
/** PARAMETERS: (*)= return-paramter                 **/
/**          none                                    **/
/** RETURNS:                                         **/
/**      random number                               **/
/*****************************************************/
double reject_gaussian()
```

Remarks: For simplicity, you can use the `drand48()` function from the standard C library to generate random numbers distributed according to $U(0, 1)$. To test your function, you can make a histogram pdf and compare with the pdf of the Gaussian distribution Eq. (8.36).

(5) **Variance of data sample**

Design, implement and test a function, which calculates the variance $s^2$ of a sample of data points. Use directly Eq. (8.55), i.e. do *not* use an equivalent form of Eq. (8.21), since this form is more susceptible to rounding errors. The function prototype reads as follows:

| SOLUTION SOURCE CODE |
|---|
| DIR: **randomness** |
| FILE(S): **variance.c** |

```
/********************* variance() ******************/
/** Calculates the variance of n data points        **/
/** PARAMETERS: (*)= return-parameter                **/
/**        n: number of data points                  **/
/**        x: array with data                        **/
/** RETURNS:                                          **/
/**      variance                                     **/
/*****************************************************/
double variance(int n, double *x)
```

Remark: The so-called corrected double-pass algorithm [Chan et al. (1983)] aims at further reducing the rounding error. It is based on the equation

$$s^2 = \frac{1}{n}\left[\sum_{i=0}^{n-1}(x-\overline{x})^2 - \frac{1}{n}\left(\sum_{i=0}^{n-1}(x_i-\overline{x})\right)^2\right].$$

The second square would be zero for exact arithmetic and accounts for rounding errors occurring in the calculation. It becomes important in particular if the expectation value is large. Perform experiments for generating Gaussian distributed number with $\sigma^2 = 1$ and $\mu = 10^{14}$, without and with the correction.

(6) **Bootstrap**

Design, implement and test a function, which uses bootstrapping to calculate the confidence interval at significance level $\alpha$ given in Eq. (8.70).

| SOLUTION SOURCE CODE |
| :--- |
| DIR: randomness |
| FILE(S): |
| bootstrap_ci.c |

The function prototype reads as follows:

```
/**************** bootstrap_ci() *******************/
/** Calculates a confidence interval by 'n_resample' **/
/** times resampling the given sample points        **/
/** and each time evaluation the estimator 'f'       **/
/** PARAMETERS: (*)= return-parameter                **/
/**          n: number of data points               **/
/**          x: array with data                      **/
/**  n_resample: number of bootstrap iterations      **/
/**      alpha: confidence level                     **/
/**          f: function (pointer) = estimator       **/
/**     (*) low: (p. to) lower boundary of conf. int.**/
/**    (*) high: (p. to) upper boundary of conf. int.**/
/** RETURNS:                                         **/
/**     (nothing)                                    **/
/*****************************************************/
void bootstrap_ci(int n, double *x, int n_resample,
                  double alpha, double (*f)(int, double *),
                  double *low, double *high)
```

Hints: Use the function `bootstrap_variance()` as example. To get the entries at the positions defined via Eq. (8.70), you can sort the bootstrap sample first using `qsort()`, see Sec. 7.1.

You can test your function by using the provided main file `bootstrap_test.c`, the auxiliary files `mean.c` and `variance.c` and by compiling with `cc -o bt bootstrap_test.c bootstrap_ci.c mean.c variance.c -lm -DSOLUTION`. Note that the macro definition -DSOLUTION makes the `main()` function to call `bootstrap_ci()` instead of `bootstrap_variance()`.

(7) **Plotting data**

Plot the data file `FTpdf.dat` using *xmgrace*. The file contains a histogram pdf generated for the Fisher-Tippett distribution. The file format is 1st column: bin number, 2nd: bin midpoint, 3rd: pdf value, 4th: error bar. Use

| SOLUTION SOURCE CODE |
| :--- |
| DIR: randomness |
| FILE(S): FTplot.agr |

the "block data" format to read the files (columns 2,3,4). Create a plot with inset. The main plot should show the histogram pdf with error bars and logarithmically scaled $y$ axis, the inset should show the data with linear axes. Describe the plot using a text label placed in the plot. Choose label sizes, line width and other styles suitably. Store the result as .agr file and export it to a postscript (eps) file.

The result should look similar to:

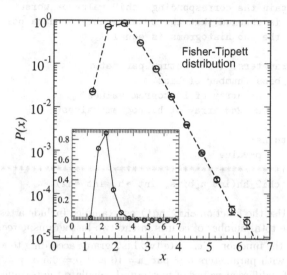

(8) **Chi-squared test**

Design, implement and test a function, which calculates the $\chi^2$ test statistics for two histograms $\{h_i\}, \{\hat{h}_i\}$ according Eq. (8.73). The function should return the p-value, i.e. the

| SOLUTION SOURCE CODE |
| --- |
| DIR: **randomness** |
| FILE(S): **chi2hh.c** |

cumulative probability ("p-value") that a value of $\chi^2$ or larger is obtained under the assumption that the two histograms were obtained by sampling from the same (discrete) random variable.

The function prototype reads as follows:

```
/********************* chi2_hh() *********************/
/** For chi^2 test: comparison of two histograms     **/
/** to probabilities: Probability to                  **/
/** obtain the corresponding  chi2 value or worse.    **/
/** It is assumed that the total number of data points**/
/*+ in the two histograms is equal !                  **/
/**                                                   **/
/** Parameters: (*) = return parameter                **/
/**   n_bins: number of bins                          **/
/**        h: array of histogram values               **/
/**        h2: 2nd array of histogram values          **/
/**                                                   **/
/** Returns:                                          **/
/**        p-value                                    **/
/****************************************************/
double chi2_hh(int n_bins, int *h, int *h2)
```

Hints: Use the function `chi2_hd()` as example. Include a test, which verifies that the total number of counts in the two histograms agree.

To test the function: Generate two histograms according to a binomial distribution with parameters $n = $ **par_n**$= 10$ and $p = 0.5$ or $p = $ **par_p**. Perform a loop for different values of **par_p** and calculate the p-value each time using the `gsl_cdf_chisq_Q()` function of the *GNU scientific library* (GSL) (see Sec. 7.3).

(9) **Linear correlation coefficient**

Design, implement and test a function, which calculates the linear correlation coefficient $r$ to measure the strength of a correlation for a sample $\{(x_0, y_0), (x_1, y_1), \ldots, (x_{n-1}, y_{n-1})\}$. The function prototype reads as follows:

> SOLUTION SOURCE CODE
> DIR: **randomness**
> FILE(S): `lcc.c`

```
/************************* lcc() *******************/
/** Calculates the linear correlation coefficient    **/
/**                                                   **/
/** Parameters: (*) = return parameter                **/
/**     n: number of data points                      **/
/**     x: first element of sample set                **/
/**     y: second element of sample set               **/
/**                                                   **/
/** Returns:                                          **/
/**     r                                             **/
/****************************************************/
double lcc(int n, double *x, double *y)
```

Remark: Write a `main()` function which generates a sample in the following way: the $x_i$ numbers are generated from a standard Gaussian distribution $N(0, 1)$ while each $y_i$ number is drawn from a Gaussian distribution with expectation value $\kappa x_i$ (variance 1). Study the result for different values of $\kappa$ and $n$.

(10) **Least-squares fitting**

Copy the program from exercise (3) to a new program and change it such that numbers for a shifted Fisher-Tippett with parameters $\lambda$ and peak position $x_0$ are generated. The numbers should be stored in a histogram and a histogram pdf should be written to the standard output.

| SOLUTION SOURCE CODE |
| --- |
| DIR: **randomness** |
| FILE(S): **fitFT.gp** |
| **fisher_tippett2.c** |

- Choose the histogram parameters (range, bin range) such that the histograms match the generated data well.
- Run the program to generate $n = 10^5$ numbers for parameters $x_0 = 2.0$ and $\lambda = 3.0$. Pipe the histogram pdf to a file (e.g. using > **ft.dat** at the end of the call).
- Plot the result using *gnuplot*.
- Define the pdf for the Fisher-Tippett distribution in *gnuplot* and fit the function to the data with $x_0$ and $\lambda$ as adjustable parameters. Choose a suitable range for the fit.
- Plot the data together with the fitted function.
- How does the result compare to the maximum-likelihood fit presented in Sec. 8.6.1?
- Does the fit (in particular for $\lambda$) get better if you increase the number of sample points to $10^6$?

Hints: The shift is implemented by just adding $x_0$ to the generated random number. Use either the histograms from Chap. 5, or implement a "poor-mans histogram" via an array **hist** ( see also in the **main()** function of the **reject.c** program partly presented in Sec. 8.2.4).

# Chapter 9

# Information Retrieval, Publishing and Presentations

When you need the information given in this chapter, you have advanced very far, congratulations! You have almost completed a full cycle of a scientific computer simulation project. The results of your simulations are analyzed and now you want to prepare a publication or give a talk.

In this chapter, some basic information about preparing your own presentations and publications is given. Since it fits best here, it is also explained how to search for literature and other science-related information, although you need to do this at all stages of a project.

The tools described in this section, should allow you to solve all technical problems occurring in the process of preparing a publication (a "paper"). Once you have prepared the paper, usually together with some coauthors, you should give it to at least one other person, who should read it carefully. Probably, he/she will find some errors or indicate passages which might be difficult to understand or misleading. You should always take such comments very seriously, because the average reader knows much less about your problem than you do.

When all necessary changes have been performed, and you and other readers are satisfied with the publication, you can submit it to a scientific journal. You should choose a journal which suits the content of your paper. Where actually to submit, you should discuss with experienced researchers, often your coauthors. It is not possible to give general advice on this issue. Nevertheless, technically the submission can be performed electronically over the Internet for almost all journals. Submitting one paper to several journals in parallel is not allowed. However, you should consider submitting also to a *preprint server* [arXiv] to make your results quickly available to the science community. Most journals allow that you make a preliminary version available in advance. Read the submission conditions carefully! You can find an overview of the publisher copyright policies for most important scientific journals in the data base *Romeo* [Romeo].

## 9.1 Searching for Literature

Before starting a project, having the goal to contribute to the science community and even to publish your results, you should be aware of what exists already. This prevents you from redoing something which has been done before by someone else. Furthermore, knowing previous results and many simulation techniques allows you to conduct your own research projects much better. Unfortunately, much information cannot be found in textbooks. The information is scattered over the world and changes continuously. On any working day, the scientific output is much larger than one scientist can read in his whole life. Thus, you must start to look systematically for literature. With modern techniques like the Internet this can be achieved very quickly. Within this section, it is assumed that you are familiar with the Internet in general and are able to use a browser. Several sources of information are stated in the following list.

- **Your local (university) library**
  Although the amount of literature is limited due to space constraints, you should always check your local library for suitable books concerning your area of research. Many old issues of scientific journals are not yet available through the Internet either. Thus, you may have to copy some articles in the library.
- **Scientific journals**
  Journals are the most important resources of information in science. Most of them can be accessed conveniently via the Internet, at least if your university or institute has subscribed to them. This is the primary source of information. You should consult the most important journals in your field regularly to see what is going on.
- **Preprint server**
  In the time of the Internet, speed of publication becomes increasingly important. Meanwhile, many researchers put their publications on the *Los Alamos Preprint server* [arXiv], where they become available worldwide mostly 72 (usually 24) hours after submission. The database is free of charge and can be accessed from almost everywhere via a browser. The preprint database is divided into several sections such as physics,

mathematics or computer science, and corresponding subsections. Similar to a conventional literature database, you can search the database, eventually restricted to a section, for author names, publication years or keywords in the title/abstract. After you have found an interesting article, you can download it and print it immediately. File formats are *postscript* and *pdf*. The submission should be usually in TeX/LaTeX (see Sec. 9.3).

Please note that there is no editorial processing at all, that means you do not have any guarantee on the quality of a paper. If you like, you can submit a poem describing the beauty of your garden. Nevertheless, the aim of the server is to make important scientific results available very quickly. Thus, before submitting an article, you should be sure that it is correct and interesting, otherwise you might get a poor reputation.

The preprint server also offers access via email. It is possible to subscribe to a certain subject. Then every working day you will receive a list of all new papers which have been submitted. This is a very convenient way of keeping track of recent developments. But be careful, not everyone submits to the preprint server. Hence, you still have to read scientific journals regularly.

- **Literature databases**

In case you want to obtain a list of all articles written by a specific author or of all articles on a certain subject, you should consult a literature database. Unfortunately, the access to these data bases is usually not free of charge. But usually your library should provide access to the most important ones via the Internet. If your library/university does not offer an access to the data base you are interested in, you should complain.

There are many specialized data bases. In technical sciences and engineering, the *INSPEC* [INSPEC] data base is the appropriate source of information. *INSPEC* frequently surveys almost all scientific journals in the areas of physics, electronics and computers. For each paper that appears, all bibliographic information along with the abstract are stored. You can search the data base, for example, for author names, keywords (in the abstract or title), publication years or journals. Via *INSPEC* it is possible to keep track of recent developments in a certain field.

Depending on your field of research, a different data base might suit you better. You should consult the web page of your library to find out which of them you can access. Modern scientific work is not possible without regularly checking literature data bases.

- **Citation data bases**

  In every scientific paper, some other articles are cited. Sometimes it is interesting to get the reverse information, i.e. to obtain all papers which are citing a given article A. This can be useful, if one wants to learn about the most recent developments which are triggered by article A. In that case you have to access a *citation index*. For science, probably the most important is the *Science Citation Index* (SCI) which can be accessed via the *Web of Science* [Web of Science]. There is also a version for social sciences. You have to ask your system administrator or your librarian, whether and how you can access it from your site.

  The *American Physical Society* (APS) [APS] also includes links to citing articles with the online versions of recent papers. If the citing article is available via the APS as well, you can immediately access the article from the Internet. This works not only for citing papers, but also for cited articles.

- **Papercore summary database**

  *Papercore* [Papercore] is an online data base for summaries of scientific journals, see screenshot in Fig. 9.1. The data base also contains review-type documents which introduce a scientific field and collect links to corresponding summaries. The summaries go much beyond abstracts, since they are about 1/10 of the length of the corresponding papers. They should contain all necessary information one would remember long-time after thoroughly reading a paper. In particular the summaries may contain formulas, figures and links to other summaries, as well as the full bibliographic information of the papers. Thus, when reading summaries, a scientist can save a lot of time, e.g. when getting an overview over a field. Also writing summaries is very beneficial, because it forces the summary author to understand a paper much deeper compared to just reading it. This is particular useful for PhD students, who in this way learn their field as well as scientific writing much better. The author of this book himself has summarized more than 100 papers and also the students of his group have been benefiting much from contributing to *Papercore*.

  *Papercore* can be accessed online from everywhere and is free of charge. Everybody can immediately read summaries. After free registration, everybody can submit summaries and alter already existing summaries, similar to the Wikipedia encyclopedia. An important difference is that the summaries can be submitted conveniently as LaTeX files. One can

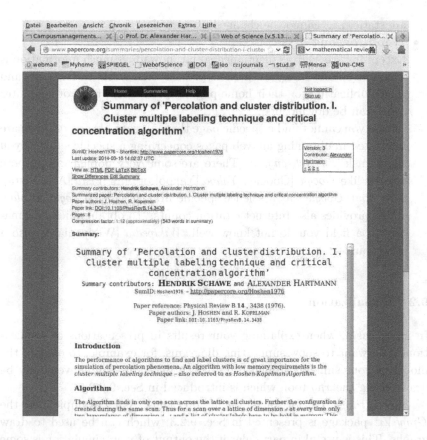

Fig. 9.1  A screenshot of the *Papercore* database. The screen shows the (beginning of a) summary of a famous paper about an percolation algorithm, widely used in statistical physics.

download a special `summary.cls` class, which allows the summary author to compile the summary on his own computer before finally submitting.

Note that in the mathematical community such summaries are well established, since they are contained in the online-accessible *Mathematical Reviews* (MR) database, which is operated by the American Mathematical Society. MR was founded in 1940 as a printed summary journal by Otto Neugebauer, who was a mathematical historian who emigrated in 1933 from Germany. In contrast to *Papercore*, the access [Math. Rev.] to MR is not free of charge, summaries can be written only by researchers who are addressed by the editors, and all summaries are static, i.e., cannot be changed afterwards.

- **Web browsing**

  Except for the sources mentioned so far, nowadays much information is available online. Many researchers present their work, their results and their publications on their home pages. Quite often, talks or computer codes can be downloaded.

  In case you cannot find a specific page through your library, or if you are interested in obtaining all web pages concerning a specific subject, you should ask a *search engine*. There are some very popular all-purpose engines like *Google* [Google], *Yahoo* [Yahoo] or *Alta Vista* [Alta Vista]. Note that *Google* has a science-specific branch called *Google scholar*, which provides also Internet citation counts. To obtain quick information in a field you do not know well, *Wikipedia* [Wikipedia] is also a good source.

## 9.2 Visualization

In many cases, when explaining your results in presentations or publications, you want to show supporting diagrams, for example, to explain the model or your simulation algorithm. Such diagrams can conveniently be drawn using the *xfig* tool, which is introduced in Sec. 9.2.2.

Special tools exist for some drawing problems. As an example, here the *GraphViz* package is presented in Sec. 9.2.3, which can be used to draw graphs. This is useful in particular if the output of your simulation is some graph which should be printed nicely.

In many cases, three-dimensional situations are to be displayed. For this purpose, *xfig* is not powerful enough. Here, the *Povray* package is quickly presented in Sec. 9.2.4.

### 9.2.1 *Presentation-ready figures using* gnuplot

Using *gnuplot* as it comes with a command line interface, is good for visualizing data quickly. Nevertheless, the plots will not look very nice immediately, e.g. the fonts are very small, the lables will be very simple. With a bit of effort, it is possible to make plots such that they can

| GET SOURCE CODE |
| --- |
| DIR: literature |
| FILE(S): |
| plot_PW_integrand.gp |
| integrand.dat |
| B.dat |

be used in scientific publications. Next, the most important commands and options are shown via one specific example. Here, the plot consist of a main plot (showing file `integrand.dat`) and an inset (`B.dat`). The resulting plot is shown in Fig. 9.2. The *gnuplot* script `plot_PW_integrand.gp` generating this plot reads as follows:

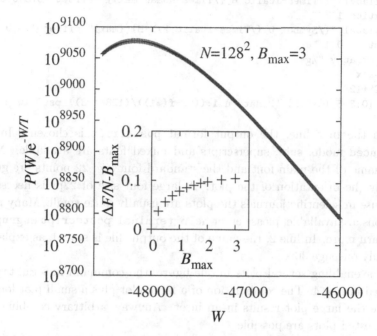

Fig. 9.2  Example plot generated using *gnuplot*, containing big fonts, different font types, subscripts, superscripts, additional labels and an inset.

```
1  set terminal postscript enhanced "Times-roman" 22 portrait
2  set output "PW_integrand.eps"
3
4  set multiplot
5
6  set size square 1
7  set nokey
8  set label "{/Times-Italic N}=128^2,{/Times-Italic B}_{max}=3"\
9        at -47500, 9060
10 set xtics 1000
11 set xlabel "{/Times-Italic W}"
12 set format y "10^{%4.0f}"
13 set ylabel "{/Times-Italic P(W)}e^{/Times-Italic -W/T}"\
14       offset 1,0
```

```
15   plot [-48600:-46000] [8700:9100] "integrand.dat" u 1:($2/log(10.0))
16
17   set size square 0.53
18   set origin 0.22, 0.175
19   set xlabel "{/Times-Italic B_{/Times-Roman max}}" offset -0.3,0.6
20   set xtics 1
21   set ylabel "{/Symbol D}{/Times-Italic F/N-B}_{max}" offset 1.5,0
22   set ytics 0.1
23   set format y "%g"
24   f(x)=a*x
25   a=-128*128
26   plot [0:3.5][0:0.2] "B.dat" u 1:(($2-f($1))/(128*128)) ps 2 lw 2
```

In the first line, the output format **postscript** is chosen. In the
**enhanced** mode, sub-, superscripts and mixed fonts are available. Next,
the name of the main font and the standard font size 22 points are given,
finally the orientation of the plot. A large font size of 22pt is necessary,
because in scientific journals the plots are usually quite small. Many more
options are available, please enter **help terminal postscript** in gnuplot
to learn more. In line 2, the name of the output file is stated, as explained
already on page 308.

For enabling several plots in one figure, the command **set multiplot**
is used in line 4. The combination of a large plot plus a small plot located
inside the large plot results in an inset. Anyway, arbitrary combinations
and nested plots are possible.

The first plot starts in line 6. The plot is squared and has scale factor 1
(including labels and margins). In line 6 it is stated that no key is shown.
Instead, in line 8, directly a label is set. Some parts of the label are printed
with mixed fonts. Here also a superscript $128^2$ and a subscript $B_{max}$ are
generated. The position of the label is determined via **at** $-47500,9060$, i.e.,
in coordinates of the data space. Note that one could also state a unique
font valid for the complete label, including a size, via the **font** option of
the **set label** command, e.g., via including **font "/Times-Italic,22"**.
More information you will obtain via entering **help set label**.

In line 10 of the script, the spacing between the tics of the $x$-axis is set,
followed by the label for this axis in line 11. For the $y$-axis, the tic labels
have a special format. One can state, more or less, arbitrary formats,
similar to the formats of the **fprint()** C command. All possible format
options can be obtained via entering **help set format**. Here, see line 12,
10 to the power of the actual $y$ value is printed. The reason is that in the
file to be plotted the log values of the actual $y$ values are shown, since the

$y$ values are really huge. Note that %f is a floating point number with 4 digits, none after the decimal point. The label for the $y$ axis, see lines 13 and 14, make use of mixed fonts and subscripts again. Note that the label is slightly shifted to the right via the offset 1,0 option, to reduce the space between the label and the axis.

The actual plotting is issued in line 15. Here, first the ranges to be plotted in $x$ and $y$ direction, are specified, respectively. The file to be plotted is stated. Note that the data in the second column was calculated as natural logarithms, while the axis labels are given to with respect to the base 10. For this reason the data is divided by log(10).

In line 17, the inset is started. It should be squared size, but smaller. The position of the lower left is stated in canvas coordinates in line 18. In lines 19 to 22 the labels and the tics spacing for $x$ and $y$ axes are specified, similar to the main plot, again with slight shifts for better visibility. The format of the $y$ axis is reset to the standard value, indicated by the format %g. The data is plotted relative to the function f(x), which is specified in lines 24 and 25. The actual plot is performed in line 26. Note that here large symbols ("points size" ps 2) with thicker lines ("line width" lw 2) are used. There are many other options to changes the style of the symbols or lines, like color ("line color" lc), or symbol type ("point type" pt). Please refer to help style for a complete list of options.

### 9.2.2 *Drawing figures using xfig*

Most scientific texts do not only contain text, formulas and data plots, but also schematic figures showing the models, algorithms or devices covered in the publication. A very convenient but also easy-to-use tool to create such figures is *xfig*. It is a window-based vector-oriented drawing program. Among its features are the creation of simple objects like lines, arrows, polylines, splines, arcs as well as rectangles, circles and other closed, possibly filled, areas. Furthermore, you can create text strings or include arbitrary (*eps*, *Jpeg*, ... ) picture files. You may place the objects on different layers which allows complex sceneries to be created. Different simple objects can be combined into more complex objects. For editing you can move, copy, delete, rotate or scale objects. To give you an impression of what *xfig* looks like, in Fig. 9.3 a screen-shot is shown, displaying *xfig* with the picture that is shown in Fig. 1.1. Again, for further help, please consult the online help function or the *man* pages.

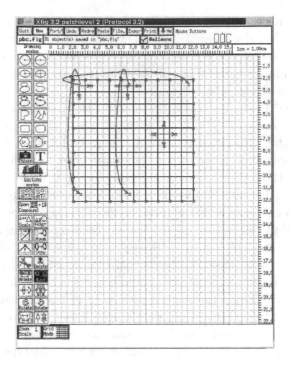

Fig. 9.3   A sample screen-shot showing the *xfig* program.

The figures can be saved in the internal fig format, and exported in several file formats such as (encapsulated) *postscript*, LATEX, *Jpeg*, *Tiff* or bitmap. The *xfig* program can be called in a way that it produces just an output file with a given fig input file. This is very convenient for larger projects where some small picture objects are contained in other pictures and for changing the appearance of the small objects in all other files. With the help of the *make* program, pretty large projects can be realized.

Last but not least, please note that *xfig* is vector-oriented, but not pixel-oriented. Therefore, you cannot treat pictures like jpg files (e.g. photos) and apply operations like smoothing, sharpening or filtering. For these purposes, the package *gimp* is suitable. It is freely available again from GNU [Loukides and Oram (1996)].

### 9.2.3   *Drawing graphs*

As we have seen in Sec. 6.8, many simulations are based on graph representations. In these cases, it is very useful to visualize graphs for debugging purposes and also to present results. Here, the *GraphViz* package [GraphViz] is explained. It works basically in the following way:

(1) One generates an ASCII file, which describes the graph and its properties. This is the so-called *dot file*.
(2) A *filter program* is used to generate an image from the dot file.

The dot file is human readable and editable. This has some advantages: You can generate graphs either using a normal text editor or you can write simple C functions to output a graph in *dot* format, which is stored using your own data structures within your simulation package. This will be described below for the graph data structures used in Sec. 6.8. There is also the interactive program *dotty*, which allows graphs to loaded, displayed, modified and saved in the *dot* format. First, we explain the general properties of the *dot* file and of the commands to generate an image from a *dot* file. Here, only short ready-to-run examples are given. Some details can be found in the man page when entering `man dot` on a Unix system or, more complete and up-to-date, at Ref. [GraphViz].

The undirected graph shown in Fig. 6.24 can be described by the following dot file, called `testgraph.dot`:

```
1   graph test
2   {
3       0
4       1
5       2
6       3
7       4
8       5
9       0 -- 1
10      1 -- 3
11      1 -- 4
12      2 -- 4
13  }
```

On line 1, `graph` shows that the file describes an undirected graph, `test` being the ID of the graph. An ID can be a string composed of alphabetic characters, digits or underscores. Also more complex IDs are possible, in particular if they are double-quoted strings "..." or html strings < ... >. Note that the key words `strict`, `graph`, `digraph`, `node` and `edge` are not allowed as IDs. Between the braces (line 2 and 14) the graph is described, this is the so-called *statement* part. For this example, first (lines 5-10) comes a list of node IDs (*node statements*). These IDs follow the same rules as the ID of the graph. Then (lines 11-14) comes a list of edges (*edge statements*), which are composed of two IDs joint by the *edge operator* --. Several statements per line are possible, if they are separated by spaces or tabs or semicolons.

To generate an image, say a postscript file from the *dot* file, one uses a *filter* program, e.g. dot. It can be called using a command line like

```
dot -Tps testgraph.dot > testgraph.eps
```

The output of dot is always to stdout. Therefore, it is redirected to a file testgraph.eps here. The option -Tps means that the output format is postscript. Other formats are given below. The resulting postscript file is shown in Fig. 9.4.

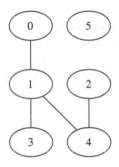

Fig. 9.4   The graph of Fig. 6.24, as plotted by the dot program.

Also lists of edges of the form node1 -- node2 -- node3 -- node4 can be given. Note that the order of nodes and edges is arbitrary within the statement list. Consequently, one can mix or alternate node and edge statements. In particular, for nodes appearing in edges, it is not necessary to include them explicitly as node statements, unless one wants to state additional attributes. These optional attributes can be written in [. . .] brackets behind the nodes and edges using a comma-separated list of attribute assignments of the form <*attribute*>=<*value*>. There are many different attributes for shape, color, (minimum) size, fonts, positions etc. We do not list them here for conciseness, but give just an example below. Please refer to the man page for details. Here we state only that attributes can be set also globally for all nodes or all edges by writing node [<*attribute list*>] or edge [<*attribute list*>]. Attributes which apply to the full graph are just given without brackets in the *<attribute list>* form. Here we show an example which uses some attributes:

```
1   graph test2
2   {
3     node [shape=box]
4     0 [label=Milk, style=bold ]
5     Butter [shape=house,height=2, color=blue]
6     Cheese [fontname="Palatino-Italic"]
7     Wheat -- Sandwich
8     Cheese -- 0 -- Butter Butter -- Cake [label = bakery]
9     Butter -- Sandwich [style = dashed]
10    Cheese -- Sandwich
11  }
```

Line 3 contains a global node attribute. Lines 4–6 contain node statements with some attributes. Note that the **shape=house** assignment overrides the global **shape=box** just for the node Butter. Lines 7–10 contain edge statements with some attributes. The resulting image made by dot is shown in Fig. 9.5.

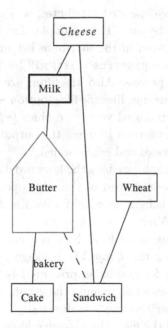

Fig. 9.5  Another graph defined by a plot file with several attributes (see text) as plotted by the dot program.

Any filter program must have some algorithm included, which places the nodes in the image. The program could, for example, try to minimize the number of crossing edges. The filter dot used above works very well in particular for graphs which can be drawn as hierarchies, such as trees. Other filters currently available are

- fdp and neato
  Different "spring models" are used to generate pretty images. The springs virtually connect nodes and edges to keep them apart. The filters relax the system of springs such that the overall stress is minimized.
- twopi
  All nodes are arranged using a radial layout. One node is placed in the center of the image, all other nodes on sequences of concentric circles around the center.
- circo
  The filter generates a circular layout where nodes belonging to the same biconnected component[1] are drawn on the same circle.

There are also filter-depended attributes, which control the algorithm calculating the graph layout. For example, for fdb and neato, nodes where the position is given in the attribute list including a final '!', like in pos="3,2!", or where pin=true is set, will be kept in the initial positions during the layout process. Also, the filters are able to generate output formats other than postscript, like *xfig* files (option -Tfig), bitmap graphics (-Tpng and -Tgif), structured vector graphics (-Tsvg and -Tsvgz), and some more. If no output format is given, the output is again in *dot* format, but with positions of nodes and edges added.

To plot a directed graph, one uses the keyword digraph instead of graph and the *edge operator* -> instead of the edge operator --. Furthermore, one can define and use subgraphs within a *dot* file; for details again see the documentation [GraphViz].

Finally, we show an example of how you can write your own graph drawing functions using the *GraphViz* package. The function is based on the data structures for graphs as presented is Sec. 6.8. The function gs_dot_graph() receives as arguments the graph, a file stream and a variable which tells whether the graph is directed. The function outputs the graph in *dot* format to the file. The C source looks as follows:

---

[1] A biconnected component is a maximal set of nodes, where two independent paths within the biconnected component exist between each pair of nodes of the set.

```
1   void gs_dot_graph(gs_graph_t *g, FILE *file, int dir)
2   {
3       int n;
4       elem_t *elem;
5
6       if(dir)
7         fprintf(file, "digraph test {\n");
8       else
9         fprintf(file, "graph test {\n");
10      fprintf(file,
11          "node [height=0.2,width=0.2,label=\"\",fixedsize=true]\n");
12
13      for(n=0; n<g->num_nodes; n++)            /* go through all node */
14      {
15        fprintf(file, "%d\n", n);
16        elem = g->node[n].neighbors;
17        while(elem != NULL)                    /* loop over all neighbors */
18        {
19          if(dir)
20            fprintf(file, "%d -> %d\n", n, elem->info);
21          else
22            if(n <= elem->info)        /* print each pair only once */
23              fprintf(file, "%d -- %d\n", n, elem->info);
24          elem = elem->next;                   /* next neighbor */
25        }
26      }
27      fprintf(file, "}\n");
28  }
```

Depending on the variable dir, a directed or an undirected graph is printed (lines 6–9). Here, the graph is always called "test". If you prefer to have a variable name, you have to include the graph name as parameter (or include it in the data structure of the graph). Next (line 10), some global node properties are set. In lines 12–26 the graph is printed by iterating over all nodes. First the node is printed (line 15), then all neighbors are visited (lines 16–25) and the corresponding directed or undirected edge is printed. Finally (line 27), the closing bracket of the *dot* file is written.

You could also in your C program automatically convert the *dot* file to an image file by using the system command to call, for example, dot with the name of the written file as a parameter, as exemplified in the following lines (assuming that name contains the prefix of the filename and g the graph):

```
1    char filename[1000], command[1000];
2    FILE *file;
3
4    sprintf(filename, "%s.dot", name);
5    file = fopen(filename, "w");
6    gs_dot_graph(g, file, 0);
7    fclose(file);
8    sprintf(command, "dot -Tps %s.dot > %s.ps", name, name);
9    system(command);
```

### 9.2.4  *Three-dimensional figures with* Povray

It is also possible to draw three-dimensional figures with *xfig*, but there is no special support for this task. This means, *xfig* has only a two-dimensional coordinate system. A very convenient and powerful tool for making three-dimensional figures is *Povray* (Persistence Of Vision RAYtraycer). Here, again, only a short example is given; for a detailed documentation please refer to the home page [Povray], where the program can be downloaded for many operating systems free of charge.

*Povray* is, as can be realized from its name, a *raytracer*. This means, you present a scene consisting of several objects to the program. These objects have characteristics like color, reflectivity or transparency. Furthermore, the position of one or several light sources and a virtual camera have to be defined. The output of a raytracer is a photo-realistic picture of the scene, seen through the camera. The name "raytracer" originates from the fact that the program creates a picture by starting several rays of light at the light sources and traces their way through the scene, where they may be absorbed, reflected or refracted, until they hit the camera, disappear into infinity or become too weak. Hence, the creation of a picture may take a while, depending on the complexity of the scene.

A scene is described in a human readable file. It can be entered with any text editor. But for more complex scenes, special editors exist, which allow a scene to be created interactively.

| GET SOURCE CODE |
| --- |
| DIR: literature |
| FILE(S): test1.pov |

Also several tools for making animations are available on the Internet. Here, a simple example is given. The scene consists of three spheres connected by two cylinders, forming a molecule. Furthermore, a light source, a camera, an infinite plane and the background color are defined. Please note that a sphere is defined by its center and a radius and a cylinder by two end points and a radius. Additionally, the corresponding color information has to be included for all objects. Here, the center sphere is slightly transparent. The scene description file `test1.pov` reads as follows:

```
#include "colors.inc"

background { color White }

sphere {  <10, 2, 0>, 2
    pigment { Blue } }

cylinder { <10, 2, 0>,  <0, 2, 10>, 0.7
    pigment { color Red } }

sphere {  <0, 2, 10>, 4
    pigment { Green transmit 0.4} }

cylinder { <0, 2, 10>,  <-10, 2, 0>, 0.7
    pigment { Red } }

sphere {  <-10, 2, 0>, 2
    pigment { Blue } }

plane { <0, 1, 0>, -5
    pigment { checker color White, color Black}}

light_source { <10, 30, -3> color White}

camera {location <0, 8, -20>
    look_at  <0, 2, 10>
    aperture 0.4}
```

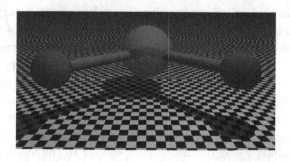

Fig. 9.6  A sample scene created with *Povray*.

The creation of the picture is started by calling (here on a Linux system via command line) `povray +Itest1.pov +P`. The program can have a different name on your system, e.g. `x-povray`. Note that there is no space between the `+I` (input) option and the name of the file describing the scene. An image will appear on your screen. The pause option `+P` makes *Povray* to wait for a mouse klick in the image to quit. In addition to showing the image on the screen, a file in the portable network graphic format `test1.png` will be created, when using *Povray* 3.7 (`.bmp` for version 3.6.2). The resulting picture is shown in Fig. 9.6. Please note the shadows on the plane.

Also other names for the output file can be given with the `+O` option, e.g. `+Oimage` will let *Povray* generate `image.png`. A summary of these and other important options, often followed by a value or another argument without a space, is as follows

- `+I<File>` input file `<File>`
- `+O<File>` output file `<File>` (without appendix `.png`)
- `+P` pause: wait for click before finishing
- `+Q<Quality>` quality of output, the higher the better
- `+W` width of image
- `+H` height of image
- `+f<format>` use other output file format, e.g. jpg (`<format>`=j), ppm (`<format>`=p), bmp (`<format>`=b).

*Povray* is really powerful. You can create almost arbitrarily shaped objects, combine them into complex objects and impose many transformations. Also special effects like blurring or fog are available. All features of *Povray* are described in a 400-page manual. The use of *Povray* is widespread in the artists' community. For scientists, it is very convenient as well, because you can easily convert configuration files of molecules or three-dimensional domains of magnetic systems into nice-looking perspective pictures. This can be accomplished by writing a small program which reads, for example, your configuration file containing a list of positions of atoms and a list of links, and puts for every atom a sphere and for every link a cylinder into a *Povray* scene file. Finally the program must add suitably chosen light sources and a camera. Then, a three-dimensional pictures is created by calling *Povray*. This is quite similar to the automatic creation of *dot* files, as explained in Sec. 9.2.3.

Here, we consider the example of a three dimensional box with atoms, which are going to be represented as spheres. The actual atom data is stored in an arry where the elements are of type atom_t which is defined as follows:

| GET SOURCE CODE |
| --- |
| DIR: `literature`<br>FILE(S): `atoms.c` |

```
/* stores data of one atom: */
typedef struct
{
  double      m;        /* mass of atom */
  double      sigma;    /* 'size' of atom for LJ */
  double      epsilon;  /* LJ energy parameter */
  double      *x;       /* position of atom */
  double      *v;       /* velocity of atom */
  double      *f;       /* force on atom */
} atom_t;
```

Note that not all data, like the force array f, is necessary for just generating a configuration. Nevertheless, here we state the actual data structure which was used to perform *Molecular dynamics simulations*, i.e. the integration of Newton's equations of motion. There is also the data structure glas_system_t which holds global information of the model. Here we only need the entries dim, which holds the dimension (here 3) of the system, the entries l[3], l[1], l[2] which store the lateral sizes and N which holds the total number of atoms. The program atoms.c contains a function glas_setup() which takes a pointer to a structure of type glas_system_t, initializes everything, generates a random configuration and returns a pointer to it.

Below, a function is shown which generates a *Povray* file from the current configuration:

```
1  void atoms_plot_cfg(atoms_system_t *system, atom_t *atom)
2  {
3    char filename[1000];
4    FILE *povfile;
5    int t, d;                                    /* loop counters */
6    double *r;                                   /* position */
7
8    r = (double *) malloc(system->dim*sizeof(double));
9    sprintf(filename, "cfg.pov");
10   povfile = fopen(filename, "w");              /* open file */
11   fprintf(povfile, "#include \"colors.inc\"\n" /* header etc */
```

```
12              "#include \"shapes.inc\"\n\n");
13      fprintf(povfile, "background { color Yellow }\n\n");
14      fprintf(povfile, "camera {\n  location <%f, %f, %f>\n",
15              0.5*system->l[0], -1.5*system->l[1], 0.5*system->l[2]);
16      fprintf(povfile, "  sky <0,0,1>\n");
17      fprintf(povfile, "  look_at  <%f, %f, %f>\n}\n\n",
18              0.5*system->l[0], 0.5*system->l[1], 0.5*system->l[2]);
19      fprintf(povfile,
20              "  light_source { <%f, %f, %f> color White}\n\n",
21              0.5*system->l[0], -0.5*system->l[1], 1.5*system->l[2]);
22      fprintf(povfile,
23              "  light_source { <%f, %f, %f> color White}\n\n",
24              -0.5*system->l[0], -0.5*system->l[1], 1.5*system->l[2]);
25
26      fprintf(povfile,                         /* print system boundaries */
27              "cylinder{ <%f, %f, %f>, <%f, %f, %f>, 0.25\n"
28              "   pigment { Red } }\n",
29              0.0, 0.0, 0.0,
30              system->l[0], 0.0, 0.0);
31      /* continued for 11 other boundary 'bars', see atoms. c */
32
```

The function takes the global system data and an array to atom data as arguments (line 1). In lines 2–6, the local variables are declared. In line 8, memory for an auxiliary vector is allocated, which is used to realize periodic boundary conditions. A file named, for simplicity, cfg.pov is opended (lines 9–10). Some standard definitions are made (lines 11–13). A camera is put a bit away from the system such that it is directed at the center of the box (lines 14–17). Two light sources ar introduced such that they are positioned above the camera and left of the camera (lines 19–24). Furthermore, 12 small cylinders are included such that they mark the boundaries of the system cube (from line 26 on). In the second part of the function, the atoms are included in the *Povray* file:

```
87      for(t=0; t<system->N; t++)                    /* print atoms */
88      {
89        for(d=0; d<system->dim; d++)
90        {
91          r[d] = atom[t].x[d];
92          if(r[d] < 0)                              /* fold positions into box */
93              r[d] += floor(-r[d]/system->l[d]+1)*system->l[d];
94          if(r[d] > system->l[d])
95              r[d] -= floor(r[d]/system->l[d])*system->l[d];
96        }
```

```
97      if(system->dim == 3)
98        fprintf(povfile,
99              "sphere { <%f,%f,%f>, %f\n    pigment { Blue }}\n",
100             r[0], r[1], r[2], 2.0*atom[t].sigma);
101       else if(system->dim == 2)
102         fprintf(povfile,
103             "sphere { <%f,%f,%f>, %f\n    pigment { Blue }}\n",
104             r[0], r[1], 0.0, 2.0*atom[t].sigma);
105     }
106     fclose(povfile);
107     free(r);
108   }
```

The main loop runs over all atoms, see line 97. For all atoms, the positions have to be subjected to periodic boundary conditions, i.e., they are folded back into the box (lines 89–96). Finally, the atom positons are converted to positions of balls in the *Povray* file (lines 97–104). Finally the output file is closed (line 106) and the memory used for the variable r is released.

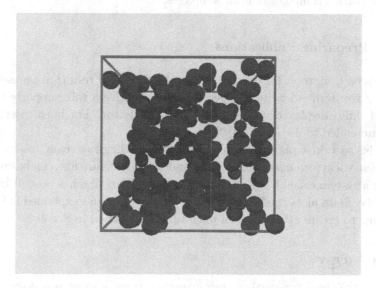

Fig. 9.7  A sample configuration of atoms, randomly placed. The configuration is converted to a *Povray* file using the function atom_plot_cfg() and printed into an image file using *Povray*.

The above sample program can be easily extend to draw, e.g. molecules, by adding small cylinders which connect spheres corresponding to atoms joint by chemical bounds.

After one ore several files have been generated, they can be transformed using *Povray* into figure files, like in the above example. In Fig. 9.7 a sample configuration generated by using this function is shown.

If many subsequent configurations are converted into images, they can be merged into a video. The most simple way to do this is to use the `convert` tool available under Linux and create an *animated gif*, e.g.

```
convert -delay 20 frame*.png -loop 0 animated.gif
```

The `-delay` option makes `convert` to produce a sequence of images, the number (here 20) denotes the pause between two images in milliseconds. Here, the set of images is stored in the files `frame*.png`. The `-loop` option states how often the movie is to be repeated, a zero means an infinite number of times. The final argument is the name of the output file. An animated gif can be easily viewed by loading it into a browser or by including it like static gif images in html webpages.

## 9.3 Preparing Publications

In this section, tools for two types of presenting your results are presented: Either you want to write a paper/report or to give a talk supported by a set of slides displayed via a laptop and a projector. For both cases, it is recommended to use TEX/LATEX.

This section explains how manuscripts, including raw texts, tables, lists, mathematical formulas, bibliography and external figure files can be created using a system called LATEX. How to make the figure files has been discussed already: Data plots can be produced using the programs explained in Chap. 8. How to create other types of diagrams is explained in Sec. 9.2

### 9.3.1 *LATEX*

TEX/LATEX is a typesetting system rather than a word processor. The basic program is TEX; LATEX is an extension to facilitate the application. In many areas of science, in particular those where many formulas occur, the combination of TEX and LATEX is a widespread standard. Nevertheless, even for publications in humanities, LATEX is the most professional tool.

When electronically submitting an article to a scientific journal, LaTeX has to be used in many cases. This book was completely written in LaTeX.

Unlike the conventional office packages, with LaTeX you do not see *immediately* the text in the form it will be printed, i.e. LaTeX is not a WYSIWYG ("What you see is what you get") program. The text is entered in a conventional text editor (like *Emacs*) and all formatting is done via special commands. An introduction to the LaTeX language can be found, for example, in Refs. [Lamport and Bibby (1994); TUG]. Although you have to learn some special commands, using LaTeX has several advantages:

- The quality of the typesetting is excellent. It is much better than self-made formats. You do not have to care about the layout. But still, you are free to change everything according to your requirements.
- Large projects do not give rise to any problems, in contrast to many commercial office programs. When treating a LaTeX text, your computer will never complain when your text is more than 500 pages or contains many huge post-script figures.
- Typesetting of formulas is very convenient and fast. You do not have to care about sizes of indices appearing themselves in indices etc. Furthermore, in case you want to replace all $\alpha$ in your formulas with $\beta$, for example, this can be done with a conventional replace, by replacing all \alpha strings by \beta strings. For the case of an office system, please do not ask how to do this conveniently.
- There are many additional packages for enhanced styles such as letters, transparencies or books. The *Bibtex* package is very convenient, which allows the user to build up large literature data bases conveniently.
- Since you can use a conventional editor, the writing process is very fast. You do not have to wait for a huge packet to come up.
- On the other hand, if you still prefer a WYSIWYG ("what you see is what you get") system, there is a program called *lyx* [Lyx] which operates like a conventional word processor but creates LaTeX files as output. Nevertheless, once you get used to LaTeX, you will never want to write a publication with something else, unless someone points a gun at your head and forces you to use an office package.

Since LaTeX is a type setting language, you have to compile your text to create the actual output. Now, an example is given of what a LaTeX text looks like and how it can be compiled. This example will give you a first impression of how the system operates. After that, some important elements like fonts, chapters, tables, lists, mathematical formulas, figures and the bibliography are presented using some examples from *this* book. Furthermore, for these style elements, the most important options and variants are given, such that 95% of all scientific writing is covered. For a complete reference, please consult the literature mentioned above.

### 9.3.1.1 *Getting started*

The following file `example.tex` produces a text with different fonts and a formula. We first show the complete file, then we explain the meaning of the different elements and show how to get the actual typeset document from it.

| GET SOURCE CODE |
| --- |
| DIR: `literature` <br> FILE(S): `example.tex` |

```
\documentclass[12pt]{article}
\begin{document}
This is just a small sample text. You can write some words
\emph{emphasized}\/, or in {\textbf bold face}. Also different
{\small sizes} are possible.

An empty line generates a new paragraph. \LaTeX\  is very convenient
for writing formulae, e.g.
\begin{equation}
M_i(t) = \frac{1}{L^3} \int_V x_i \rho(\vec{x},t) d^3\vec{x}
\end{equation}
\end{document}
}
```

The first line introduces the type of the text, here `article`, which is the standard, and the font size. You should note that all TEX commands begin with a backslash ($\backslash$), in case you want to write a backslash in your text see Sec. 9.3.1.2. There are few exceptions like the dollar symbol '$' which switches the mathematical in-line mode on and off. Arguments to commands are given in { } braces, like the type of the text for the `\documentclass` command, which should always be the first command in your file. Optional arguments are given in [ ] braces, such as the font size here (the default is `10pt`). Note that the { } are also used to group words together. Sometimes it is necessary to group an empty word, e.g. if a command specifies a special character and you want that immediately another letter follows. An example is the word "Straße", which has to be set as `Stra\ss{}e` to tell the LATEXcompiler that not the command `\sse` is meant, which does not exist, anyway.[2]

The actual text is written between the lines starting with `\begin{document}` and ending with `\end{document}`. In the example you see some commands such as `\emph`, which switches highlighting on, `\textbf`, which switches to bold font, or `\small`, which switches to a small font, see also Sec. 9.3.1.2. Changes of fonts or sizes are only effective within one group of words, or until the choice is overwritten by a different one.

Mathematical formulas can be written, for example, in between `\begin{equation}` and `\end{equation}`. This is a so-called *environment*. There are many other environments, see below. The name of the environment is always an argument to the `\begin` and `\end`. LATEX will create in this case a formula which is separated from the text and contains an automatically created equation number. To learn how to address these numbers via *labels*, see below. If you want a separate formula without equation number, use `\begin{equation*}` and `\end{equation*}`. For inline formulas, write the formula inside a pair of dollar symbols `$  $`. For mathematical formulas, a huge number of commands exists. Below, the most important elements like Greek letters (`\alpha`), operators (`+`, `\Rightarrow`), subscripts (`x_i`), fractions (`\frac`), integrals (`\int`), functions (`\sin`), vectors (`\vec`), and multi-line formulas (`\begin{eqnarray}` ... `\end{eqnarray}`) will be explained.

The text can be compiled by entering

```
latex example
```

in a shell (the suffix `.tex` can be omitted). This is the command for UNIX, but LATEX exists for all operating systems. There exist also development environments, which combine editor, latex compilation and display of the final document. Examples for these environments are *Kile*, which is included in most Linux distributions, or *TeXworks*, which is available for all big operating systems. Please consult the documentation of your local installation.

The output of the compiling process is the file `example.dvi`, where "dvi" means "device independent". The `.dvi` file can be inspected on screen by a viewer via entering `xdvi example.dvi`. Alternatively, it can be converted into a *postscript* file via typing `dvips -o example.ps example.dvi` or to

---

[2]In this case LATEX would complain.

a *pdf* file via typing `dvipdf example.dvi`. In the first case, the `-o` option is necessary on many systems, because otherwise the file is directly printed. For the `dvipdf` command, always the corresponding *pdf* file (here: `example.pdf`) is generated. The result will look like this:

> This is just a small sample text. You can write some words *emphasized*, or in **bold face**. Also different sizes are possible.
> An empty line generates a new paragraph. LATEX is very convenient for writing formulae, e.g.

$$M_i(t) = \frac{1}{L^3} \int_V x_i \rho(\vec{x}, t) d^3 \vec{x} \qquad (9.1)$$

LATEX takes care of the full formatting business for you. This means, you do not have to care about spacings, justification, hyphenation, font sizes, numbering of pages, sections and so on. In particular the way you format your source `.tex` file has not much influence on the final output. You could have, e.g. in each line just one word, or you can put many words in one line of the source file. Important are spaces and tabs, which separate words. The number of spaces or tabs between two words has no influence on the output. Furthermore, as mentioned in the first example, one or several empty lines in the source file start a new paragraph. This usually means that the first line of the new paragraph in the output file is a little bit indented. This can be avoided by including globally `\parindent0pt` in the beginning of the source file, or by using `\noindent` to avoid indentation just for the next paragraph. Starting new lines without starting a new paragraph can be forced by using `\\` or by using `\newline`. This is also used to indicate new lines in tables etc., see below. Page breaks can be forced using `\newpage` or `\clearpage`.

Hyphenation works well, but not always, in particular if language-specific symbols appear. You can help LATEX by providing possible (or wanted) hyphenation positions via inserting `\-` in the corresponding positions as, e.g. in `si\-mu\-la\-tion`. For each word you can decide yourself what position you want to specify. If a word is not hyphenated, it will appear normal. Otherwise, it will be hyphenated at one the given positions, the one which created the best layout. You can also state a list of words with possible hyphenation positons (each indicated just by a −) at the begin of the source file via the `\hyphenation` command. If a word in the list contains no possible hyphenation point, it will never be hyphenated in the resulting document. Also you can prevent words or sentenced from being hyphenated or split over several lines by using the \mbox command with the text as argument.

The appearance of the final document is controlled by many parameters, like for spaces between lines, width of the text, size of indices in formulas, etc. Depending on the document and on the font size, default values exists, which you can alter, if you like. This will not be covered here, because for most cases the default values are sufficient and give perfectly looking results. Note that most scientific journals offer special packages, which will format your paper draft such that it looks like a printed paper in the journal. Just look at the homepage of the journal where you intend to submit your paper.

You can add comments to a LaTeX source file via the % symbol: in a line everything including and after the symbol will be ignored by the compilation.[3] For example, for scientific journals you can often download journal-taylored classes and corresponding templates for manuscripts. The templates show how the special class commands are used, e.g. to specify title, authors, affiliations, abstracts and keywords. Within the templates, comments are used to explain the meanings of these commands.

Although LaTeX contains already a lot of design elements, there are many extensions available, called *packages*. They can be made available in the source file via the \usepackage command with the name of the package given as argument in { ... } brackets. Any \usepackage must appear after the \documentclass command and before \begin{document}, i.e. in the *preamble*. Often used packages are graphicx for including figures, see Sec. 9.3.1.7, the babel package for language specific settings, like special hyphenation rules, or the amsmath package of the American Mathematical Society including many additional math commands. Important is the inputenc package, which enables the user to specify the coding system such as to write special characters of their language directly in a source file. For example

```
\usepackage[latin1]{inputenc}
```

selects the *iso-latin-1* coding system. Another common coding system is *utf8*. Other packages which are important in the scientific context are geometry, which implements different paper layouts like DinA4, ifthen, which allows for conditional compiling, makeidx, which enables the user

---

[3]This is a rare example where the actual formatting of the source file has an influence on the output.

to create an index, and `fp` which makes simple floating-point calculations available in LATEX. Please refer to the documentation supplied with these packages.

Next, we introduce the most important standard LATEX elements using examples actually from this book. Each LATEX command usually offers many options and variants, which cannot be shown completely in this little chapter. Please consult the LATEX documentation for details.

### 9.3.1.2  *Fonts, special characters and symbols*

All text can be set in different fonts. The simplest way is to change the font via `\bf` to **bold**, via `\it` to *italic* or via `\sf` to sans-serif. The change of the font is active until it is changed again, or inside a {...} group. There are alternative versions, where the text to be shown in a different font has to be given as argument, like `\texttt{typewriter}` will result in `typewriter`. Similarly, `\textsf`, `\textbf`, `\textit` can be used. Even more commands like `\textsc` for SMALL CAPS exist, please consult the detailed LATEX documentation.

Also different font-sizes, relative to the chosen font size are available, again valid until the end of a group or until another font size is chosen:

| | |
|---|---|
| `\tiny` | tiny |
| `\scriptsize` | very small |
| `\footnotesize` | quite small |
| `\small` | small |
| `\normalsize` | normal |
| `\large` | a bit larger |
| `\Large` | even larger |
| `\LARGE` | very large |

Often used, there is also the command `\emph` to *emphasize* the text given as argument, and `\underline` to underline the text given as argument.

LATEX knows four different types of dashes. The first three types appear as - (called "hyphen"), -- ("en-dash"), and --- ("em-dash") in the source file. The first form is for combined words like in "direct-simulation approach". The second is used to specify ranges like "page 21–45". The third is used to create long dashes like in "to be — or not to be?". The forth form is the minus sign in mathematical formulas, also appearing as a single dash - in the source file.

There are several symbols which have special meanings in LATEX:

```
#   &   $   {   }   %   ^   _   ~   \
```

If you actually want these symbols to appear in your text, you have to write, respectively

```
\#  \&  \$  \{  \}  \%  \^  \_  \~{}  \textbackslash
```

For creating quotes it is not allowed you use the " character. Instead, you have to use `` for opening and '' for closing quotation marks. For example, ``This is a quote'' will generate "this is a quote". These characters can be used also in single-character quotes, , e.g. to quote 'c'.

An ellipsis ... is not produced by writing three dots, this would look like ..., instead the command \ldots should be used.

As mentioned above, LATEXautomatically creates spaces as necessary, i.e. to align with the right border, it may distribute extra space over the line. If you want to prevent this at a certain position, i.e. to keep the space short, use a backslash in front of a space. This also works when using a tilde ~, but in addition prevents a line break at this position. Usually, the space after the end of a sentence is longer. But when the last letter before a period is an upper case letter, LATEX assumes that it is an abbreviation. In this case you can use \@ before the period if it is actual the end of a sentence.

Sometimes you want to give extra spaces. For this purpose a couple of commands are defined: tiny, small, medium and large spaces are added by \, (backslash and comma), \; , \quad, \qquad, respectively. This is in particular useful for mathematical formulas, where usually no extra spaces are put by LATEX. The command \hfill means *horizontal filling*. It will cause everything written after it, till the end of the printed line, to be formatted such that it is aligned with the right border of the page. In the same way \vfill will shift everything after it to the bottom of the current page, which may have to be given, e.g. using \newpage. Arbitrary horizontal spaces can be created using the command \hspace*. The length of the space has to be supplied as argument in { } brackets either in real units, e.g. 1.3cm, or in units of a given length, e.g. 0.5\linewidth. Note that you can define your own lengths, see Sec. 9.3.1.9. The * indicates that the space always is to be included, even if it appears at the beginning of a line. Without the * the space is only created inside a line. For vertical spaces, there exist the corresponding commands \vspace and \vspace*.

In many languages letter with different accents exist, like in the French, or with umlauts, like in German, or other special characters. Here, the most important accents are listed acting on the letter 'o', some of them work for other letters as well:

| ò | \'o | ó | \'o | ȯ | \. o | õ | \~o |
|---|-----|---|-----|---|------|---|-----|
| ô | \^o | ŏ | \u o | ǒ | \v o | o͡o | \t oo |
| ö | \"o | ő | \H o | o̧ | \d o | ß | \ss |
| ō | \=o | o̩ | \b o | ç | \c c | o̧ | \c o |
| œ | \oe | Œ | \OE | ø | \o | Ø | \O |
| æ | \ae | Æ | \AE | å | \aa | Å | \AA |
| ł | \l | Ł | \L | ı | \i | ȷ | \j |
| i | !' | ¿ | ?' | | | | |

Note that many more language-specific adjustments are available, like hyphenation rules, other alphabets, right-to-left writing, etc. For more information, please refer to the documentation of the `babel` package.

### 9.3.1.3 *Chapters, sections and footnotes*

You can structure your documents using chapters (for the class type `book`), sections, subsections etc. For example, Sec. 8.3 (see page 283) was started using

```
\section{Basic data analysis\label{sec:statistics}}
```

while the following subsection (page 284) was started with

```
\subsection{Estimators}
```

In the `article` class, also \subsubsection, \paragraph and \subparagraph are available. LATEX does the numbering of the sections, subsections etc. automatically for you. You can reference these numbers if you attach *labels* with arbitrary names using the \label command and the name (here `sec:statistics`) as argument. Then you can reference the section everywhere in the document with the \ref command, i.e. \ref{sec:statistics} in this case. In the printed document, you will see just the section number. You can also reference the number of the page (in print) where the label occurs via \pageref, i.e.

\pageref{sec:statistics} here. Labels are also frequently used for equation numbers. If you place a \label command inside an equation environment, the printed value of the label will be the equation number. Note that LATEX needs usually at least two runs to write the labels correctly. Within the first run, it learns the corresponding values of the labels, in the next run, it can use the values to include them in the printed document.

Also very convenient is that LATEX will keep track of chapters, sections and subsection in an ordered way, such that simply by including the command \tableofcontents, a full table of contents will appear in the output document.

Finally, footnotes can be generated using the command \footnote which the footnote text as argument. The command must be placed in the source file at the position where the footnote should be referenced. The actual placement of the footnote, starting on the same page as the reference point, will be done automatically by LATEX.

### 9.3.1.4 *Lists*

You can structure your documents even more by using lists. A list without numbers is created by an itemize environment. The different *items* are indicated by bullets (and other symbols if several lists are nested). For example, the list starting on page 313 is created using

```
\begin{itemize}
\item \verb!\\! prints a backslash.
\item \verb!\0! selects the {\rm Roman} font, which is also
the default font. A font
is active until a new one is chosen.
\item \verb!\1! selects the {\it italic} font, used in equations.
```

... (left out) ...

```
\end{itemize}
```

Each list element is preceded by a \item command. Another important list environment is enumerate, which puts numbers instead of bullets. For nested enumerate environments, this results in a sub numbering with letters. Since this does not appear elsewhere this book, we give a special example here:

```
\begin{enumerate}
\item Write program
\begin{enumerate}
```

```
\item Design data types and functions
\item Code in C
\item Test and debug
\end{enumerate}
\item Perform simulations and analyse data
\begin{enumerate}
\item For standard case, to compare with analytics
\item For the two-dimensional case
\item For the three-dimensional case.
\end{enumerate}
\end{enumerate}
```

This will result in:

(1) Write program

    (a) Design data types and functions

    (b) Code in C

    (c) Test and debug

(2) Perform simulations and analyse data

    (a) For standard case, to compare with analytics

    (b) For the two-dimensional case

    (c) For the three-dimensional case.

Arbitrary lists can be defined using the `list` environment; please have a look at the LaTeX documentation.

The `\verb` command used in the above list allows an arbitrary string to be printed in typewriter font as it stands. Here, the string to be printed is not given as an argument, but is embraced in ! !. In this way arbitrary symbols can be contained in the string, also '{' and '}'. Always the first symbol after the `\verb` is taken as bracketing symbol here. Hence, one could write also `\verb?\\?` in the above example. The string is not allowed to extend beyond the limits of one line. For longer texts which should be printed verbatim, there is the `verbatim` environment. Thus, you have to embrace the text by `\begin{verbatim}` and `\end{verbatim}`, which is used in this book, for example, to display the C source codes.[4]

---

[4]The line numbers which are frequently shown next to the source code can be generated using the `lineno` package, which has to be included using the `\usepackage` command.

### 9.3.1.5 *Mathematical formulas*

The biggest advantage over conventional office packages shows LaTeX when setting mathematical formulas. It is much simpler to enter them using a standard editor (once you know the few most important LaTeXcommands) and the resulting typesetting looks just professional. As mentioned above formulas can appear within the main text (inside a pair $ ...$ of dollar symbols) or as separated equations with equations numbers (\begin{equation}...\end{equation}) or without (\begin{equation*} ...\end{equation*}). For the former one, the short form \[ ...\]) exists, for the latter one $$ ...$$. Note that many formulas will be typeset "tighter" inside a line as compared to a separate formula:

```
A formula printed in a line appears like
$\lim_{n\to\infty} \sum_{k=0}^n \frac{1}{n!} = e$
while as a separate formula it appears as
$$
\lim_{n\to\infty} \sum_{k=0}^n \frac{1}{n!} = e
$$
```

will result in

A formula printed in a line appears like $\lim_{n\to\infty} \sum_{k=0}^n \frac{1}{n!} = e$ while as a separate formula it appears as

$$\lim_{n\to\infty} \sum_{k=0}^n \frac{1}{n!} = e$$

Here, \sum and \lim are mathematical *operators*, where upper and lower bounds or limits can be stated via the underscore _ or the caret ^ character, respectively. Please note that by default only one symbol is taken as bound or limit. If a more complex string should appear, it has to be presented as { ...} group. There are several other important operators which can be used in this, in particular for integrals (\int) or products (prod):

| $\sum$ | \sum | $\prod$ | \prod | $\coprod$ | \coprod |
|---|---|---|---|---|---|
| $\bigcap$ | \bigcap | $\bigcup$ | \bigcup | $\biguplus$ | \biguplus |
| $\bigwedge$ | \bigwedge | $\bigvee$ | \bigvee | $\bigsqcup$ | \bigsqcup |
| $\bigoplus$ | \bigoplus | $\bigotimes$ | \bigotimes | $\bigodot$ | \bigodot |
| $\int$ | \int | $\oint$ | \oint | | |

The underscore and the caret are most often used to denote subscripts or superscripts. LATEX takes are of the correct font sizes even for nested sub- and superscripts, as, e.g.,

```
$$
A_y^k = \prod_{i=1}^n B_{y,z_i}^k
$$
```

will be set as

$$A_y^k = \prod_{i=1}^n B_{y,z_i}^k$$

As in the above formula, variables are always denoted by single letters. This means a sequence of several letters like in `$velocity$` will be always interpreted as several variables in a row, i.e., printed as $velocity$, which contains slightly more spaces between the letters as compared to italics printing of *velocity*. To increase the space of available variables, Greek letters are used exhaustively in mathematical formulas. The following tables shows the available letters:

| $\alpha$ | \alpha | $\beta$ | \beta | $\gamma$ | \gamma |
|---|---|---|---|---|---|
| $\delta$ | \delta | $\epsilon$ | \epsilon | $\varepsilon$ | \varepsilon |
| $\zeta$ | \zeta | $\eta$ | \eta | $\theta$ | \theta |
| $\vartheta$ | \vartheta | $\iota$ | \iota | $\kappa$ | \kappa |
| $\lambda$ | \lambda | $\mu$ | \mu | $\nu$ | \nu |
| $\xi$ | \xi | $o$ | o | $\pi$ | \pi |
| $\varpi$ | \varpi | $\rho$ | \rho | $\varrho$ | \varrho |
| $\sigma$ | \sigma | $\varsigma$ | \varsigma | $\tau$ | \tau |
| $\upsilon$ | \upsilon | $\phi$ | \phi | $\varphi$ | \varphi |
| $\chi$ | \chi | $\psi$ | \psi | $\omega$ | \omega |

Furthermore, letters can be decorated with several addition symbols like tildes, dots, vectors etc:

Clearly, also standard binary operator symbols like '+', '−' and / or relational symbols like '=' will be used often and can be directly entered into equations. Furthermore, many less common operators are available as the following table shows:

| Γ | \Gamma | Δ | \Delta | Θ | \Theta |
|---|--------|---|--------|---|--------|
| Λ | \Lambda | Ξ | \Xi | Π | \Pi |
| Σ | \Sigma | Υ | \Upsilon | Φ | \Phi |
| Ψ | \Psi | Ω | \Omega | | |

| $\hat{a}$ | \hat{a} | $\breve{a}$ | \breve{a} | $\check{a}$ | \check{a} |
|-----------|---------|-------------|-----------|-------------|-----------|
| $\dot{a}$ | \dot{a} | $\ddot{a}$ | \ddot{a} | $\mathring{a}$ | \mathring{a} |
| $\bar{a}$ | \bar{a} | $\vec{a}$ | \vec{a} | $\underline{a}$ | \underline{a} |
| $\grave{a}$ | \grave{a} | $\acute{a}$ | \acute{a} | $\widehat{aaa}$ | \widehat{aaa} |
| | | $\tilde{a}$ | \tilde{a} | $\widetilde{aaa}$ | \widetilde{aaa} |

| + | + | − | - | · | \cdot |
|---|---|---|---|---|-------|
| × | \times | * | \ast | ⋆ | \star |
| / | / | ÷ | \div | \ | \setminus |
| ± | \pm | ∓ | \mp | ⨿ | \amalg |
| ⋄ | \diamond | ◁ | \lhd | ◂ | \triangleleft |
| ◇ | \Diamond | ▷ | \rhd | ▸ | \triangleright |
| □ | \Box | ⊴ | \unlhd | △ | \bigtriangleup |
| ≀ | \wr | ⊵ | \unrhd | ▽ | \bigtriangledown |
| ∘ | \circ | • | \bullet | ◯ | \bigcirc |
| ⊕ | \oplus | ⊖ | \ominus | ⊗ | \otimes |
| † | \dagger | ‡ | \ddagger | ⊘ | \oslash |
| ∩ | \cap | ∪ | \cup | ⊎ | \uplus |
| ∨ | \vee | ∧ | \wedge | | |
| ⊓ | \sqcap | ⊔ | \sqcup | | |

In the *amsmath* package many more (less common) operator symbols are available. Please refer to the corresponding documentation. In standard LaTeX predefined binary relations are listed in the following table:

| $=$ | `=` | $<$ | `<` | $>$ | `>` |
|---|---|---|---|---|---|
| $\le$ | `\le` | $\ge$ | `\ge` | $\sim$ | `\sim` |
| $\ll$ | `\ll` | $\gg$ | `\gg` | $\doteq$ | `\doteq` |
| $\simeq$ | `\simeq` | $\approx$ | `\approx` | $\asymp$ | `\asymp` |
| $\subset$ | `\subset` | $\supset$ | `\supset` | $\smile$ | `\smile` |
| $\subseteq$ | `\subseteq` | $\supseteq$ | `\supseteq` | $\frown$ | `\frown` |
| $\sqsubset$ | `\sqsubset` | $\sqsupset$ | `\sqsupset` | $\cong$ | `\cong` |
| $\sqsubseteq$ | `\sqsubseteq` | $\sqsupseteq$ | `\sqsupseteq` | $\equiv$ | `\equiv` |
| $\in$ | `\in` | $\ni$ | `\ni` | $\propto$ | `\propto` |
| $\vdash$ | `\vdash` | $\dashv$ | `\dashv` | $\perp$ | `\perp` |
| $\parallel$ | `\parallel` | $\mid$ | `\mid` | $\models$ | `\models` |
| $\prec$ | `\prec` | $\succ$ | `\succ` | $\bowtie$ | `\bowtie` |
| $\preceq$ | `\preceq` | $\succeq$ | `\succeq` | | |

Each binary relation command can be preceeded by `\not` which will result in a negated symbol, e.g. `\not=` will result in $\ne$. For some of the negated relation operators, special commands are defined, like `\neq`. Sometimes the special form will appear differently like `\notin` appearing as $\notin$ compared to `\not\in` appearing as $\not\in$. Sometimes it is necessary to write two symbols on top of each other, here the command `\stackrel` can be used. For example, `$a^2\stackrel{(*)}{=}b^2+c^2$` will be shown as $a^2 \stackrel{(*)}{=} b^2 + c^2$.

Furthermore, LaTeX offers a variety of arrow symbols, as collected in the following table:

| $\leftarrow$ | `\leftarrow` or `\gets` | $\rightarrow$ | `\rightarrow` or `\to` |
|---|---|---|---|
| $\longleftarrow$ | `\longleftarrow` | $\longrightarrow$ | `\longrightarrow` |
| $\leftrightarrow$ | `\leftrightarrow` | $\longleftrightarrow$ | `\longleftrightarrow` |
| $\Leftarrow$ | `\Leftarrow` | $\Rightarrow$ | `\Rightarrow$` |
| $\Longleftarrow$ | `\Longleftarrow` | $\Longrightarrow$ | `\Longrightarrow` |
| $\mapsto$ | `\mapsto` | $\longmapsto$ | `\longmapsto` |
| $\hookleftarrow$ | `\hookleftarrow` | $\hookrightarrow$ | `\hookrightarrow` |
| $\leftharpoonup$ | `\leftharpoonup` | $\rightharpoonup$ | `\rightharpoonup` |
| $\leftharpoondown$ | `\leftharpoondown` | $\rightharpoondown$ | `\rightharpoondown` |

| ⇌ | \rightleftharpoons | ⟺ | \iff |
|---|---|---|---|
| ↑ | \uparrow | ↓ | \downarrow |
| ⇑ | \Uparrow | ⇓ | \Downarrow |
| ↕ | \updownarrow | ⇕ | \Updownarrow |
| ↗ | \nearrow | ↘ | \searrow |
| ↙ | \swarrow | ↖ | \nwarrow |
| ↝ | \leadsto | | |

LATEXcontains also a variety of other symbols:

| ∀ | \forall | ∃ | \exists | ∂ | \partial |
|---|---|---|---|---|---|
| ℜ | \Re | ℑ | \Im | ı | \imath |
| ℏ | \hbar | ȷ | \jmath | ℓ | \ell |
| ⋯ | \cdots | … | \dots | ⋮ | \vdots |
| ⋱ | \ddots | ℵ | \aleph | ℘ | \wp |
| ∇ | \nabla | ∅ | \emptyset | ∞ | \infty |
| ′ | , | ′ | \prime | ¬ | \neg |
| △ | \triangle | ∠ | \angle | √ | \surd |
| ⊥ | \bot | ⊤ | \top | ◇ | \diamondsuit |
| ♡ | \heartsuit | ♣ | \clubsuit | ♠ | \spadesuit |
| ♭ | \flat | ♮ | \natural | ♯ | \sharp |

Often used important elements are fractions (\frac), and roots (\sqrt). The first commands takes two arguments, the second command one, plus an optional value [n] for the n'th root. Both commands can be nested, as the following example shows:

```
\begin{equation*}
\sqrt[3]{\frac{a+\sqrt{b+c}}{\frac{1}{x^2}+\sin(x)}}
\end{equation*}
```

This result appears as follows:

$$\sqrt[3]{\frac{a + \sqrt{b + c}}{\frac{1}{x^2} + \sin(x)}}$$

Binomials can be used in the following rather nonstandard[5] way:

```
\begin{equation*}
{r \choose k-1}+ {r \choose k} = {r+1 \choose k}\,.
\end{equation*}
```

result ins

$$\binom{r}{k-1} + \binom{r}{k} = \binom{r+1}{k}.$$

There exists also the command \atop which works in the same way but omits the brackets

In the above example for fractions also the function \sin is used. Note that function names are not printed in *italic* but in Roman font. Within LATEX a couple of functions are predefined as the following list shows:

| | | | | | |
|---|---|---|---|---|---|
| \arccos | \arcsin | \arctan | \arg | \cos | \cosh |
| \cot | \coth | \csc | \deg | \det | \dim |
| \exp | \gdc | \hom | \inf | \ker | \lg |
| \lim | \liminf | \limsup | \ln | \log | \max |
| \min | \Pr | \sec | \sin | \sinh | \sup |
| \tan | \tanh | | | | |

You can declare your own functions using the command

```
\DeclareMathOperator!{}\verb!{\operator}{}!\verb!{text}
```

which takes two argument, the name of your operator and the text which shall be printed. The \DeclareMathOperator can be used only in the preamble of your source file before the begin{document}.

Complicated formulas can be structured using brackets and other delimiters. Vertical arrows can be used as delimiters. Apart from those, LATEX knows the following brackets and delimiters:

---

[5]The two "arguments" are not given after the command but before and after it.

| ( | ( | ) | ) | ⌊ | \lfloor | ⌋ | \rfloor |
|---|---|---|---|---|---------|---|---------|
| [ | [ | ] | ] | ⌈ | \lceil | ⌉ | \rceil |
| { | \{ | } | \} | ⟨ | \langle | ⟩ | \rangle |
| \| | \| | ‖ | \\| | | | | |

By default, the delimiters are printed in the standard font size. If larger elements are to be put inside delimiters, one can precede the opening delimiters by `\left` and the closing delimiter by `\right` as in

```
\begin{equation*)
\left(z+\frac{x+y}{x-y}\right)\,.
\end{equation*}
```

which results in

$$\left(z+\frac{x+y}{x-y}\right).$$

In combination with the `array` environment and two large delimiters, matrices or determinants can be created, see Sec. 9.3.1.6. Note that the left and right delimiters do not have to match, so mixed pairs are possible. One can omit even one of the two delimiters by writing a dot instead of a bracket, see the example for using an *array* on page 429. Instead of using `\left` and `\right` one can specify larger delimiters explicitly using commands such as `\big`, `\Big`, `\bigg`, and `\Bigg` in front of the delimiter.

Another way to group parts of the forumla are lines above (`\overline`) below (`\underline`) as well as curly brackets above (`overbrace`) and below (`underbrace`) the formula. The following example

```
\begin{equation*}
\overbrace{\overline{(x-\overline x)^2}}^{\mbox{variance}}
= \underbrace{\overline{x^2}}_{\mbox{mean(squared)}} -
\overbrace{{\overline x}^2}^{\mbox{square(mean)}}
\end{equation*}
```

will result in:

$$\overbrace{\overline{(x-\overline{x})^2}}^{\text{variance}} = \underbrace{\overline{x^2}}_{\text{mean(squared)}} - \overbrace{{\overline{x}}^2}^{\text{square(mean)}}$$

Note that you can use the braces also without the additional part which is printed above or under the braces, respectively. The `\underline` command is also allowed outside mathematical formulas.

Also different fonts are available for formulas. Here in particular the caligraphic font, e.g.

```
\begin{equation*}
\mathcal{A,B,\ldots,Y,Z}\,.
\end{equation*}
```

will result in

$$\mathcal{A}, \mathcal{B}, \ldots, \mathcal{Y}, \mathcal{Z}.$$

Also standard text, e.g. using the Roman font, can be used in formulas. Here the `\mbox` command with the wanted text as argument can be used. Alternative, one can switch via `\textrm` by hand to the Roman font. Note that the sizes of the fonts when used as indices will differ:

```
\begin{equation*}
x_{\textrm{critical}} \quad  x_{\mbox{critical}}\,.
\end{equation*}
```

results in

$$x_{\text{critical}} \quad x_{\text{critical}}.$$

Even more fonts can be used when the *amsmath* (or just the *amsfonts*) package is used. Here, the Gothic font `mathfrak` is available. Very useful is the blackboard font, which allows to write, e.g.

```
\begin{equation*}
\mathbb{C},\; \mathbb{N},\; \mathbb{R},\; \mathbb{Z}\,.
\end{equation*}
```

which yields

$$\mathbb{C}, \mathbb{N}, \mathbb{R}, \mathbb{Z}.$$

The equations considered so far contain one single line. Formulas which extend over several lines can be generated using the `\begin{eqnarray}` ... `\end{eqnarray}` environment, or the corresponding * version without equation numbers. Each line consist of three parts, separated by & symbols. A line is terminated by `\\`. The first part will be printed right aligned, the second part centered and the third part left aligned. For example the formula shown on page 290 was generated using

```
\begin{eqnarray}
\sigma^2_l & = & \frac{ns^2}{\chi^2(1-\alpha/2, n-1)} \nonumber \\
\sigma^2_u & = & \frac{ns^2}{\chi^2(\alpha/2, n-1)}\,.
\end{eqnarray}
```

Note the \nonumber prevents an equation number being printed in the corresponding line of the formula. Finally, it should be emphasized that the American Mathematical Society's LaTeX add-on *amsmath* offers several additional types of multi-line formula environments.

### 9.3.1.6 *Minipages, tables and arrays*

Even more structure can be created by subdividing paragraphs horizontally. For this purpose, the minipage environment can be used. Inside a minipage, one can put almost everything which can be put onto a normal page, also nested minipages. From outside, a minipage is treated like a *single* symbol, but a possibly large symbol. As example, we show the source for one item of the list on page 15, where a small paragraph and a little table are shown next to each other:

```
\begin{minipage}[t]{0.65\textwidth}
Calculates a bitwise OR  of the two operands, defined as
shown in the table on the right: The result is 1 if $a$ OR $b$
are 1. Hence, for the numbers 201  and
158  one will obtain the result 223 (binary \verb!11011111!).
\end{minipage}
\hfill
\begin{minipage}[t]{0.2\textwidth}
\vspace*{-1mm}
\begin{tabular}{cc|c}
$a$ & $b$ & $a$\verb!|!b\\\hline
0 & 0 & 0\\
0 & 1 & 1\\
1 & 0 & 1\\
1 & 1 & 1
\end{tabular}
\end{minipage}
```

A minipage environment receives always one argument, its width. You can give the width relative to some predefined *lengths*, such as \textwidth in this example. Another important length is the \columnwidth, which is for a two column format, quite common in scientific journals, about half of the textwidth. Alternatively one can give lengths also in real units, like 3cm, 30mm or 15pt, where pt is the size of a point.

Here, the minipages get the optional argument [t], which means that the minipages are aligned at the top. Other possibilities are [b] and [c] for alignment at bottom and center, respectively.

In the second minipage, a *table* is used, which is also a widespread LATEX element. Note that the full table is shifted upwards by \vspace*{-1mm}. Also the command \hfill is used, see Sec. 9.3.1.2. A table is defined using the tabular environment. It expects as arguments the format of the table. For each column of the table a format has to be given. In this case, there are two columns with *centered* content (indicated by the letter 'c'), followed by a vertical line ('|'), followed by another centered line. Other formats are left ('l') and right ('r') justified columns. A column which always looks the same can be given via @{ content }, where content has to be substituted by the actual content. This can be also a space, e.g. @{\hspace*{5mm}}. If you want to have a frame around your table, you have should also use vertical lines at the beginning and at the end of the format line. As for the minipages, one can give options [t], [b], or [c] for alignment of tables at top, bottom or center, respectively. Such an option, as usual, can be written between the tabular and the {...} argument.

The actual entries of the table come next. The entries for different columns of each row are separated by '&' symbols. Here, where we have three columns, we need two '&' symbols per row. A row is finished by \\. If you want to have a horizontal line under a row, use \hline. For the above mentioned frame around the table, you need a horizontal line before the first and after the last line. Note that each table row of the c, r, or l format is restricted to one line on the printed page. To overcome this you can put a minipage in an entry, which, as mentioned above is treated by LATEX as a single symbol. Since this is a bit cumbersome, the tabular environment offer as alternative the paragraph column format. This is indicated by the format letter p where the width of the column has to be specified as argument, as in p{0.2\textwidth}. The entry will be formatted possibly over several lines to fit into the given width. Furthermore, it is also possible to create entries which span several columns in a row. For this purpose, use a \multicolumn entry, which takes three arguments, the number of columns, the format of the column (as for the full table, e.g. |c|) and the actual content of the entry. The following example table (not occurring in this book) provides an example for the use of the paragraph and multi column formats, as well how to create a frame around the table:

```
\begin{quote}
\begin{tabular}{|l|cccp{0.3\textwidth}|}
\hline
System & $T^{\star}$ & $\rho$ & $\mu$  & comment \\
\hline
A & 32.7K & 35g/cm$^3$ & 12 A/V & \\
B & 12.5K & 48g/cm$^3$ & 9.3 A/V &
       only measured under very high pressure\\
C & 24.2K & 18g/cm$^3$ & 17.3 A/V & \\
D & $\sim$ 40K & \multicolumn{2}{c}{not measured} &
from literature \\ \hline
\end{tabular}
\end{quote}
```

When latexing this table, the result looks like this:

| System | $T^{\star}$ | $\rho$ | $\mu$ | comment |
|--------|------|------|------|---------|
| A | 32.7K | 35g/cm$^3$ | 12 A/V | |
| B | 12.5K | 48g/cm$^3$ | 9.3 A/V | only measured under very high pressure |
| C | 24.2K | 18g/cm$^3$ | 17.3 A/V | |
| D | $\sim$ 40K | not measured | | from literature |

Within the mathematical formulas, like in the **equation** environment, one can use a similar environment called **array**. For example, Eq. (6.1) on page 163 was created using:

```
\begin{equation}
n! = \left\{\begin{array}{ll} 1 &  \mbox{if } n=0 \mbox{ or } n=1\\
                n \times (n-1)! & \mbox{else}\\
    \end{array} \right.
\end{equation}
```

The \mbox command is used (see alsp page 426) to display a normal text string within a formula, such that the string is not formatted like a formula. For the formatting, the array is treated also like one single symbol. Using the \left\{, a curly bracket (open to the right) is created, where the size is adjusted such that everything next to the bracket until the closing \right is embraced. Note that the '.' in \right. means that the closing bracket is "invisible", but you can use a closing bracket instead. For example, if you use a \left( and a \right) enclosing an **array** environment, you get a matrix. When using `\left|` ... `\right|` one can get a determinant, as the following examples shows:

```
\begin{equation*}
\left|
  \begin{array}{ccc}
    a_1 & b_1 & c_1 \\
    a_2 & b_2 & c_2 \\
    a_3 & c_3 & c_3
  \end{array}
\right|\,.
\end{equation*}
```

will be typeset as

$$\begin{vmatrix} a_1 & b_1 & c_1 \\ a_2 & b_2 & c_2 \\ a_3 & c_3 & c_3 \end{vmatrix}.$$

You might have noticed that many tables or figures presented in this book are centered left-right in the middle of the page. This can be done with any text or more complex LATEX-objects like minipages, figures (see below) or arrays via the **center** environment. For example

```
\begin{center}
This is an example of\\
a centered text.
\end{center}
```

will result in

This is an example of<br>
a centered text.

### 9.3.1.7 *Figures*

To include figure files in your document, such as plots generated with *xmgrace* or images drawn using *xfig*, you can use the `\includegraphics` command provided by the **graphicx** package. To make this package available to your LATEX file you have to write in the preamble, i.e., between the `\documentclass` and `\begin{document}` commands:

```
\usepackage{graphicx}
```

The figures you want to include can be available most conveniently in the encapsulated postscript (`.eps`) format. When using just `\includegraphics`, the figure will appear right where the command appears, even if the figure does not fit on the page. It is usually better to use the `figure` environment, which lets the figures "float" and places them "close" to the current position. For example, Fig. 8.36 (page 329 of this book) is generated using:

```
\begin{figure}[!ht]
  \begin{center}
    \includegraphics[width=0.8\textwidth]{pic_random/ks_example.eps}
  \end{center}
  \caption{Kolmogorov-Smirnov test: A sample distribution function
    (solid line) is compared to a given  probability distribution
    function (dashed line). The sample  statistics $d_{\max}$  is
    the maximum difference between the two functions.
    \label{fig:KSExample}}
\end{figure}
```

The `figure` environment carries the optional argument `[!ht]`, which tells LaTeX that the figure should be placed preferentially *here* (`h`). If this is not possible, it should be placed at the top (`t`) of the next possible page. Also 'b' for *bottom* is often used. The figure appears centered on the page, because it is put inside a `center` environment. The caption of the figure is given via the `\caption` command. Here, also a label is given, which enables us to reference the figure number via `\ref` or the page where the figure appears via `\pageref`.

### 9.3.1.8 *Bibliography*

Next, we show how a bibliography is generated. There are two versions: One can either put all cited literature into the document, or one can use the *ibtex* package which uses an external ASCII file as literature database, where the cited literature is automatically picked from and sorted.

First, we explain how the literature is included directly with the document. Not surprisingly, the `thebibliography` environment is used for this. The bibliography of the current book reads as follows

```
\begin{thebibliography}{}

\bibitem[Abramson and Yung (1986)]{abramson1986}
  Abramson, B. and Yung, M. (1986).
```

Construction through decomposition: a divide-and-conquer algorithm for
the N-queens problem, {\it CM '86: Proceedings of 1986 ACM Fall joint
computer conference}\/, pp 620--628
(IEEE Computer Society Press, Los Alamitos)

\bibitem[Abramson and Yung (1989)]{abramson1989}
 Abramson, B. and Yung, M. (1989).
Divide and conquer under global constraints: A solution to the
N-queens problem, {\it Journal of Parallel and Distributed
Computing}\/ {\bf 6}, pp 649--662

... (left out) ...

\end{thebibliography}

The argument of the \begin{thebibliography} is empty here. Usually, it should contain a sample of what the citation labels look like. By
default, labels are enumerated 1,2,3 .... In this case, one could write {11}
as argument, if you have less than 100 entries in the bibliography, because
this creates two-digit labels.

Each entry is given after a \bibitem command. If you want to use
your individual labels, you can state them as optional arguments in [   ]
brackets, like here. Each \bibitem carries as argument a *marker*, which
can be used via the \cite command to refer to the publication, for example
\cite{abramson1989}, see page 174 of this book.

Next, it is shown how *Bibtex* can be used to maintain a literature
database with just a small effort. Here you have one or several separate
Bibtex files, ending with .bib, e.g. mycitations.bib, containing the bibliographic entries. The main idea is that these entries contain all possibly
necessary information, see below. At the position where the bibliography is
to be placed, one writes the \bibliography command and gives the names
of the bibtexfiles (without the ending .bib) as argument in {...} braces,
e.g.

\bibliography{mycitations}

The way the bibliograph and the citations actually look like is determined in the manuscript via a \bibliographystyle command, with the
name of the style as argument in {...} braces. Standard styles of Bibtex
are plain (alphabetical order in the bibliography, citations as numbers),
unsrt (like plain but order of occurrence), alpha (like plain, citations
as author name plus year) and abbrv (like plain but very compact bibli-

ography). There exists many more style files, in particular many journals offer custom-made style files which make the bibliography look exactly like in the journal, i.e., order of entries, fonts, journal articles with or without titles, etc. Please have a look at the web page of the journal you would like to publish in.

Within the main body of the text, the references are again given via the \cite command, as when using the thebibliography environment. To compile the bibliography into your article, say simulation.tex, you have to latex your manuscript first, then you call (e.g. in a shell) bibtex simulation (without the suffix .tex), followed by (usually) two more latex runs, i.e. you can enter in your shell:

```
latex simulation
bibtex simulation
latex simulation
latex simulation
```

When using a LATEX environment, e.g. *Kile* or *TeXworks*, Bibtex can be called via corresponding buttons or menus.

Using Bibtex has three advantages compared to explicitly giving the bibliography:

- Bibtex picks automatically those entries from the Bibtex files which are cited, the others are not included in the bibliography. Note that you can force an entry to be included in the bibliography via the \nocite command placed somewhere in the LATEX file.
- Bibtex puts the citations in the right order, e.g. order of occurrence or alphabetical order.
- The way the citations and the bibliographic looks like is completely determined by the style file. You do not have to reformat if you write a manuscript for a different journal.

To understand the format of the Bibtexfile, we start with an example with two entries:

```
@Book{practical_guide2009,
    author =     {A. K. Hartmann},
    title =      {{Practical Guide to Computer Simulations}},
    publisher =  {World Scientific},
    address =    {Singapore},
    year =       {2008},
}
```

```
@Article{aspect-ratio2002,
    author =    {A. K. Hartmann and A. J. Bray and A. C. Carter and
                 M. A. Moore and A. P. Young},
    title =     {The stiffness exponent of two-dimensional {I}sing
                 spin glasses for non-periodic boundary conditions
                 using aspect-ratio scaling},
    journal =   {Phys. Rev. B},
    volume =    {66},
    pages =     {224401},
    year =      {2002},
}
```

The Bibtex file is a pure ASCII file. Each entry starts with an @ character. Next follows the *type* of the entry. For scientific publications the types Book, InBook, Article, Proceedings, InProceedings, Unpublished and Misc are most important. The type can be written in any mixture of upper and lower case letters (not case sensitive). The actual entries is embraced by { and }. Next comes the citation *key*, which is to be used in the \cite command, followed by a comma. Then comes a list of comma separated *tags*. Each tag consists of a tag name, followed by a = symbol, followed by the actual tag text embraced either by a { } or a " " pair. For scientific publications, important tag names are author (or editor), title, and year. For the author tag, names can be given in the format *first names* followed by *family name*, or *family name* followed by a comma followed by the *first names*. Names of several authors are separated by an and, respectively. Note that one can form groups via { } braces, e.g. when names consist of several words. These braces can also be used to prevent one or several upper case letter in the title tag to be converted to lower case letters, which might be the default for some style files. For the aspect-ratio2002 entry, this is used to make sure that the name "Ising" starts with an upper case letter.

For the type article the tags journal, volume, and pages are important. For the type Book you also need the tag publisher, and optional the tag address. In principle arbitrary tags are possible. Some journal bibliography styles recognize tags like article-number, url (giving a link to the article) or doi (digital object identifier). Tags with unknown names (depending on the style) are ignored by Bibtex. Thus, you have to have a look at the style file of the journal you want to submit to. Note that many journals offer on their web page also the possibility to download a Bibtex entry for each of the papers you would like to cite. This saves you a bit of time and avoids mistakes in the bibliographic information.

### 9.3.1.9  *Self-defined commands and lengths*

Finally, we note that you can create your own commands or lengths or your own environments. Here, we show one example for a self-defined command, which can be used in math mode:

```
\newcommand{\pd}[2]{\frac{\partial #1}{\partial #2}}
```

The new command is called \pd, as shown by the first argument of \newcommand. The new command itself has two arguments, indicated by the option [2]. The arguments can be addressed using #1 and #2 in the command definition, which comes last. Here, the \pd command is used to construct a formula for a partial derivative, using the two arguments. For example, Eq. (8.95) on page 365 was generated using:

```
\begin{equation}
\pd{l(\myvec{\theta})}{\theta_k} \stackrel{!}{=} 0 \quad
(k=1,\ldots,n_{\rm p})
\end{equation}
```

A new length can be introduced anywhere in the source file via using the \newlength command with the name of the new length, including a leading backslash, as argument, e.g.

```
\newlength{\fieldwidth}
```

The length, a self-defined or a LaTeX-predefined one, can be set using \setlength with the name of the length (with backslash) and the actual length as arguments. The latter one can be given in real units or in multiples of an existing length:

```
\setlength{\fieldwidth}{0.3\columnwidth}
```

Alternatively, one can set the length to the width of a text element using \settowidth with the name of the length (with backslash) and the text element as arguments, e.g.

```
\settowidth{\fieldwidth}{simulation tool}
```

will assign to \fieldwidth the length the expression "simulation tool" will have in the final document. Similar commands for setting lengths are \settoheight and \settodepth. Furthermore, you can alter lengths via the command \addtolength. For example, to half the self-defined length you can use:

```
\addtolength{\fieldwidth}{-0.5\fieldwidth}
```

### 9.3.1.10  *More information*

These examples should be sufficient to give you an impression of the philosophy of LaTeX.   The amount of information given here should be even enough to enable you to write papers or theses with some additional help from online resources.   Comprehensive instructions are beyond the scope of this section, so please consult the literature [Lamport and Bibby (1994); TUG].   Note that many extensions are available in addition to the standard LaTeX, for example, to have text floating around small figures (\usepackage{wrapfig}), to display algorithms (\usepackage{algorithms}), to have more flexible display of formulas (\usepackage{amsmath}) or to include line numbers (\usepackage{lineno}). All of these packages and many more, including the required documentation, can be found on the *Comprehensive TeX Archive Network* [CTAN].

Under UNIX/Linux, the spell checker *ispell* is available. It allows a simple spell check to be performed. The tool is built on a dictionary, i.e. a huge list of known words. The program scans any given text and also a special LaTeX mode is available. Every time a word occurs, which is not contained in the list, *ispell* stops. Should similar words exist in the list, they are suggested. Now the user has to decide whether the word should be replaced, changed, accepted or even added to the dictionary. The whole text is treated in this way. Please note that many mistakes cannot be found in this way, especially when a misspelled word is equal to another word in the dictionary. However, at least *ispell* finds many spelling mistakes quickly and conveniently, so you should use the tool.

### 9.3.2  *Beamer class*

With the LaTeX package it is also possible to prepare slides for presentations in high quality. This is particularly easy using the *beamer* class [Beamer]. Here, a short introduction is given. We will create two sample slides, which are actually taken, with small modifications, from a talk of the author of this book. These slides feature the most basic *beamer* class elements, such that you can start preparing your own slides quickly. The *beamer* class has everything adjusted already such that it results in nice-looking slides, when showing the resulting *pdf* file using *acroread*. You do not have to worry about slide sizes, fonts or spacings.

To use the class, you have to set the class to `beamer` in the corresponding LATEX source file. Furthermore, you should use the packages `graphicx` to include graphics, `babel` and

| GET SOURCE CODE |
|---|
| DIR: `literature` |
| FILE(S): `testtalk.tex` |

`inputenc` for input coding, as well as `times` and `fontenc` for font coding. This leads to the following:

```
\documentclass{beamer}
\usepackage{graphicx}
\usepackage[english]{babel}
\usepackage[latin1]{inputenc}
\usepackage{times}
\usepackage[T1]{fontenc}
```

To generate a *pdf* file from the input file, you should use the `pdflatex` command on a Unix-like system, i.e. you just type `pdflatex testtalk` in this case, or the corresponding command for a different operating system. This directly generates the final output file `testtalk.pdf`, which can be presented on a laptop using *acroread* in the full screen mode.

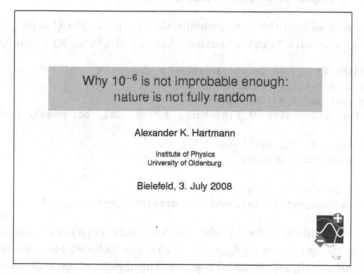

Fig. 9.8 The title page of the test presentation using LATEX together with the *beamer* class. Note that the thin black frame is not part of the presentation. It is just shown here to indicate the limits of the slide.

Next, in the source file, you have to choose the style of presentation. For this purpose, there are some predefined *themes* and *color themes*. Here, we use the theme `default` and the color theme `crane`:

```
\usetheme{default}
\usecolortheme{crane}
```

The default theme is rather puristic, see Figs. 9.8 and 9.9. Other themes tend to fill the screen. For example, `Luebeck` provides additional boxes, which show the name of the talk, the name of the author, the name of the current section and of the current subsection on each slide. This leaves less space for the actual information and the name of the speaker should be known by the second slide. Thus, a more puristic theme is recommended here. Anyway, among the predefined themes, there are, for example, `Bergen` , `Marburg`, `Berkeley`, and many more. To name just a few of the color themes: `beetle`, `fly` and `whale`. For details, please consult the documentation and try which themes suit your style best.

Within the default settings, all slides include some navigation symbols in the lower right. If you do not like this, and if you prefer to have page numbers instead, you should use

```
\setbeamertemplate{navigation symbols}{}
\setbeamertemplate{footline}[frame number]
```

The first slide of the talk is usually the title page. For this purpose you can use the macros \title, \author, \date and \logo, for example:

```
\title{Why $10^{-6}$ is not improbable enough:\\
nature is not fully  random}
\author{Alexander K. Hartmann}
\institute[University of Oldenburg] % (optional, but mostly needed)
{
   Institute of  Physics\\
   University of Oldenburg
}
\date{Bielefeld, 3. July 2008}
\logo{\includegraphics[width=0.11\textwidth]{mylogo.pdf}}
```

Here, a figure file is included using the \includegraphics command, provided by the `graphicx` package. Note that the included file must be basically a *pdf* as well, for technical reasons. This means, if you only have other file formats available, you must first convert the figures, for example using `epstopdf` in case you have encapsulated postscript (`.eps`) figures.

A slide within a *beamer* source file is defined by a *frame* environment. Inside the \begin{frame} and \end{frame} brackets in principle any valid

LATEX code can be written. For the title page, this is particularly simple, because one just has to write `\titlepage`. The remaining part of the source file looks as follows:

```
\begin{document}
\begin{frame}
  \titlepage
\end{frame}

\logo{}  % show logo only on first page

\section{Group}
\subsection{Overview}
\include{slide_group/group}

\end{document}
```

Note that the logo will be shown by default on *every* slide, unless you change the logo after the title slide to the empty logo, as it is done here. To structure the presentation, one uses the standard LATEX commands `\section` and `\subsection`. It is recommended to create for each slide a corresponding subdirectory, where all source material is collected. This makes it easier to exchange complete slides between talks. This will happen once you start to present your results on different occasions. Here, the second slide (see Fig. 9.9) is stored in the subdirectory `slide_group` in the file `group.tex`.

This slide is also implemented using the `frame` environment. Now a title of the slide is given:

```
\begin{frame}
  \frametitle{Computational Physics Group}
```

The top of the slide is composed of a centered text in blue and a centered picture:

```
\centerline{
  \textcolor{blue}{``Complex behavior of disordered systems''}}
\centerline{
  \includegraphics[width=0.6\textwidth]{slide_group/physics_cs}}
```

Most of the slide is implemented using standard LATEX elements, in particular you can tune your presentation a lot using the `minipage` environment. Please have a look at the source file. Nevertheless, the *beamer* class offers some additional features, which allow the slides to be changed

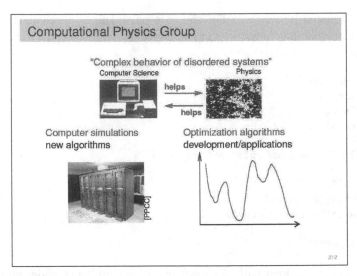

Fig. 9.9   The second slide of the test presentation using LaTeX together with the *beamer* class. Note that this slide changes dynamically. The current situation is the first stage of this slide. The next one (see text) will show a different figure in the lower right corner, after you hit the space bar on your computer when viewing the file.

dynamically. Here, we mention in particular the `overprint` environment. In our example, this looks like

```
\begin{overprint}
\onslide<1>\centerline{
\includegraphics[width=0.8\textwidth,height=0.6\textwidth]
                {slide_group/opt_problem}}
\onslide<2>
\centerline{\includegraphics[width=0.9\textwidth,height=0.5\textwidth]
{slide_group/function2dB}}
\end{overprint}
```

Note that some of the arguments to *beamer*-class commands are given in angular brackets < >, like in this example: As indicated by the `onslide<1>` command, initially the figure file `opt_problem.pdf` will be shown on the slide.[6] When you hit the space bar on your laptop during the presentation, next, the figure file `function2db.pdf` will be shown at the same place. Technically, the `testtalk.pdf` file just contains two different slides, which are identical except the areas tagged by the `\onslide` commands. You

---

[6]Note that the suffix `.pdf` can be omitted, because only *pdf* files are allowed.

will notice this immediately if you look at `testtalk.pdf` in non-full-screen mode, but when using full-screen mode, it looks like the slides change dynamically. Note that the numbers given along the `\onslide` commands are always only for the current frame. Thus, for the next frame, it starts with 1 again. Furthermore, you can give also ranges such as `onslide<1-3>` or `onslide<2->`. Thus presentations can be created, which change dynamically in a quite complex way. You can, for example, explain how your simulation algorithm works, by showing how it performs on a small sample. For this purpose, you prepare different figures showing the sample at different stages of the algorithm and use the `overprint` environment as shown above. To avoid that the figures "wiggle" during the presentation, you should make sure that all figure files which are shown on the same position have exactly the same size.[7]

You should know that you can give these ranges also along with standard `\item` commands in an `itemize` environment, such as `\item<1>` `...item<2> ...\item<3->`. This allows a slide to be unfolded stepwise, one or several items at a time.[8] Finally, please do not forget to close each frame environment by `\end{frame}`.

These examples should be enough to enable you to create nice-looking standard presentations rather quickly. To learn more about other features of the *beamer* class and to become an expert, please consult the *beamer* manual.

---

[7]If you prepare the figures using *xfig*, you can, for example, use a small box, which has in all figures the same size and is drawn using the background color around the actual content of the figures, respectively.

[8]Please do not exaggerate. The author has experienced presentations that were unfolded almost word by word, which was quite annoying.

# Appendix A

# Supplementary Materials

From the website: http://www.worldscientific.com/r/9019-supp, you will find all example programs and scripts used throughout this book, as well as the solutions to exercises.

We assume that you have a standard C compiler available. In any case, all programs run under a Linux environment, which is available free of charge. For some applications, you need further tools. All tools used in this book can be obtained free of charge.

The content of the supplementary materials is as follows:

The directory **programs** contains the source codes of the C/C++ programs and some scripts, as well as the solutions to exercises. For each chapter, there is a corresponding subdirectory, in the order of appearance in the book:

```
c-programming
scripts
se
debugging
oop
algorithms
libraries
randomness
literature
```

Each directory contains a README file, which shortly describes the contend of the programs and how to compile and run them.

# Bibliography

Abramson, B. and Yung, M. (1986). Construction through decomposition: a divide-and-conquer algorithm for the N-queens problem, *CM '86: Proceedings of 1986 ACM Fall joint computer conference*, pp. 620–628. (IEEE Computer Society Press, Los Alamitos)

Abramson, B. and Yung, M. (1989). Divide and conquer under global constraints: A solution to the N-queens problem, *Journal of Parallel and Distributed Computing* **6**, pp. 649–662.

Aho, A. V., Hopcroft, J. E., and Ullman, J. D. (1974). *The Design and Analysis of Computer Algorithms*, (Addison-Wesley, Reading (MA)).

Albert, R. and Barabási, A.-L. (2002). *Statistical mechanics of complex networks*, Rev. Mod. Phys. **74**, pp. 47–97.

Allen, M. P., and Tildesley, D. J. (1989). *Computer Simulation of Liquids*, (Oxford University Press, Oxford).

*Alta Vista*, search engine, see http://www.altavista.com/.

APS, *American Physical Society*, journals see http://publish.aps.org/.

*arXiv*, preprint server, see http://arxiv.org/.

Bauke, H. (2007). Parameter estimation for power-law distributions by maximum likelihood methods, *Eur. Phys. J. B* **58**, pp. 167–173.

*Beamer* class, a LaTeX package; written by Till Tantau, see http://latex-beamer.sourceforge.net/.

Becker, P. (2007). *The C++ Standard Library Extensions*, (Addison-Wesley Longman, Amsterdam).

Binder, K. (1981). Finite size scaling analysis of ising model block distribution functions, *Z. Phys. B* **43**, pp. 119–140.

Binder, K. and Heermann, D. W. (1988). *Monte Carlo Simulations in Statistical Physics*, (Springer, Heidelberg).

Bolobas, B. (1998). *Modern Graph Theory*, (Springer, New York).

*boost* collection of libraries; available, including documentation, at http://www.boost.org/.

Cardy, J. (1996). *Scaling and Renormalization in Statistical Physics*, (Cambridge University Press, Cambridge).

Chan, T. F., Golub, G.H., and LeVeque, R. J. (1983). Algorithm for Computing the Sample Variance: Analysis and Recommendations, *Amer. Statist.* **37**, pp. 242–247.

Claiborne, J. D. (1990). *Mathematical Preliminaries for Computer Networking*, (Wiley, New York).

Computational Provenance, special issue of *Computing in Science & Engineering* **10** (3), pp. 3–52.

Cormen, T. H., Clifford, S., Leiserson, C. E., and Rivest, R. L. (2001). *Introduction to Algorithms*, (MIT Press).

*Comprehensive TeX Archive Network*: http://www.ctan.org/.

Dekking, F. M., Kraaikamp, C., Lopuhaä, H. P., and Meester, L. E. (2005). *A Modern Introduction to Probability and Statistics*, (Springer, London).

Devroye, L. (1986). *Non-Uniform Random Variate Generation*, (Springer, London).

Dhar, A. (2001). Heat Conduction in a One-dimensional Gas of Elastically Colliding Particles of Unequal Masses, *Phys. Rev. Lett.* **86**, pp. 3554–3557.

"Diehard" test provided by George Marsaglia, see source code at
http://www.stat.fsu.edu/pub/diehard/.

Efron, B. (1979). Bootstrap methods: another look at the jacknife, *Ann. Statist.* **7**, pp. 1–26.

Efron, B. and Tibshirani, R. J. (1994). *An Introduction to the Bootstrap*, (Chapman & Hall/CRC, Boca Raton).

Fernandez, J. F. and Criado, C. (1999). Algorithm for normal random numbers, *Phys. Rev. E* **60**, pp. 3361–3365.

Ferrenberg, A. M., Landau, D. P. and Wong, Y. J. (1992). Monte Carlo Simulations: Hidden Errors from "Good" Random Number Generators, *Phys. Rev. Lett.* **69**, pp. 3382–3384.

Flyvbjerg, H. (1998). Error Estimates on Averages of Correlated Data, in: Kertész, J. and Kondor, I. (Eds.), *Advances in Computer Simulation*, (Springer, Heidelberg), pp. 88–103.

Fukui, K. and Todo, S. (2009). Order-N cluster Monte Carlo method for spin systems with long-range interactions *J. Comp. Phys.* **228**, pp. 2629–2642

Galassi M. et al (2006). *GNU Scientific Library Reference Manual*, (Network Theory Ltd, Bristol), see also http://www.gnu.org/software/gsl/.

N.D. Gagunashvili, Chi-Square Tests for Comparing Weighted Histograms, *Nucl. Instrum. Meth.* A **614**, 287–296 (2010)

Ghezzi C., Jazayeri, M. and Mandrioli, D. (1991). *Fundamentals of Software Engineering*, (Prentice Hall, London).

*Google* search engine, see http://www.google.com/.

*GraphViz* graph drawing package, see http://www.graphviz.org/.

Grassberger, P., Nadler, W. and Yang, L. (2002). Heat conduction and entropy production in a one-dimensional hard-particle gas, *Phys. Rev. Lett.* **89**, 180601, pp. 1–4.

Haile, J. M. (1992). *Molecular Dynamics Simulations: Elementary Methods*, (Wiley, New York).

Hartmann, A. K. (1999). Ground-state behavior of the 3d ±J random-bond Ising model, *Phys. Rev. B* **59**, pp. 3617–3623.

Hartmann, A. K. and Rieger, H. (2001). *Optimization Algorithms in Physics*, (Wiley-VCH, Weinheim).

Heck, A. (1996). *Introduction to Maple*, (Springer, New York).

*HotBits* webpage: here you can order files with random numbers which are generated from radioactive decay, see `http://www.fourmilab.ch/hotbits/`.

Hucht, A. (2003). The program *fsscale*, see `http://www.thp.uni-duisburg.de/fsscale/`.

*INSPEC*, literature data base, see `http://www.inspec.org/publish/inspec/`.

Jain, A. K. and Dubes R. C. (1988), *Algorithms for Clustering Data*, (Prentice-Hall, Englewood Cliffs, USA).

*JAVA* programming language, see `http://www.java.com/`.

Johnsonbaugh, R. and Kalin, M. (1994). *Object Oriented Programming in C++*, (Macmillan, London).

Josuttis, N. M. (1999). *The Standard C++ Library*, (Addison-Wesley, Boston).

Karlsson, B. (2005). *Beyond the C++ Standard Library. An Introduction to Boost*, (Addison-Wesley Longman, Amsterdam).

Kernighan, B. W. and Pike, R. (1999). *The Practice of Programming*, (Addisin-Wesley, Boston).

Kernighan, B. W. and Ritchie, D. M. (1988). *The C Programming Language*, (Prentice Hall, London).

Lamport, L. and Bibby, D. (1994). *LaTeX : A Documentation Preparation System User's Guide and Reference Manual*, (Addison Wesley, Reading (MA)).

Landau, D.P. and Binder, K. (2000). *A Guide to Monte Carlo Simulations in Statistical Physics*, (Cambridge University Press, Cambridge (UK)).

Lefebvre L. (2006). *Applied Probability and Statistics*, (Springer, New York).

Lewis, H. R. and Papadimitriou, C. H. (1981). *Elements of the Theory of Computation*, (Prentice Hall, London).

Lui, J. S. (2008). *Monte Carlo Strategies in Scientific Computing*, (Springer, Heidelberg).

Loukides, M. and Oram, A. (1996). *Programming with GNU Software*, (O'Reilly, London); see also `http://www.gnu.org/manual`.

Lüscher, M. (1994). A portable high-quality random number generator for lattice field-theory simulations, *Comput. Phys. Commun.* **79**, pp. 100–110.

*Lyx*, document processor based on LᴬTᴇX, see `http://www.lyx.org/`.

*Mathematical Reviews* is a database for summaries and reviews of mathematical papers and books, see `http://www.ams.org/publications/math-reviews/math-reviews`.

Matsumoto, M. and Nishimura, T. (1998). Mersenne twister: a 623-dimensionally equidistributed uniform pseudo-random number generator, *ACM Transactions on Modeling and Computer Simulation* **8**, pp. 3–30.

Mehlhorn, K. and Näher, St. (1999). *The LEDA Platform of Combinatorial and Geometric Computing* (Cambridge University Press, Cambridge); see also `http://www.mpi-sb.mpg.de/LEDA/leda.html`.

Meyers, S. (2005). *Effective C++: 55 Specific Ways to Improve Your Programs and Designs*, (Addison-Wesley, Reading (MA)).

Morgan, B. J. T. (1987). *Elements of Simulation*, (Cambridge University Press, Cambridge).

Newman, M. E. J. and Barkema, G. T. (1999). *Monte Carlo Methods in Statistical Physics*, (Clarendon Press, Oxford).

Newman, M. E. J. (2003) *The Structure and Function of Complex Networks*, SIAM Review **45**, pp. 167–256.

Newman, M. E. J., Barabasi, A.-L., and Watts, D. (2006). *The Structure and Dynamics of Networks*, ( Princeton University Press, Princeton).

*Papercore* is a free and open online database for summaries of scientific papers, see www.papercore.org.

Phillips, J. (1987) . *The Nag Library: A Beginner's Guide* (Oxford University Press, Oxford); see also http://www.nag.com.

Oram, A. and Talbott, S. (1991). *Managing Projects With Make*, (O'Reilly, London).

*PhysNet*, the Physics Departments and Documents Network, see http://physnet.uni-oldenburg.de/PhysNet/physnet.html.

The *Python* programming language can be downloaded for all major operating systems from https://www.python.org/, including extended documentation and tutorials.

"Diehard" test provided by George Marsaglia, see source code at http://www.stat.fsu.edu/pub/diehard/.

Press, W. H., Teukolsky, S. A., Vetterling, W. T., and Flannery, B. P. (1995). *Numerical Recipes in C*, (Cambridge University Press, Cambridge).

*Povray, Persistence of Vision* raytracer, see http://www.povray.org/.

*Quantis*: A hardware true random number generator. It is based on the quantum mechanical process of photon scattering. It can be connected to a computer via USB port or PCI slot. More information can be found at http://www.idquantique.com/products/quantis.htm.

R is a software package for statistical computing, freely available at http://www.r-project.org/.

Rapaport, D. C. (1995). *The Art of Molecular Dynamics Simulations*, (Cambridge University Press, Cambridge).

Robert, C. P. and Casella, G. (2004). *Monte Carlo Statistical Methods*, (Springer, Berlin)

Robinson, M. T. and Torrens, I. M. (1974). Computer simulation of atomic displacement cascades in solids in the binary collision approximation, *Phys. Rev. B* **9**, pp. 5008–5024.

*Romeo*, a database for publisher copyright policies and self-archiving, see http://www.sherpa.ac.uk/romeo/.

Rumbaugh, J., Blaha, M., Premerlani, W. , Eddy, F., and Lorensen, W. (1991). *Object-Oriented Modeling and Design*, (Prentice Hall, London).

Scott, D. W. (1979). On optimal and data-based histograms, *Biometrica* **66**, pp. 605–610.

Sedgewick, R., (1990). *Algorithms in C*, (Addison-Wesley, Reading (MA)).

Skansholm, J. (1997). *C++ from the Beginning*, (Addison-Wesley, Reading (MA)).

Sommerville, I. ( 1989). *Software Engineering*, (Addison-Wesley, Reading (MA)).

Sosic, R. and Gu, J. (2001). Fast Search Algorithms for the N-Queens Problem, *IEEE Trans. on Systems, Man, and Cybernetics* **21**, pp. 1572–1576.

*Standard Template Library*, see
`http://www.sgi.com/tech/stl/download.html`.

Stroustrup, B. (2000). *The C++ Programming Language*, (Addison-Wesley Longman, Amsterdam).

Sutter, H. (1999). *Exceptional C++: 47 Engineering Puzzles, Programming Problems, and Solutions*, (Addison-Wesley Longman, Amsterdam).

*Subversion* version control system, see `http://subversion.tigris.org/`.

Swamy, M. N. S. and Thulasiraman, K. (1991). *Graphs, Networks and Algorithms*, (Wiley, New York).

Texinfo system, see `http://www.gnu.org`. For some tools there is a *texinfo file*. To read it, call the editor 'emacs' and type <crtl>+'h' and then 'i' to start the texinfo mode.

TUG, TEXUser Group, see `http://www.tug.org/`.

*Valgrind* memory checker; more information, including a user manual, can be obtained from `http://www.valgrind.org/`.

Vattulainen, I., Ala-Nissila, T. and Kankaala, K. (1994). Physica Test for Random Numbers in Simulations, *Phys. Rev. Lett.* **73**, pp. 2513–2516.

Walker, A.J. (1997). An efficient method for generating discrete random variables with general distributions, *ACM Trans. Math. Softw.* **3**, pp. 253

Web of Science, see `http://www.isiwebofknowledge.com/`

*Westphal Electronic* (`http://www.westphal-electronic.com/`) sells divices which produce true random numbers based on thermal noise in Z diodes. They can be connected to a computer via USB port or bluetooth.

*Wikipedia* is a free online encyclopedia, currently containing more than 2.5 million articles, see `http://www.wikipedia.org/`.

Wilson, M. (2007). *Extended STL*, (Addison-Wesley Longman, Amsterdam).

Wolfsheimer, S., Hartmann, A. K., Rabus, R., and Nuel, G. (2012). Computing posterior probabilities for score based alignments using ppALIGN, *Stat. Appl. Gen. Mol. Biol.* **11**, issue 4, article 1

*Xmgrace* (X Motiv GRaphing, Advanced Computation and Exploration of data), see `http://http://plasma-gate.weizmann.ac.il/Grace/`.

*Yahoo* search engine, see `http://www.yahoo.com/`.

Ziff, R. M. (1998). Four-tap shift-register-sequence random-number generators, *Computers in Physics* **12**, pp. 385–392.

*zlib* compression library, see `http://www.zlib.net/`.

# Index

Printed in the United States
By Bookmasters